Judging Medicine

Judging Medicine

George J. Annas

Humana Press • Clifton, New Jersey

Library of Congress Cataloging-in-Publication Data

Annas, George J.
  Judging medicine.

    Includes index
    1. Medical laws and legislation--United States.
    I. Title.
    KF3821.Z75A56   1988   344. 73'041   88-9024
    ISBN   0-89603-132-3   347. 30441

2, 636

To Mary

Foreword

In the early 1970s, well before the field of bioethics had established itself in medicine or anywhere else, the Hastings Center organized a small meeting of law school professors. The question we put to them was: what could or should be done to stimulate legal interest in the field? The answer we got was a wise one. We should do nothing to forcefeed the interest. It should simply be allowed to develop on its own, by the ordinary route of attracting a following because of its inherent importance.

That is just what happened, and one of the first young legal scholars drawn to what remains (oddly enough) a relatively small field was George Annas. The idea of a column on law and ethics for the *Hastings Center Report* was not by 1976 a particularly bold one. It had been clear to us from the outset of the Center in 1969, and the establishment of the *Report* in 1971, that the rapidly emerging moral problems in medicine and biology would have enormous legal and policy implications. Even so, we were hardly prepared for the large and steady number of cases that were to come before the courts during the 1970s and that were to continue unabated in the 1980s. But our concern about a column on the subject was of a more pedestrian kind. Could such a column remain interesting month after month? Could its author write in a way accessible to a wide variety of people and fields? Could it be lively and strong in its opinions and yet judicious and thoughtful in its reasoning? Could it avoid becoming predictable?

That was a heavy set of demands. George Annas met them all. More than that, over the years his column has remained one of the most popular of the various features of the *Report*. It is read and respected by his fellow legal scholars and by those philosophers, physicians, and others who on occasion express some reservations about lawyers as acceptable members of the human race. Annas has made them think twice about that bias. He has shown himself sensitive both to the clinical setting and nuances of the many ethical problems in medicine and to their moral dimensions and complexity.

In his Preface to this volume, Annas notes the pervasiveness of the view that "conflicts in medicine are too difficult to ever resolve in a pluralistic society," a view he sharply challenges. I very much agree with his judgment on that most basic of points. Far from it being true that we cannot find good moral answers to anything, it is more often the case that we do, in however halting and confused a fashion, come to some agreement as a society; and that agreement is often based on solid moral principles. It is just hard work, very hard work, requiring seriousness of purpose and persistence of effort. More than that, it seems clear (at least to me) that any so-called morality that is nothing more than a set of decisionmaking procedures—a morality of process rather than of substance—is a betrayal of the high calling of ethics, which is to tell us something about right and wrong, good and evil, not simply about useful processes to help us keep peace with each other.

George Annas knows this and it shows in his work. He fully understands that law and morality are not identical, however much they may overlap. He no less understands that each has something to give to the other, that their relationship, though often troubled, can and should be a creative one. In the law, we often see the working out of our social and political values; we see implications we might not otherwise have seen. In ethics, we no less see the way in which our legal practices and institutions are themselves a source of values and moral principles.

Yet while the line between law and ethics is a fine one, their differences are important also. More people than ever seem to confuse law and ethics, as if what is beyond the pale of law or regulation is beyond moral judgment altogether. One senses in this attitude, more implied than articulated, a residue of that view of morality that relegated it to subjectivism or emotivism. Although that position, an outgrowth of the positivism that had a powerful sway from the 1930s through at least the 1960s, has had few serious followers of late, it left a powerful mark on the culture. If morality is nothing more than an expression of taste, bias, or personal predilections, then law becomes the only viable bulwark against social chaos. The result is to place too high a premium on law and too low a value on ethics.

This is not a mistake that George Annas makes. He shows the law in all of its frailty and his frequent assaults on the reasoning of judges is sufficient in itself to show his sensitivity to the limits of the law as a guide to moral conduct. At the same time, he allows his own strong moral convictions to come through; he does not take refuge in the uncertainty or obscurity of the law to evade moral judgment. The result, time and again, is a telling commentary on difficult moral and legal problems of contemporary biomedicine, one illuminated by telling legal insight and braced by moral vigor.

Daniel Callahan
The Hastings Center

Preface

Judging medicine requires criteria. The criteria used in these pages is that of law, with a special emphasis on liberty and social justice. This book deals with the premier issues of medical law and bioethics as those fields have explosively developed over the past decade. Medical law is to law what astrophysics is to physics; a general body of knowledge applied to a specific human endeavor. What makes medical law the most compelling and intellectually satisfying of the legal specialties is that it directly impacts on the health, life, and well being of real people. Its most legitimate goal is to define legal principles that can enable medical practice and scientific advance to enhance basic human rights and values.

Law tends to concentrate on rights, ethics, on values. Law tells us what we must do; ethics what we ought to do. The disciplines are nonetheless far from distinct: law needs an ethical base to be useful to human beings, and medical law, like medical ethics, demands an understanding of the world of medicine and medical care for its practitioners to be helpful to physicians and their patients. And the major principles underlying both medical law and medical ethics are the same: self-determination, beneficence (or at least nonmaleficence), and justice as fairness.

In his masterpiece, *The Great Gatsby*, Fitzgerald tells us why a civil rights lawyer might find medical law so attractive: "There is no difference between men, in intelligence or race, so profound as the difference between the sick and the well." Helping the sick to exercise their rights in the medical care area is perhaps the most challenging of all civil rights endeavors. Adding to this challenge is the inherently fascinating relationship between two of society's most powerful professions. Hemingway, struck by the seeming loyalty of these professionals to their peers, once opined that "writers should stick together like doctors, lawyers, and wolves." He meant the comparison with wolves as a compliment, insisting that they "do not hunt each other... or prey on each other." This still seems true *within* the two professions, but physicians often view themselves as prey for the legal profession, and the relationship between the professions has been unnecessarily combative and overtly hostile, at least in the courtroom. Courts have responded with varying degrees of success to the quandaries raised by modern medicine, resolving some, defining some, blundering through some, and disgracefully washing their hands of others. As many of these essays illustrate, judges are not more intelligent, more thoughtful, or more compassionate than other professionals as a class. That is not why they have been so often called upon to resolve these quandaries. They have had to resolve conflicts because that is their social role: publicly resolving disputes, based on legal principles is what judges are for. The case-based, common law approach of deciding real disputes on the basis of legal principles and precedents has proven very serviceable in exposing the underlying issues and gradually developing rules that have the potential for universal application. Legislatures and regulatory agencies also have their social roles. The law's struggle to engage, control, or accomodate medicine through litigation, legislation, and regulation is the subject of the sixty-one essays that comprise this book.

Almost all of these essays originally appeared in the *Hastings Center Report*, generally acknowledged as the world's leading bioethics publication. The *Report* appears bimonthly, and I have written for it in every issue but one since June, 1976. I first suggested a law feature for the *Report* to Dan Callahan in 1975 because it was my observation that many bioethicists (and physicians) feared and mistrusted the law, and in almost all cases that attitude was based on factual ignorance and a misunderstanding of the law's role in society. It was my hope that an accurate portrayal of the major medical law issues and cases of our day could enhance dialog and contribute to the resolution of those issues that were resolvable. But as much as exposition and analysis, constructive criticism and patient-centered advocacy is at the core of these essays. They should be of interest to health care professionals, lawyers, law makers, judges, bioethicists, and everyone interested in the way in which law interacts with medical practice, and the effects of this interaction on real people.

The essays are arranged in groups that, with the exception of the first (patient rights) and last two (government regulation, and transplants and implants), roughly correspond to the human life cycle. Each of the book's nine sections opens with a brief introduction that puts its essays into legal and historical context. These sections stand on their own and can be read independently. The essays themselves focus on matters of intense public concern manifested in courtroom confrontations, legislature proposals, or regulatory schemes. Many were the subject of lectures and talks I have been privileged to give health professionals and legal groups in more than thirty states and eight countries over the past decade. Others derive from my experiences as a member of various state regulatory agencies, work on appellate briefs, and preparation of Congressional testimony. To create a whole that is more than the sum of its parts, all of the essays have also been updated (to Jan. 1988) and cross-referenced.

There is no natural endpoint to this collection. During the time that the book has been in press, for example, another eight essays have appeared in the *Hastings Center Report*. These have included commentary on the judicial decisions in the cases of Baby M, Nancy Jobes, and "A.C.," and thoughts on separating Siamese twins, using anencephalic newborns as organ donors, and patenting "new" life forms.

As this book goes to press it is vogue in medical ethics circles to declare that conflicts in medicine are too difficult to ever resolve in a pluralistic society. Since they are essentially indeterminate, the argument goes, rather than concentrate on developing substantive rules for decisionmaking, we should instead concentrate on developing an acceptable process for decisionmaking. The implication is that we cannot ever hope to find the "right" answer to any (or most) of these issues. I remain unpersuaded. Granted that the task of defining substantive principles and rules that we can live with is a complex and frustrating one, it nonetheless strikes me as one that is worthy of law and social policy, and the only one that can hope to prevent medical law (and medical ethics) from becoming more than simply the arbitrary exercise of power.

There are many people who contributed to this work. My friend and colleague, Leonard Glantz, deserves special thanks for his uncompromising intellectual insight and uncommon generosity in reviewing and criticizing almost every one of these essays. Without his challenging analyses, these essays would not deserve publication. My only other critic to consistently match his courage and tenacity has been my wife, Mary Annas. Of course, the editors of the *Hastings Center Report* must be given credit for their constuctive editorial suggestions, for prodding me to produce on time, and for often changing the titles (some of which I have restored). Margaret O'Brien Steinfels was the editor until 1980, and Carol Levine until mid-1987. The *Report's* Joyce Bermel and Nancy deKoven have also helped keep the enterprise on track and on time. Tom Lanigan first suggested publishing these essays as a collection in 1980 (when less than half of them existed), and gets much of the credit for its ultimate shape. Production itself owes much to the word processing skills of Mary Lou Hannigan and Meg Picariello, and the proofreading of Ms. Hannigan and Martha Neary. In the mid-1970s, my colleagues at the Boston University Law School's Center for Law and Health Sciences (in addition to Leonard Glantz) Jay Healey and Barbara Katz, worked with me on almost every project, and helped define the field of health law (of which medical law is a subspecialty). Many of the ideas for and in these columns were also influenced by my health law colleagues Alex Capron, Jane Greenlaw, Fran Miller, John Robertson, Dick Scott, Margo Somerville, and Ken Wing; my bioethics colleagues (in addition to Dan Callahan) Jane Boyajian, Art Caplan, Tris Englehardt, Peter Singer, Judy Swazey, and Bob Veatch; as well as many medical colleagues including David Allen, Ward Casscells, Ned Cassem, Sherman Elias, Will Gaylin, Michael Grodin, Jay Katz, Mel Levine, Elliot Sagall, David Todres, and Herman Wigodsky. Since I moved to Boston University Medical School in 1978 (and its School of Public Health), I have been unfailingly supported by Norman Scotch, the Director of the School of Public Health, an academic role model for nourishing scholarship and creativity. Partial support for some of these essays came from Concern for Dying, and the Rubin Family Fund, for which I am grateful.

George J. Annas
Boston, 1988

Contents

Reproductive Liberty

Medical Practice

The Mentally Retarded and Mentally Ill Patient

Death, Dying and Refusing Medical Treatment

Government Regulation

Transplants and Implants

Patients Rights

Introduction

Opening this collection of essays with a section concentrating on patients rights seems natural for two reasons. The first is that patients rights is what "judging medicine" is all about. Medicine has to do with health, healing, and comfort: the health of the patient, healing the patient, and comforting the patient. The law has little to say directly about any of these, but has much to say about the centrality of the patient's own view of health, healing, and comfort. Because the doctor–patient relationship is inherently one-sided, the law's direction over the past decade has been to attempt to give the patient more authority over decisions that directly affect the patient's mind and body.

Movement in this direction has occurred by enunciating the doctrine of informed consent, giving the patients the right to refuse treatment in almost all circumstances, and attempting to humanize the hospital environment through various strategies like statements of patients rights. The law works best in the medical area when it promotes the rights of patients in a way that enhances their autonomy in the doctor–patient relationship, and supports the supposition that decisions made in this confidential, voluntary and informed relationship can be carried out without governmental or institutional interference (at least so long as they do not violate important public policy). In this sense, patients rights become central to all medical decisionmaking,

1

and enunciating patients rights in the hospital setting becomes a matter of critical importance to both physicians and patients.

The second reason to begin this book with patients rights is more personal. Most of my initial work in the field of law and medicine was concentrated on patients rights, and my first book, *The Rights of Hospital Patients* (1975), was devoted exclusively to this topic. This book was written before I began to write regularly for the *Hastings Center Report,* and thus the columns for that publication do not exhibit my interests and views on this subject. The opening essay in this chapter reflects my mid-1970s views, and reflects my opinion of the proper role of the law in medical practice.

Literature from around the world informs us that many physicians cannot understand why patients seem to take autonomy so seriously, when the physicians *know* that fighting death and teaching others to fight death is so much more important. In Aleksandr Solzhenitsyn's *The Cancer Ward*, for example, a patient, Oleg Kostoglotov, discovers he has been receiving hormone therapy for cancer without his knowledge:

> By some right—it doesn't occur to them to question the right —they are deciding for me, without me, on a terrible treatment, hormone therapy. This is like a red-hot iron which, once it touches you, leaves you maimed for life. *But it appears so ordinary in the ordinary life of the hospital!*

Oleg understands fully that his physicians believe the treatment is absolutely necessary if he is to survive. But he also understands much more fully than they that there are fates worse than death:

> What, after all, is the highest price one should pay for life? *How much should one pay, how much is too much?...*—to save one's life at the cost of surrendering everything that gives it color, flavor, and sparkle. To get a life of digestion, breathing, muscular and mental activity, and nothing more? To become a walking husk of a man—isn't that an exorbitant price?

Most Americans believe they should be involved in medical treatment decisions, and at least retain the right to refuse *any* medical intervention. Nonetheless, we do not realistically expect these rights to be honored in the hospital. Larry McMurtry's description of Emma's feelings on being hospitalized for cancer in his *Terms of Endearment* conveys an almost universal resignation:

From that day, that moment almost, she felt her life pass from her own hands and the erring but personal hands of those who loved her into the hands of strangers—and not even doctors, really, but technicians: nurses, attendants, laboratories, chemicals, machines.

This is powerful imagery, powerful primarily because of its accuracy. Does it really have to be like this? Must doctors continue to train doctors to ignore the person and concentrate exclusively on the body? Must we really choose between expert technicians and humane care—takers? Can we really expect no better when we deliver ourselves "into the hands of strangers"?

The following essays explore this question. The law is important. But patients rights are often more a matter of informal agreement, hospital policy, and the ethical sensibilities of providers, than of explicit legal rule. Indeed, the only nonstatutory legal right to medical care that continues to exist in this country is the right of a person to treatment for an emergency condition in a hospital emergency room. Whether and to what extent this emergency care duty should be expanded is also dealt with in this opening section.

The opening essay, "The Hospital: A Human Rights Wasteland," originally appeared in *The Civil Liberties Review* (Fall, 1974). "The Emergency Stowaway" was published in *Law, Medicine and Health Care* eight years later, in February, 1982, and reflects my basic views on patients rights. All of the remaining essays were originally published in the *Hastings Center Report,* the breast cancer piece in April, 1980, the Good Samaritan piece in April, 1978, the emergency room piece in August, 1985, and the delivery room essay in December, 1981.

The Hospital
A Human Rights Wasteland

Civil libertarians have little difficulty appreciating the plight of prisoners or mental patients. But tell the average civil libertarian that there are significant and unnecessary restrictions on the individual rights and liberties of patients in general hospitals, and you are likely to encounter a blank stare. There are a number of reasons for this. One is the general misconception that the problems are minor, or that certain temporary restrictions on individual liberty are essential if hospitals are to treat sick people properly. An unconscious desire not to perceive ourselves at risk may be another reason; we seldom seriously think we will ever be either prisoners or mental patients. But almost all of us have been hospital patients at least once, and each of us will be a hospital patient an average of seven times during our life. By not dealing with the issue, perhaps we are seeking to avoid thinking about our own future hospitalization, an event which is almost always traumatic and undesired.

Other than the fact that the average length of a hospital stay is less than a week, there are probably more similarities than differences among prisoners, mental patients, and general hospital patients. Although no social stigmatization attaches to admission, the patient almost never comes to the hospital voluntarily. Some outside force, usually an illness or injury, has made admission mandatory. Unless the patient enters through the emergency room, a doctor is likely to have ordered the admission and chosen the particular hospital. Upon arriving, the patient is made to sign a variety of forms which generally are explained only with the assurance that they are "routine." The patient is then separated from accompanying friends or relatives and escorted to a preassigned room. The patient's clothes are replaced with a johnnie, a one-piece garment designed for the convenience of hospital staff to make testing and treating the patient easier. The patient is given a plastic wristband with a number written on it, a

4

number that becomes more important than the patient's name. The patient is confined to a bed, and may even have to await permission to use the toilet. Medication and food may be prescribed without consulting the patient. Nurses, students, aides, and physicians may enter the patient's room without knocking and submit the patient to all manner of examination and treatment without explanation. Moreover, all of this will be carefully recorded in a written record that the patient generally is not allowed to see, but which is available to almost anyone on the hospital staff, and may be seen as well by medical researchers and insurance companies. Unless the patient is in an expensive private room, visiting hours are usually restricted.

The experience tends to intimidate and disorient, and discourage any assertion of individual rights. While medical care in the past was a one-to-one relationship involving only the patient and physician, in the modern hospital the patient is treated by a team in a complex, unfamiliar setting. Because of this, the institution–patient relationship becomes at least as important as the doctor–patient relationship.

Those interested in the rights of women, children, and the poor should also be able to identify with many problems in the hospital. Patients are asked to behave like children, investing in the physician and staff the faith they once had in their parents. Women are often treated as more neurotic and emotional than men, and thus frequently deprived of information concerning treatment alternatives and possible complications because this information might "upset" them unduly.* To suit their convenience, hospital staffs often separate children from parents, ignoring the child's physical and emotional needs. Finally, the source of payment can determine the quality and type of care, as well as the manner in which it is administered. And hospitals, which have become major financial enterprises, may view their primary functions as research and education (or even as pleasing their shareholders) rather than patient care.

Why hasn't the voice of the consumer risen up in effective protest to demand changes in the more dehumanizing practices of hospitals? First, the average length of stay of about a week makes formation of an in-patient consumer group impossible. Second, most patients in hospitals are simply too sick to assert their rights.

They may even willingly forego them if they believe this will speed their treatment and return home. Third, healthy individuals do not see the issue as one which is as vital to their lives as others which affect

---

*See, e.g., "Breast Cancer: The Treatment of Choice," at p. 36.

them daily, such as food, housing, education, and racial discrimination. Finally, there is the great difficulty outsiders have in understanding what the institution called a hospital is all about.

Modern hospitals are a product of only the past three or four decades. Before the turn of this century, it was unlikely that going to a hospital (or almshouse or pesthouse as they were then often termed) would do any good. Most people went to them only because they were poor and dying, or had an incurable condition. The only recorded hospital patients rights measure before this century was instituted in France in 1793 by the National Convention of the French Revolution. It decreed that there should be only one patient in a bed (as opposed to the usual of two to eight), and that beds should be at least three feet apart. At least some progress has been made since then.

As drugs and surgery replaced the purging and bleeding of the nineteenth and early twentieth centuries, and as medical education and specialty medicine progressed, the hospital gradually became the center of medical research. States and localities founded hospitals for their citizens, following the lead of private charities and nonprofit corporations. Perhaps the most significant development was the teaching hospital. Usually a large, general hospital in which medical students, interns, and residents are trained under the direction of experienced physicians on the faculty of a medical school, it is here that most medical research takes place, most medical advances are initiated, and where the latest medical techniques are available.

It is also likely that the least attention is paid to patients rights in teaching hospitals. But rights problems exist in all hospitals. Mitchel Rabkin, the physician-director of Boston's Beth Israel Hospital, has described the hospital patient's dilemma in the following terms:

> Having evolved from a shelter for wayfarers and then a place for the sick poor to die in a therapeutic community where scientific medicine can bring about diagnosis and healing, today's hospital stands increasingly to become a jungle, whose pathways to the uninitiated are poorly marked and fraught with danger, the same pathways which, as they become overgrown by the teeming hospital environment, may be poorly perceived and even invisible from above.

The purpose of this essay is to help chart a path through this hospital jungle, to identify those areas and activities of the hospital that place the patient rights most at risk, and to suggest ways to enhance the rights of patients and the humanness of the hospital environment, and thus improve the quality of health care delivered.

The argument for enhanced patients rights is based on two fundamental premises: (1) The American medical consumer possesses certain interests, many of which may properly be described as rights, that are not automatically checked at the hospital admissions desk along with the patient's other valuables, and (2) Most health care facilities fail to recognize the existence of these interests and rights, fail to provide for their protection or assertation, and frequently limit their exercise without recourse. The following sections review some of the more important predicaments faced by hospital patients.

Areas of Risk in the Hospital

The Emergency Room

More than twice as many patients as are admitted to hospitals are seen in emergency rooms. As a consequence, emergency rooms have become the primary source of medical care for communities surrounding major hospitals, especially for the poor. Since so many people depend on them for their health care, the type of treatment they afford patients is critical. Problems can arise from a refusal to examine or treat an individual, a requirement of a cash down payment prior to examination, long and harmful delays before examination, transfer to another institution, treatment and discussion of a patient in full view or hearing of others, or inability to understand the patient's language.

Probably not atypical is a case that occurred at a major Boston teaching hospital. A psychiatrist was called in to the emergency ward to see a Puerto Rican woman whose stomach had just been pumped out. He was the first person on the scene able to speak Spanish. The woman told him that she had received some very distressing news at home, had taken two Alka-Seltzers, and had come to the hospital to talk with a doctor. The emergency ward staff had assumed that she was an overdose case because, they explained, "most Puerto Ricans who demonstrate symptoms like those shown by the woman have overdosed." Though the woman's life was not endangered, the example is important because hospitals frequently fail to provide an interpreter despite the presence of a large, foreign speaking population in their area, and may make ethnically stereotyped diagnoses. Had this hospital been properly concerned with informed consent, it would have recognized its obligation to communicate to the patient the medical staff's assumptions about her condition, which she would then have promptly contradicted, refusing her consent to a stomach-pumping.

More serious are instances in which hospitals refuse to provide even emergency treatment, in direct contravention of the law, to patients who have no cash or insurance. The father of a ten-year-old child related the following story to Senator Edward Kennedy's Subcommittee on Health. His son, Paul, had a seizure at home and passed out. He rushed the boy to the nearest hospital, a private institution (his son had been receiving treatment at the county hospital, a considerable distance away). When they arrived, he was subjected to an interview about his finances and insurance. The staff refused to examine his son until he had answered such questions as: "Do you own your own home?" "Who is your employer?" and "How long have you worked there?" The interviewer also refused to call the county hospital. In frustration the father left the emergency room and drove the long distance to the county hospital. In the course of his trip he passed several other private hospitals, but was afraid to stop for fear of receiving another interrogation but no treatment. His son died within an hour after he arrived at the county hospital. Prompt attention would have saved the boy's life.

Cases abound of hospitals refusing to see emergency patients who have neither insurance nor cash with them, hospitals transferring emergency patients to county or municipal hospitals because of their apparent inability to pay, and patients dying in emergency rooms while they are waiting to be seen by a nurse or physician.

There is some evidence that the emergency room situation is improving under pressure from the courts (which have required hospitals with emergency facilities to treat all emergency cases that present themselves for treatment), federal Hill-Burton regulations for providing funds to hospitals, Medicare and Medicaid regulations regarding emergency services, and the promulgation of emergency room standards by the Joint Commission on Accreditation of Hospitals. Few would disagree, however, that much remains to be done before patients needing immediate treatment can present themselves confidently at any emergency ward.

Consent to Treatment

In his often-quoted (but less often observed) essay *On Liberty,* John Stuart Mill wrote:

> The only purpose for which power can be rightfully exercised over any member of a civilized community, against his will, is to prevent harm to others. His own good, either physical or moral, is not a sufficient warrant.

This principle stands in sharp contrast to the aphorism, "The doctor knows best." The doctor might, but it is the doctor's legal duty to explain in lay terms the treatment he proposes, the risks of death and serious side effects, problems of recuperation, probabilities of success, and alternatives, *and* to obtain the patient's informed, competent, and voluntary consent before proceeding with the treatment. In order to preserve the patient's right of self-determination, it is essential that the patient make the decision regarding treatment, not the physician. As a leading law textbook expresses it: "Individual freedom here is guaranteed only if people are given the right to make choices which would generally be regarded as foolish."

The principle may seem almost self-evident, but it is difficult for many physicians to comprehend. Some argue, for example, that they can get patients to consent to anything, depending on the way they phrase their description of it. Patients can, of course, be deceived by coloring or distorting the truth. In California, for example, an orthopedic surgeon, John Nork, admitted that he had persuaded over 30 patients to submit to laminectomies (surgery on an intervertebral disc) that they did not need. Many of the conditions of these patients were exacerbated by the surgery, and some were handicapped for life because of the physician's ineptitude in performing the surgery. The power of the physician is illustrated by the fact that one of the patients he permanently injured testified on his behalf at a malpractice trial.

Malpractice case law is filled with instances of physicians failing to disclose information concerning suggested surgery to patients. Examples include doctors failing to mention the risk of paralysis in a laminectomy, failure to mention any risks in an ulcer operation, failure to mention any possible harmful effects of cobalt treatment for cancer, failure to disclose reasonable alternative treatments,* failure

See, e.g., "Breast Cancer: The Treatment of Choice," at p. 36.

to explain a procedure in words the patient can understand, obtaining consent while the patient is under the influence of drugs or alcohol, and failure to explain to the patient that the recommended procedure can actually make the patient's condition worse.

Providing patients with complete and accurate information concerning their conditions and the treatment alternatives can also directly improve the quality of care by reducing the number of unnecessary operations performed. Insistence on consultation with a specialist in internal medicine would also significantly reduce this number. As to the hazards to patients from fully disclosing information about their illnesses, surveys have found that although some patients are upset at learning the risks of operations, most prefer to know and either are unaffected emotionally by the information or feel more comfortable after receiving it.

Access to Medical Records

It is fairly standard procedure for hospitals never to permit patients to see their records. Although this is usually justified by the rationale that patients cannot understand records and will only be distressed by them, this paternalism is as misplaced as it is unjustifiable.

The record is about the patient, and while the hospital may own the paper on which it is printed, the patient has the most vital interest concerning its content. The real reasons that records are not routinely given to patients seem to be that: (1) Physicians don't want to take the time to explain their contents to patients; and (2) Physicians often write impressions of the patient in the record that may be seen by them as useful, but which may be insulting to the patient and have no basis in fact.

Both these reasons are illustrated by a recent case that arose at another major Boston teaching hospital. Massachusetts has a statute that gives patients a legal right to see their records while they are in the hospital, and after they are discharged, and to obtain a copy at "reasonable cost." A former Medicaid patient made a request to the hospital for a copy of his lengthy medical record. The person in charge, who knew that the individual was indigent, said that a copy could be had at a cost of $40.00. Unable to raise this much money, the former patient asked simply to see his records. This request was refused. The patient then sought legal help. An attorney obtained an agreement that the patient's former physician would sit down with him and go through his record. At this meeting the patient asked to go through the entire record carefully, a procedure that could have taken six to eight hours. The hospital then decided to give him a copy of the record to study at home. As he read it, he found a number of statements by physicians, unsupported by any data or psychiatric evaluation, to the effect that he was crazy or "nuts," or that his illnesses were psychosomatic.*

It is essential that patients know what the record contains not only so they can decide where to go for future care (it is unlikely a patient would receive as serious attention in a hospital which had a record calling him "nuts" as in one which had not previously treated him), but also so they can decide whether or not to permit copies of medical records to be examined by prospective employers or insurance companies. Moreover, knowledge of what one's medical record contains should be a prerequisite to giving informed consent to specific hospital treatments.**

Medical Experimentation

Whenever physicians depart from standard medical practice to obtain new knowledge, they are engaging in human experimentation. Studies have shown that the "ignorant, the poor, and the ethnically despised" are the most frequently used subjects in human experimentation. In the Tuskeegee syphilis study, to choose one egregious example, 600 black men with syphilis had effective treatment withheld from them to permit public health officials to study the natural course of the disease. In a Brazilian heart transplant case, surgeons even withheld from the recipient the fact that he was going to receive a transplant, explaining: "If a man is incapable of understanding an operation he vitally needs, there is no choice but to proceed...the patient was psychologically better off not knowing and worrying about the risks." The recipient first learned that he had received a heart transplant when he saw himself on television a week later.

---

*Legal issues involving mentally retarded and mentally ill patients are dealt with at p. 217 *et seq.*

**See the following essay for a summary of studies on patient access to medical records, at p. 30.

A leading medical commentator, Franz Ingelfinger, has noted that although "he is far from exercising free power of choice...the most used and useful of all experimental subjects is the patient with disease." Other unusually horrible examples include the terminally ill patients in New York's Jewish Chronic Disease Hospital who were injected with live cancer cells without their knowledge, to determine if and how quickly the body would reject the foreign cancer cells, and use of retarded children at New York's Willowbrook Hospital who were intentionally infected with hepatitis to test the effectiveness of an experimental vaccine. Most abuses, however, are far less dramatic.

Much experimentation simply involves reviewing the medical records of patients who have a certain disease, doing additional tests on tissue (like blood or skin) that would have been removed from the patient in any event, or drawing an extra syringe of blood from a patient who would not otherwise have had it drawn. All these studies, however, raise questions of confidentiality, privacy, and self-determination. Medical researchers sometimes seem to take for granted that in experiments involving minimal risks, there is no need to obtain a patient's express consent.*

It was not until 1966 that the National Institutes of Health, which fund much of the medical research that is performed in this country, issued guidelines for researchers that included having the research design, the risks and benefits, and the informed consent procedure reviewed in advance and approved by a hospital-based review committee. As interesting as what these guidelines contain, is what they do not contain. For example, NIH makes no provisions to defray the medical bills or loss of earnings of someone injured in an experiment unless the injury resulted from negligence. Usually, the experimental subjects, those least able to bear the risk, are left to their own devices.

Reports on research suggest that most patients do not understand in any clear way the distinction between treatment and experimentation, and that institutional review is often a rubber stamp for experimentation. Therefore, while not a legal requirement, the provision for indemnification for injury of the research subject should be a *sine qua non* of ethical research.**

---

* See "Transplants and Implants" for a discussion of informed consent to highly invasive and risky experimentation at p. 363 *et seq.*

** See pages 325–339 for a discussion of federal commissions set up to improve oversight of human experimentation.

The Terminally Ill Patient

The patients perhaps most at risk of losing all their human rights in the hospital are those who are dying or terminally ill. The first right lost is the right to the truth. Studies have indicated that most doctors never tell their patients they are dying, and over 90% seldom tell them.* Instead doctors usually tell the family, depriving the patient simultaneously of another right: the right of confidentiality. In addition, dying patients are likely to receive less attention than other patients from medical personnel, and are more likely than other patients to be used as the subjects of medical experimentation. Since they are not told that they are considered terminal, informed consent is never obtained for their "treatments."

Many of the problems of the dying are related to society's discomfort with the elderly (nursing homes, for example, present most of the same problems as hospitals extended over an average length of stay of three years). Though many of the elderly would prefer to die a "natural death," without the massive intervention of medical equipment and procedures that might prolong their agony for days or even months, hospitals are often very reluctant to follow their wishes. Some fear malpractice suits; others place the wishes of the surviving family above those of the patients; still others consider a patient who no longer desires to live incompetent to make treatment decisions or to refuse treatment.

Some examples may help to illustrate these points. Elisabeth Kubler-Ross relates the following story in her book *On Death and Dying*. In a patient's words:

> Nobody knows how long this can still last. I surely have always held on to hope, but this is the lowest I have ever been. The doctors have not revealed anything to me. They have not told me what they have found during my operation...Why in the world can't they talk with me? Why can't they tell you before they do a certain procedure? Why don't they let you go to the bathroom before they take you out of the room like a thing, not a person?

A terminally ill college professor was so upset about the way he was treated while being diagnosed that he wrote an article about it. He

---

*This seems to have changed over the past decade—at least physicians answer survey questions to the effect that they do now inform patients of their prognoses.

relates his feelings after having been given a battery of tests by a neurologist and three medical students:

> I got a reinforcement of the sense of not only am I a patient who is supposed to behave in a certain way, but I'm almost an object to demonstrate to people that I'm not really people any more, I'm something else. I'm a body that has some very interesting characteristics about it...I began to feel not only the fear of this unknown, dread thing that I have, that nobody knows anything about—and if they know they're not going to tell me—but an anger and a resentment of "Goddamn it, I'm a human being and I want to be treated like one!" And feeling that if I expressed anger, I could be retaliated against because I'm in a very vulnerable position.

Another case is related by a physician who observed it. The patient was a 68-year-old physician who was dying of inoperable cancer of the stomach. First his stomach was removed. Ten days later he collapsed with a pulmonary embolism that was surgically removed. He thanked the doctors for their efforts, but asked that if he had a further cardiovascular collapse that no steps be taken to prolong his life, because the pain of his cancer was now more than he could endure, and wrote a note to this effect in his own record. Nevertheless, two weeks later he suffered a series of five heart attacks, and was revived after each. Following three more weeks of pain, vomiting, and general convulsions, he died as the staff was making preparation for further "heroic" treatments. Throughout, medication did little to ease his pain.*

The Teaching Hospital

Since some of the illustrations in this essay highlight the horrible and ignore the routine, they do not reach some of the more fundamental problems in hospitals. The teaching problem, for example, has only been alluded to. In teaching hospitals the main mission is often viewed as education rather than patient care. Doctors argue both that medical students and interns need clinical experience to become good doctors, and that patients do not like to be treated or examined by students, or operated on by interns or residents. One result is that often

*Developing enforceable rights for the dying has been one of the major movements in the past decade. See "Death, Dying and Refusing Medical Treatment," for a detailed chronology and discussion of the issue at p. 259 *et seq.*

medical students are introduced as "doctor" when they are not yet MDs. A dental student has related, for example, how while in training at a major teaching hospital he was invited into an examining room by a physician and, along with three medical students, introduced to a sixteen-year-old girl as a "young doctor." Each of the four then performed a pelvic examination on the mortified and frightened patient. Although some physicians might attempt to defend this unjustifiable deception on the part of the medical students, there is absolutely no argument to make in favor of the dental student.

Another abuse is perpetrated by surgeons who assure their patients that they will be performing the surgery, when in fact a resident performs the operation and the surgeon observes and instructs. When the surgeon is not in the operating room at all, this deceptive practice is termed "ghost surgery."

Not only does this method of "teaching" involve the direct deception of the patient, it also teaches medical students that it is permissible to lie if you have a "good reason." Some operations are performed not because they are necessary, but because the intern or resident hasn't done enough of the procedures to become proficient. When a student-investigator questioned a resident as to why a particular patient was having a hysterectomy instead of a tubal ligation, for example, he was told: "We like to do a hysterectomy, it's more of a challenge...you know a well-trained chimpanzee can do a tubal ligation...and it's good experience for the junior resident...good training."

Humanizing the Hospital

From the variety and extent of the potential infringements on human rights in the hospital it should be apparent that there are no simple solutions. Also, unlike most areas in which civil liberties attorneys work, litigation probably is of relatively little significance. Most incidents become irrevocable before a hearing can be held. Furthermore, most of the law that has developed in the field has been based on malpractice litigation. Although there are important principles to be established regarding standards of care and the elements of informed consent, most of the problems in hospitals demand either new legislation or the promulgation of regulations under existing legislative authority. Even with these, however, the actions of individual patients may be of greater importance. The women's movement, for example, has already demonstrated that doctors will change their practices if their patients insist upon change.

A Patients Bill of Rights

In late 1972 the American Hospital Association issued a twelve-point Patients Bill of Rights and encouraged its 7000 member hospitals to adopt it or a similar declaration. As one could probably guess from the source, the document's provisions were a vague restatement of the law involving such concepts as informed consent and the right to refuse treatment. One commentator likened it to the fox telling the chickens what their rights were. Johnny Carson parodied the AHA effort by proposing that the following items be added to it: "A patient who is in a coma has a right not to be used as a doorjamb; no patient who has been given an autopsy may be denied the right to seek further medical consultation; and no matter what the extenuating circumstances, no patient can be forced to take a sponge bath with Janitor-in-a-Drum."

In Minnesota, a bill of rights similar to the AHA model has been enacted into law, and all health care institutions are required to post it in conspicuous places in their facilities. This trend toward publishing rights is important because it not only reminds people that they have rights, it also encourages them to assert them and to make further demands. To be really significant, however, such bills should deal with the fundamental problems that patients encounter in trying to retain self-determination and privacy in health care facilities. The following model bill of rights contains a minimal listing of the rights that should be accorded all patients both as a matter of policy by hospitals and as law by state legislatures.

The term "rights" is used in three senses: (1) rights that a citizen clearly or probably can claim as a matter of law under the Constitution, existing statutes, or judicial doctrines; (2) rights that a person probably can claim as judicially enforceable because of his or her relationship with another party, such as a doctor or hospital administrator; and (3) rights that a growing body of people believe should be recognized as the moral rights of individuals and the obligations of authorities, even though the courts would probably not yet recognize them as such.

Though some would like to see the emphasis placed on enforceable legal rights only, at this stage in the development of patient's rights such a limitation would be both conceptually and strategically unwise; humanizing the hospital will require a movement that joins the legal and moral aspects of the cause into one campaign, and buttresses the arguments from legal precedent and logic with the spirit of the moral cause. This was the manner of the civil rights and women's movements, and it would be useful for patients as well.

Where the phrase "legal right" is used in the model bill, the right is one well recognized by case law or statute. The term "right" refers to one that probably would be recognized if the case were brought to court, and "we recognize the right" refers to a statement of what "ought to be." Once these rights are recognized, some mechanism for hearing complaints and enforcing rules must also, of course, be established. If enacted into law in the future, all of these rights would then become legal rights.

The model bill is set out as it would apply to a patient's chronological relationship with the hospital: sections 1–4 for a person not hospitalized but a potential patient; 5 for emergency admission; 6–15 for in-patients; 16–22 for discharge and after discharge; and 23 relating back to all 22 rights.

A Model Patients Bill of Rights[1]

Preamble

As you enter this health care facility, it is our duty to remind you that your health care is a cooperative effort between you as a patient and the doctors and hospital staff. During your stay a patients rights advocate will be available to you. The duty of the advocate is to assist you in all the decisions you must make and in all situations in which your health and welfare are at stake. The advocate's first responsibility is to help you understand the role of all who will be working with you, and to help you understand what your rights as a patient are. Your advocate can be reached at any time of the day by dialing_____. The following is a list of your rights as a patient. Your advocate's duty is to see to it that you are afforded these rights. You should call your advocate whenever you have any questions or concerns about any of these rights.

1. The patient has a legal right to informed participation in all decisions involving the patient's health care program.

2. We recognize the right of all potential patients to know what research and experimental protocols are being used in our facility and what alternatives are available in the community.

3. The patient has a legal right to privacy regarding the source of payment for treatment and care. This right includes access to the highest degree of care without regard to the source of payment.

4. We recognize the right of a potential patient to complete and accurate information concerning medical care and procedures.

5. The patient has a legal right to prompt attention, especially in an emergency situation.

6. The patient has a legal right to a clear, concise explanation in layperson's terms of all proposed procedures, including the possibilities of any risk of mortality or serious side effects, problems related to recuperation, and probability of success, and will not be subjected to any procedure without the patient's voluntary, competent and understanding consent. The specifics of such consent shall be set out in a written consent form, signed by the patient.

7. The patient has a legal right to a clear, complete, and accurate evaluation of the patient's condition and prognosis without treatment before being asked to consent to any test or procedure.

8. We recognize the right of the patient to know the identity and professional status of all those providing service. All personnel have been instructed to introduce themselves, state their status, and explain their role in the health care of the patient. Part of this right is the right of the patient to know the identity of the physician responsible for the patient's care.

9. We recognize the right of any patient who does not speak English to have access to an interpreter.

10. The patient has a right to all the information contained in the patient's medical record while in the health care facility, and to examine the record on request.

11. We recognize the right of a patient to discuss the patient's condition with a consultant specialist, at the patient's request and expense.

12. The patient has a legal right not to have any test or procedure, designed for educational purposes rather than the patient's direct personal benefit, performed on the patient.

13. The patient has a legal right to refuse any particular drug, test, procedure, or treatment.

14. The patient has a legal right to privacy of both person and information with respect to the hospital staff, other doctors, residents, interns and medical students, researchers, nurses, other hospital personnel, and other patients.

15. We recognize the patient's right of access to people outside the health care facility by means of visitors and the telephone. Parents may stay with their children and relatives with terminally ill patients 24 hours a day.

16. The patient has a legal right to leave the health care facility regardless of the patient's physical condition or financial status, although the patient may be requested to sign a release stating that the patient is leaving against the medical judgment of patient's doctor or the hospital.

17. The patient has a right not to be transferred to another facility unless the patient has received a complete explanation of the desirability and need for the transfer, the other facility has accepted the patient for transfer, and the patient has agreed to transfer. If the patient does not agree to transfer, the patient has the right to a consultant's opinion of the desirability of transfer.

18. A patient has a right to be notified of impending discharge at least one day before it is accomplished, to insist on a consultation by an expert on the desirability of discharge, and to have a person of the patient's choice notified in advance.

19. The patient has a right, regardless of the source of payment, to examine and receive an itemized and detailed explanation of the total bill for services rendered in the facility.

20. The patient has a right to competent counseling from the hospital staff to help in obtaining financial assistance from

public or private sources to meet the expense of services received in the institution.

21. The patient has a right to timely prior notice of the termination of eligibility for reimbursement by any third-party payer for the expense of hospital care.

22. At the termination of the patient's stay at the health care facility we recognize the right of a patient to a complete copy of the information contained in the patient's medical record.

23. We recognize the right of all patients to have 24-hour-a-day access to a patients rights advocate who may act on behalf of the patient to assert or protect the rights set out in this document.

As is apparent from the preamble of this document, it is my view that a statement of rights alone is insufficient. What is needed in addition is someone whom I term an advocate, to assist patients in asserting their rights. As indicated previously, this advocate is necessary because a sick person's first concern is to regain health, and in pursuit of health patients are willing to give up rights that they otherwise would vigorously assert.

The Patients Rights Advocate[2]

Many health care facilities view the problem of patients rights and consumer demands as one of public relations. Some have therefore assigned members of their staffs to act as some form of "patient representative." In a 1973 survey, for example, 462 of the 1000 hospitals with more than 200 beds that responded said that they had at least one employee whose primary job is "to serve as management's direct representative to patients." Their duties were almost always limited to nonmedical matters, and they would accordingly be more properly denoted "management representatives." One hospital director, for example, described the type of problems he thinks are most important to his patients, and those his representatives are designed to meet: "...the burned-out lightbulb, the dirt in the corner, the afternoon nourishment missed while at X-ray, the allergy-free pillow, the patient's car left in front of the emergency unit, the airline tickets needed on discharge, and so forth." His representatives are specifically forbidden to deal with nursing or medical complaints patients might have. The philosophy behind such a system is that what the patient really cares about are basic creature comforts, not the quality of health care, the medical alternatives, and the likely outcome. Though such a stance is in many ways ludicrous on its face, the rationale often given is that such "housekeeping" problems are in fact what patients complain about most. There *are* real "housekeeping" problems, requiring not a patient's representative, but a larger and more efficient hospital staff to solve them properly. But patients don't discuss or complain about the type of care they are getting because this is frowned on and almost always meets with the response: "Ask your doctor." The average patient is lucky to see his or her doctor five minutes a day, and the doctor is likely to discourage any lengthy discussion about the patient's condition. Moreover, since most hospital bills are paid by third-party private or government insurance, patients may feel that they have nothing to bargain with because even if they are dissatisfied with the quality of care they receive, they know the bill will still be paid, and they know the hospital knows this too. Under these circumstances, to equate patients rights with adequate housekeeping in the hospital is hypocritical at best, and does nothing to help the patient maintain either self-determination or privacy.

Refocusing patients' concerns on health care, and affording patients the opportunity to exercise the rights outlined in the bill of rights, requires an advocate whose main concern is medical care and treatment and whose powers are the same as the legal powers of the patient. The advocate should be able to exercise at least the following unrestricted powers on behalf of individual patients and at their direction:

- complete access to all medical records
- the ability to call in qualified consultants
- ex-officio participation in all hospital committees responsible for monitoring the quality of health care
- the power to lodge complaints directly with the hospital's director and executive committee
- immediate access to all chiefs of service
- access to all patient support services
- the ability to delay discharges

There is no single set of qualifications for the advocate. The advocate must deal with people of varying degrees of education and ability to communicate, and of different ethnic, religious, and social backgrounds. Some knowledge of law, medicine, and psychology would appear essential, but the extent to which formal education would prepare a person for this position seems minimal. Knowledge of the community served and the language of its population will probably prove the most essential attributes of a successful advocate.

Ideally this person should be financially independent of the institution in which she or he will function. Financing could come, for example, from a state's department of consumer affairs or attorney general's office, or the US Department of Health and Human Services, which could establish a national patient advocacy office. In the health maintenance organization (HMO) framework, funding could be included in the yearly premiums, and advocates hired by consumer - dominated boards of directors. In this regard, however, experimentation is probably in order. It already seems evident, however, that patients' representatives paid by the hospital in which they work would have divided loyalties and could not take too strong a stand in favor of patients rights. Only patients should have the power to fire the advocate.

As far as I am aware there are no advocate systems currently in existence that follow this model. Initiatives from hospitals are unlikely, pressure must be applied to governmental units to sponsor such programs in the hospitals under their jurisdictions. The goal is not to so disrupt the hospital routine as to make effective care impossible, but to improve patient care by making the recognition and exercise of patients rights the rule rather than the exception.

Legislation

Access to Medical Records. One way to have patients rights legally recognized is to pass a law. Even if enacting a strong patients bill of rights is not feasible, work can progress on a piecemeal basis. Access to medical records is one example. The only legal method by which hospital patients in most states currently may get to view and copy their medical records is by filing a medical malpractice suit. Indeed, the Malpractice Commission of HEW concluded that routine denial of access to records is a primary reason for the instigation of such suits;

i.e., patients weren't sure if malpractice had occurred until they could view the records, but weren't allowed to view them until after they sued.

Massachusetts has had a law since 1945 that gives patients in and out of the hospital a legal right to see their records. A 1974 survey by the Boston University Center for Law and Health Sciences disclosed, however, that most major Boston-area hospitals routinely violated this statute by refusing to allow patients to see their records without their doctor's permission. Since the publication of that survey a number of hospitals have changed their policies to conform with the law, and the number of requests from patients for their medical records has increased.

The passage of laws will not perforce open up hospital records, but will make it extremely difficult for a doctor or hospital to deny access to patients as a routine matter and will encourage patients to make such demands. If enough patients put direct pressure on hospitals to see their records, hospitals will respond and policies will change.*

*Malpractice Litigation.*³ Some physicians and medical societies have begun to advocate the abolition of the malpractice suit, proposing that it be replaced by either binding arbitration or a no-fault insurance system. While binding arbitration is useful in the context of health maintenance organizations, in which payment is made and contracts are signed before the person is ill or injured, it is completely out of place in the more common office and hospital situations. No-fault, on the other hand, is an appropriate solution only for indemnifying the victims of medical experimentation, since most of the harm done, though foreseeable, is unavoidable. As it happens, most malpractice claims are based on the failure of the doctor or hospital to conform to good and accepted medical practice or to obtain informed consent. It is therefore ironic that while they tout no-fault as a replacement for the litigation that arises in the course of ordinary medical practice, most doctors and hospitals are unwilling to provide for the indemnification of victims of experimentation.

The malpractice suit is currently the only way patients can successfully challenge the actions of doctors and hospitals and obtain some compensation for substandard care. As noted previously, it is also the only way in many states that patients can see their medical records.

*The modern trend is to enact such legislation. New York's statute, for example, took effect in 1987.

The major problems with the system from the consumer's point of view are that it is time-consuming and that few lawyers will take small claims. These arguments make both binding arbitration and no-fault attractive *supplements* as long as strict upper limits are placed on the applicability.*

Computerization of Medical Records. Another development that patients should view with some alarm is the computerization of medical records. While there is some value for a doctor anywhere in the country or world to be able to obtain your medical record upon making the proper query to a computer, the implications for invasion of privacy are enormous.

Erroneous medical records circulated without the patient's consent can have a devastating effect on his or her ability to get a job or life insurance; they also affect the quality of the patient's future health care. It has been estimated that at least 10% of computerized medical records contain errors and that it would be financially infeasible to reduce this to below 3%. All other arguments for access to one's records aside, in the face of these figures it is imperative that patients be able to inspect and approve the entries in their medical records. When errors occur, patients should also have the right to have them corrected. Unfortunately, as important as they are, medical records generally are exempted from even the limited consumer credit information access statutes currently on the books.[4]

The Living Will. The increasing use of drastic "heroic" measures to prolong the life of the terminally ill has generated increasing public awareness of the problems of dying. A 1973 Harris poll found that 62% of Americans favored allowing the terminally ill patient to direct the doctor to "let him die rather than extend his life when no cure is in sight," and only 28% thought this practice was wrong. Legally, of course, physicians are free to follow the wishes of a patient in this matter. The problem is that many doctors continue to aggressively treat terminally ill patients because of their own beliefs and convictions or those of the patient's family.

One proposed solution is the "living will" (so-called because this "will" takes effect while the patient is still alive), which instructs the physician to take no steps to prolong the dying process. A state statute

*See "Medical Malpractice" at p. 183, for a view of the subject from a mid-1980s perspective.

is thought to be necessary by some to make a living will legally binding on doctors and hospitals, since otherwise, it is argued, such a document would violate public policy against suicide or euthanasia. I do not agree: the patient has a legal right to refuse treatment, and this clearly includes any treatment proposed by a physician who views restoration of health as an impossibility.[5]

Much more difficult questions are raised at the beginning of life, when it is proposed that treatment of severely handicapped or retarded infants be suspended and they be "permitted" to die. Such action on the part of physicians in a public facility can be a violation of the equal protection guarantee of the US Constitution. At all facilities, both doctors and consenting parents may be committing criminal acts, and if so, they should be prosecuted. The medical care system and the secrecy of the hospital must never be used to circumvent the criminal law to the detriment of a class of individuals unable to protect themselves.[*]

Conclusion

Although I have argued that hospital patients are in many ways treated like prisoners and mental patients, I do not propose either abolishing hospitals or fundamentally altering their role in the delivery of health care. What I do propose is making them more responsive to human needs and human rights—something I believe can be accomplished without a decrease in the quality of patient care, and perhaps even with a significant increase in the quality of care. While care is unlikely to be seriously affected by the patients rights movement, the manner in which it is delivered and the role of the patient in its delivery will be altered drastically. The system of keeping relevant information from the patient will be eliminated. Open discussion will replace guarded comments and outright deception. Teaching and experimentation will be acknowledged (and undoubtedly accepted) as such and not presented to the patient as treatment. Patients who wish to die in the hospital, without having their dying prolonged, will be allowed to do so.

None of these changes will take place overnight, but they need not take long in an institution as young as the modern hospital. Development of a bill of patients rights backed by an effective patient advocate system would significantly accelerate the process of change. Even without these, however, all of us, since we have frequent contact with

[*]For a more complete discussion see the three essays on the "Baby Doe" regulations at pp. 126–142.

various parts of the health care system, can begin by encouraging change from outside.

All those seriously interested in making the health care system in general and hospitals in particular more responsive to human needs should routinely do a number of things. First, whenever visiting a doctor or clinic, alone or with a friend, insist that all diagnoses be explained in language you can understand, that the proposed treatment be fully explained, including possible side effects, length, restrictions on activity, and what alternatives exist and their probable outcome. Second, routinely ask to see your medical records, to educate yourself as to what they contain, to correct errors, and to get doctors and hospitals used to the idea of being accountable to consumer-patients. Third, support movements to increase consumer information concerning the availability, cost, and quality of health care services in the community. This includes pressing for physician directories that list specialties, experience, office hours, and prices, and listings of the policies of local hospitals regarding such things as access to medical records, visiting restrictions, natural childbirth, and human experimentation. If patients get used to asserting their rights in the doctor's office and on an outpatient basis, it will be far less difficult to continue to assert them in a hostile hospital environment.

Though the concept of patients rights is novel, it is a mistake to view it as unimportant simply because the hospital stay usually is brief. In a society in which individual rights are compromised on all sides, the patients rights movement provides all of us an opportunity to assert those civil rights we espouse so eloquently in all other contexts.

The Emerging Stowaway
Patients Rights in the 1980s

At one point in Edgar Allan Poe's *Narrative of Arthur Gordon Pym of Nantucket*, Pym, who has stowed away in the hold of a whaling vessel, believes he has been abandoned and that the hold will be his tomb. He expressed sensations of "extreme horror and dismay," and "the most gloomy imaginings, in which the dreadful deaths of thirst, famine, suffocation, and premature internment, crowded in as the prominent disasters to be encountered."

It is probably uncommon for hospitalized patients to feel as gloomy as Pym. Nevertheless, installed in a strange institution, separated from friends and family, forced to wear a degrading costume, confined to a bed, and attended by a variety of strangers who may or may not keep the patient informed of what they are doing, the average patient is intimidated and disoriented. Such an environment encourages dependence and discourages the assertion of individual rights.

The primary argument against patients rights is that patients have "needs" and that defining these needs in terms of rights leads to the creation of an unhealthy adversary relationship. It is not, however, the creation of rights, but the disregard of them, that produces adversaries. When provider and patient work together in an atmosphere of mutual trust and understanding, the articulation of rights can only enhance their relationship.

Many issues, however, cannot be resolved entirely within the provider–patient relationship. Providers not only have formal relationships with their patients, but also have relationships with other providers, health care institutions, and numerous governmental agencies. A provider's relationship with these institutions and individuals is often a very complex one, and providers often find themselves con-

fused and therefore submissive in cases where they do not understand
their own rights or those of their patients.

Rights in Health Care

In most instances, both the health care provider and the patient will
be better off if the status of the law regarding both patient and provider
rights is understood, and the means of change or challenge well delin-
eated. I would go even further. An understanding of the law can be
as important to the proper care of patients as an understanding of
emergency medical procedures or proper drug dosages. But how are
rights to be understood, and how does a person know that he or she has
a "right" to something?

There is a formidable amount of literature on rights in the archives
of philosophy and jurisprudence. Rather than review it, let me note
briefly the thoughts of two relatively recent entrants who have written with great insight. The first is John Rawls. In expounding his
*Theory of Justice,*⁶ he imagines that a group of men and women come
together to form a social contract. These individuals all have ordinary
tastes, talents, ambitions, and convictions, but each is temporarily
unaware of his own personality and best interests and must agree to
the terms of the contract before his awareness of his own identity is
restored. The theory postulates that under such circumstances all will
agree to two principles: (1) each person shall enjoy the most extensive
liberty, compatible with a like liberty for others; and (2) inequalities
in wealth and power should exist only where they work to the benefit
of the worst-off members of society. One could develop an entire
system of patients rights that rests on these premises. Such a
document would be strongly pro-patient since this group is currently
the one that generally lacks rights and is always the group that will be
viewed as "worst-off" in the health care setting.

A second approach is suggested by the writings of Ronald Dworkin
in his book of essays *Taking Rights Seriously.*⁷ Dworkin notes the
great confusion in "rights language," generally created by attributing
to it different meanings in different contexts: "In most cases when we
say that someone has a right to something, we imply that it would be
wrong to interfere with his doing it, or at least that some special
grounds are needed for justifying an interference." An example is the
right to spend one's money, e.g., gambling it away, is the "right" thing
to do, or that there is nothing "wrong" with it. When we speak of pa-

tient's rights, this distinction may be critical to understanding what we are talking about. For example, a woman may have a legal right to have an abortion, but such a decision may still be considered "wrong" by her.

Dworkin argues further that there are some rights that can be said to be fundamental in the sense that the government is bound to recognize and protect them. Such rights, which we often denote as "legal rights," and less frequently as "constitutional rights," are generally spelled out in statutes and court decisions. By respecting such rights, the government guarantees to the weakest members of the society that they will not be trampled on by the strongest. In Dworkin's words:

> The bulk of the law—that part which defines and implements societal, economic, and foreign policy—cannot be neutral. It must state, in its greatest part, the majority's view of the common good. The institution of rights is therefore crucial, because it represents the majority's promise to the minorities that their dignity and equality will be respected. When the divisions among the groups are most violent, then this gesture, if law is to work, must be most sincere...taking individual rights seriously is the one feature that distinguishes law from ordered brutality.

Without going too far afield, one can apply Dworkin's notion directly to health care and note that rights can form a useful means of guaranteeing to defenseless patients that they will be treated with human dignity and respect. While the health care provider often has the power to deny certain rights almost at will, he or she does this only at the peril of jeopardizing the integrity of the health care system itself.

It must strike most as ironic that the first major health care organization to put forward a patients bill of rights was the American Hospital Association, an organization composed primarily of hospital administrators. One would not expect landlords to pen a bill of rights for tenants, police for suspects, or wardens for prisoners. Nor would one reasonably expect that the hospital administrator's view on rights for patients would be the same as either the patient's or society's. Nevertheless, physicians and nurses should be ashamed that the administrators were out in front on this issue. Even though it leaves much to be desired in terms of completeness, specificity, and enforceability the AHA Bill has tremendous symbolic value in legitimizing the notion of rights in the health care institution. On the other hand, fewer than half of all AHA-member hospitals have formally adopted even this bill, and the symbolic victory of the 1970s is currently under attack.

The Attack on Patients Rights

Physicians, who perhaps value their own professional autonomy more than any other group, nevertheless devalue it for their own patients. Instead, paternalism is the norm with the majority of physicians who believe that the health and continued life of their patients is much more important than their patients' right to self-determination. This belief system not only leads to conflicts with individual patients about their own care, but also to a general view that patients rights are a luxury item in medicine rather than a necessity.

A few examples illustrate the point. Two particular rights of patients have recently come under attack in the medical literature: access to medical records and informed consent. In an attack on "record reading," four psychiatrists at a major Boston teaching hospital, interviewed 11 out of 2500 patients at the hospital who, in a one year period, asked to see their medical records.[8] It is doubtful that anything of general importance about a patient's reactions to reading medical charts can be learned from an uncontrolled, nonblind, clinically impressionistic study of those few individuals who, for whatever reason, buck a system that routinely fails to inform them of their right of access to their records. Nonetheless, the authors' conclusion that such patients have a variety of personality defects, usually manifesting themselves in mistrust of and hostility toward the hospital staff, should not be permitted to go uncontested. In a setting where trusting patients are not routinely told of their right to access, it seems reasonable to assume that only the least trusting or most angry will ask to see their records. To locate the source of mistrust in the patient's personality or in the stress of illness and hospitalization is to forget, as Donald Lipsett perceptively suggests, that "the doctor–patient relationship cannot be understood simply in terms of the patient's side of the equation."[9] The authors thus fall into what Robert Burt of Yale Law School has referred to as "the conceptual trap of attempting to transform two-party relationships, in which mutual self-delineations are inherently confused and intertwined, by conceptually obliterating one party..."[10] Thus, it would seem that the ten women who asked to read their charts "to confirm the belief that the staff harbored negative personal attitudes toward them..." were correct in their belief; the psychiatrists labeled them "of the hysterical type with demanding, histrionic behavior and emotional over-involvement with the staff."

The authors also seem unaware of the wide variety of settings in which patients have *benefited* from routine record access; and incorrectly assert that there were no strikingly beneficial effects in the two studies they do cite. In the first study, for example, two patients expressed their unfounded fear that they had cancer only after their records were reviewed with them, and one pregnant patient noted an incorrect Rh typing that permitted Rho-Gam to be administered at the time of delivery.[11] In the other study they cite, 50 percent of the patients made some factual correction in their records.[12]

In short, the study seems to have been done and published for the primary purpose of proving that the right to access one's medical record is unimportant since it is only exercised by "mentally disturbed" people who are not improved by reading their charts. The study fails to prove this, and even if it succeeded, I would still be unwilling to deprive the other 2489 patients at that hospital of their right to access in the future.

If we believe in individual freedom and the concept of self-determination, we must give all citizens the right to make their *own* decisions and to have access to the same information that is widely available to those making decisions about them. It is as irrelevant in this connection that 2489 patients at one hospital did not ask to see their records as it is that more than 200 million Americans never have had to exercise their right to remain silent when arrested. Rights serve us all, whether we exercise them or not.

The attack on informed consent, which many physicians have long considered a "legal fiction,"[13] recently surfaced in reaction to a study on informed consent based on patients' memories which concluded that informed consent was not important. The methodology involved interviewing 200 consecutive cancer patients within 24 hours after they had signed consent forms for chemotherapy, surgery, or radiation therapy.[14] Upon questioning, most could not recall the procedure consented to, its major risks, or the alternatives to it. From this the authors concluded that the process is not working. While this may seem a reasonable conclusion (although an alternative one is simply that patients have poor recall), it turns out that the authors assumed their major premise. Approximately two-thirds of their sample group (66 percent) opted for radiation therapy. That group signed a consent form that said "the procedure, its risks and benefits and alternatives have been explained to me." Maybe they were, but maybe they weren't. The authors did not know, so their entire study was based on a premise that was unsubstantiated. Such a poorly designed study, it seems to me, could only be published if the doctors agreed so strongly with the conclusion that they did not even review the methodology.

A more interesting part of the study asked patients some general questions about informed consent. The first was "What are consent forms for?" Approximately 80 percent responded: "To protect the physician's rights." The authors were upset at this response, but the patients, of course, were correct. That *is* the primary function of *forms*. If one wants these forms to protect the patient, three simple steps are necessary: (1) the forms must be complete; (2) they must be in lay language; and (3) the patients must be given a copy and time to think over the information it contains.[15] The reason none of this is usually done is clear: informed consent is not taken seriously in the hospital setting. It is, like record access, a luxury which is secondary to caring for the medical "needs" of the patient, and besides, it really does not matter anyway because patients cannot remember anything they've been told.

Other significant findings indicate the extent to which patients understand and appreciate the consent process: 80 percent thought the forms were necessary; 76 percent thought they contained just the right amount of information; 84 percent understood all or most of the information; 75 percent thought the explanations given were important; and 90 percent said they would try to remember the information contained on the forms. To me, this suggests that the patients surveyed understood and appreciated the informed consent process much better than the researchers did. While their data is certainly not flawless, one can conclude from it just the opposite of what the researchers did: for almost all patients, informed consent is very important.

Related to this general attack on rights is an attack on the patient population itself. The notion is that the major problems with the health care delivery system are not problems with providers, but with patients. We eat too much, smoke too much, don't exercise, take too many risks, etc., and so what do we expect when we get sick? Not only must the American health care enterprise deal with a bad class of patients, but now they want some say in what kind of care is provided! As Lewis Thomas has put it in a related vein: this is becoming "folk doctrine about disease. You become ill because of not living right. If you get cancer it is, somehow or other, your own fault. If you didn't cause it by smoking or drinking, or eating the wrong things, it came from allowing yourself to persist with the wrong kind of personality in the wrong environment."[16]

This attitude would be humorous if it were not so pervasive and did not affect patient care so profoundly. Martha Lear has given us some excellent and telling examples in her deeply moving book, *Heart-*

sounds, that chronicles the final four years of her physician–husband who goes through eight operations and eleven hospitalizations during that period. Together they identify the "it's your fault ploy" which means that no matter what goes wrong in the hospital setting, it is the fault of the patient, not the health care system:

> Why did the operation take so long?
> Because you lost so much blood.
> *Not*: Because the surgeon blew it.
>
> Why do you keep making these tests?
> Because you have a very stubborn infection.
> *Not:* Because I can't diagnose your case.
>
> Why did I get sick again?
> Because you were very weak.
> *Not:* Because I did not treat you competently the first time.[17]

Dr. Lear is constantly asking himself if he treated patients that way, and usually admits that he did. He suggests that every physician spend at least a week per year in a hospital bed: "That would change some things in a hurry."

An Agenda for the 80s

Since patients *do* have rights and *do* want to exercise them, and since the major attacks on the notion of patients rights have been based on sloppy studies and false premises, the patients rights movement is likely to gain momentum. Indeed, the 1970s can be most properly viewed as a decade in which the existence of patients rights became legitimized through basic education of the health care providers. The 1980s should be a decade in which the primary thrust is to work on ways to directly enhance the status of patients in the hospital as a means of humanizing the hospital environment so that patients can have a greater voice in how they are treated.

I suggest the following five point agenda for the '80s:

Patients Rights Agenda

1. No Routine Procedures
2. Open Access to Medical Records

3. Twenty-Four-Hour-a-Day Visitor Rights
4. Full Experience Disclosure
5. Effective Patient Advocate

1. No Routine Procedures. It is common for nurses and others to respond to the question, "Why is this being done?" with, "Don't worry, it's routine." This should not be an acceptable response. No procedure should *ever* be performed on a patient because it is routine; it should only be performed because it is *specifically* indicated for a patient. Thus routine admission tests, routine use of johnnies, routine use of wheelchairs for in-hospital transportation, routine use of sleeping pills, to name a few notable examples, should be abolished. Use of these procedures means patients are treated as fungible robots rather than as individual human beings. Moreover, these procedures are often demeaning and unnecessary.

2. Open Access to Medical Records. While currently provided for by federal law and many state statutes and regulations, open access to medical records by patients remains difficult, and a patient often asserts his right to see his record at the peril of being labeled "distrustful" or a "trouble-maker." The information in the hospital chart is about the patient and properly belongs to the patient. The patient must have access to it, both to enhance his own decision-making ability and to make it clear that the hospital is an "open" institution that is not trying to hide things from the patient. Surely if hospital personnel make decisions about the patient on the basis of information in the chart, the patient also deserves access to the information.

3. Twenty-Four-Hour-a-Day Visitor Rights. One of the most important ways both to humanize the hospital and to enhance patient autonomy is to assure the patient that at least one person of the patient's choice has unlimited access to the patient at any time of the day or night. This person should also be permitted to stay with the patient during any procedure (e.g., childbirth or induction of anesthesia) as long as the person does not interfere with the care of other patients.

4. Full Experience Disclosure. The most important gain of the past decade has been the almost universal acknowledgment of the need for the patient's informed consent. Nevertheless, some information that is material to the patient's decision is still withheld: the experience of the person doing the procedure.[18] Patients have a right to know if the person asking permission to draw blood, take blood gases, do a bone marrow aspiration, or do a spinal tap has ever performed the procedure before, and if so, what the person's complication rate is. This applies not only to medical students and student nurses, but also to board certified surgeons. We all do things for the first time, and not every patient wants to take such an active role in education.

5. *An Effective Patient Advocate.* While a patients bill of rights is necessary, it is not sufficient. Rights are not self-actualizing. Patients are sick and desire relief from pain and discomfort more than they demonstrate a desire to exercise their rights; they are also anxious, and may hold back complaints for fear of retaliation. It is critical that patients have access to a person whose job it is to work *for the patient* to help the patient exercise the rights outlined in the institution's bill of rights. This person should be a member of all major hospital committees that deal with patient care, have authority to obtain medical records for patients, call consultants, launch complaints directly with all members of the hospital, medical, nursing, and administrative staff, and be able to delay discharges. While there appear to be some successful "patient representatives" hired by hospitals, it is not accurate to give them this title since they must represent their employer, the hospital. It is likely that ultimately effective representation can only be obtained by someone who is hired by a consumer group or governmental agency outside the hospital in which the representative works.

Conclusion

We have begun the long journey toward humanizing the hospital and promoting patients' self-determination. But more specific measures are needed before patients will be assured that they can effectively exercise their rights in institutional settings.

Like Poe's Arthur Gordon Pym, the notion that patients have rights has survived days of darkness, isolation, and starvation. Patients rights are now generally accepted (although sporadically attacked) and it is up to patients and providers alike to see to it that these rights become a reality for every citizen.

Breast Cancer
The Treatment of Choice

The title of Tom Wolfe's best seller, *The Right Stuff,* refers to those qualities—manliness, fearlessness, survivability, and recklessness—that characterize successful test pilots in general, and the Mercury astronauts in particular. A similar list could be compiled to characterize successful surgeons. It would likely include decisiveness, confidence, dexterity, and dedication. The list would be unlikely to include open-mindedness or a concern to present patients with all the options available for particular disease conditions.

This "self-righteous silence" has recently brought down the wrath of the women's movement on surgeons who perform breast cancer operations without informing their patients about alternatives. And in 1979 Massachusetts became the first state to enact a law requiring physicians to provide "a patient suffering from any form of breast cancer" with "complete information on all alternative treatments which are medically viable." This essay examines the law's legislative history and suggests that it is an appropriate, although unusual, response to a continuing problem.

The Wages of Authoritarianism

The story was not an uncommon one. Brett Bridge (the name has been changed) was told by her physician that she probably had breast cancer. He insisted she check into the hospital immediately for a biopsy and possible radical mastectomy. She became confused and upset, and asked that the operation be postponed for a week while she thought it over. The surgeon refused her request, and also refused either to discuss alternatives with her or to refer her to a surgeon who would. Stronger than most patients in this situation, Ms. Bridge researched her case in a medical school library, and located a surgeon who discussed alternatives with her and eventually did a lumpectomy, a less radical procedure than a mastectomy. She vowed to make sure that no other woman in Massachusetts would have to suffer similarly.

She asked Massachusetts State Senator Carol Amick to introduce legislation to require all physicians licensed in Massachusetts to swear under penalties of perjury every two years (at the time of re-registration) that "he has fully advised each patient suffering from a form of breast cancer, of all alternatives available for treatment in addition to mastectomies." Failure to so swear would be grounds for license revocation. At a public hearing the proposal won the support of the Women's Legislative Caucus and various other women's rights groups. Nevertheless, the Health Care Committee objected to placing this "additional burden" on the Board of Registration in Medicine. In a compromise, that portion of the bill was scrapped, and the disclosure requirement alone was added to a proposed Patients Bill of Rights. The entire bill was signed into law on May 23, 1979, and became effective three months later.[19]

The Medical Society Reacts

Almost as soon as the legislature reconvened in fall of 1979, the Massachusetts Medical Society attempted to get the breast cancer provision repealed. The Society's President, a surgeon at Massachusetts General Hospital, testified before the Health Care Committee on September 24, 1979. He listed five arguments against the provision: it is unclear, it interferes with the doctor-patient relationship, it will increase the cost of cancer care, alternative methods are not equally effective, and the law cannot solve scientific disputes.

He argued first that the term "complete information" is "indefinable" and asked if a physician is required to read the "entire libraries of information" available on this topic to the patient. He found the term "medically viable" unacceptable, and claimed not to know if a treatment with an 85 percent five-year survival rate qualified as medically viable, or if no treatment at all qualified. The short answer is that the law generally permits the medical profession itself to define these terms, and if this really is impossible, more explicit regulations are the answer.

Interference with the doctor–patient relationship is a catchall phrase that always seems to surface when patients rights are discussed. And here he is simply wrong; he thought that patients *must* be informed, even if they don't want to be. This belief simply shows a misunderstanding of the rationale behind the doctrine of informed consent (we promote self-determination only if we allow patients to "opt out" of the informed consent mode). Likewise, the claim that some patients cannot recall after surgery everything they learned in a prior informed consent discussion is irrelevant. The proper question is whether patients were given adequate information at the time they made their decision for surgery. The third argument, that giving patients this information will add three pre-operative days to their hospital stay and "triple" the expense of consultants, is equally unpersuasive. There is no basis for either figure, but the argument begs the question: would this added information improve patient care?

Finally, the Medical Society President asserted both that the literature is inconclusive, and that it is "inappropriate to attempt to settle a scientific controversy by law or regulation." Here he is actually making a strong argument for full disclosure. No one asks a patient to settle a scientific dispute; only to decide which of a number of apparently equally effective treatments she wants.

His argument seems to be twofold: (1) women cannot be trusted to make this type of decision; and (2) informed consent should not be applicable to breast cancer. In his first assumption he continues a tradition of some physicians who have argued against package inserts for oral contraceptives and estrogens. For example, the late Alan Guttmacher testified in 1970 before Senator Gaylord Nelson's Committee when it was considering a package insert for the pill:

> My feeling is that when you attempt to instruct the American womanhood in this, which is a pure medical matter which I am afraid she has not the background to understand, you are creating in her simply a panic reaction without much intellectual background.[20]

On informed consent the Medical Society's arguments are almost all generic, and they should not be taken any more seriously because they are resurrected in this context. The only unique characteristic of breast cancer is that there are many available alternatives of apparently equal efficacy. But this argues *for* disclosure, not against it.

The Medical Evidence

While publicly pleading ignorance, the Medical Society's President cannot be completely unaware of the revolution in the treatment of breast cancer during the past two decades. Breast cancer strikes more than 100,000 American women annually. Since 1955 the survival curves for breast cancer patients have remained almost constant, and treatment modalities have accordingly changed significantly. Though in 1950 almost all victims were treated with radical mastectomy, by 1970 the figure was 51 percent and by 1976 this method was used in only 25 percent of newly diagnosed patients. Today there seems to be no justification for routinely performing radical mastectomies, and at least in Stage I and II of the disease, no justification for performing radicals at all. In the words of the authors of the most recent review article on the subject:

> The routine use of the classical radical mastectomy or of adjuvant radiotherapy, both standard practices a decade ago, should now be abandoned; newer approaches should be considered viable alternatives in the treatment of many patients.[21]

The authors go on to describe those "viable alternatives," which include "the use of primary irradiation without mastectomy after biopsy or lumpectomy..." They also strongly recommend that many specialists become involved with the patient immediately following diagnosis, saying:

> It is no longer acceptable for one specialist to treat a patient without early consultation with the other specialists involved. With so many treatments available, the possibilities for cure are real, and there is no justification for a nihilistic approach to the treatment of breast cancer.

All this must come as a vindication of the pioneering work of surgeons like George Crile who for years have openly encouraged women to question their surgeons, and have actively involved them in the decision-making process concerning alternatives to breast cancer. Crile has long argued that since there is no scientific evidence proving one treatment better than others, the patient should choose, since it is she who has to live with the consequences. Further, in his experience he has found it impossible to predict which women will choose mastectomies, and which less disfiguring surgery.[22]

Two on-going clinical trials sponsored by the National Cancer Institute support the proposition that alternative modes of breast cancer treatment are equally effective. One involves randomizing women between a simple mastectomy and a lumpectomy followed by radiation; and the other randomizes them among (1) breast removal; (2) removal of the cancer alone; and (3) removal of the cancer followed by radiation. In the words of NCI's study director, Allen Lichter:

> If we didn't believe strongly that a lesser operation plus radiation will prove as effective as total mastectomy, we wouldn't be doing this. It would not only be immoral and unethical, it would be illegal.[23]

The same adjectives—immoral, unethical, and illegal—can be applied to withholding information about alternatives from breast cancer victims. Yet very few physicians are telling their patients about this study and letting them choose, with informed consent, whether or not to participate. As of early January, 1980, for example, only twelve women had volunteered to let the computer pick their treatment in the NCI study. Some women who have been informed about the study have refused to be randomized, preferring to make their own treatment decisions, or to let their physicians make them; but the vast majority of eligible patients have not even been given the choice.

Legislating Medical Practice

Although it is probably a mistake to legislate in particular areas of medicine regarding informed consent (since the list is unending), this legislation has had the beneficial effect of bringing into public debate the views of surgeons concerning what it is appropriate to tell patients with breast cancer. Informed consent in this area seems to have been the exception rather than the rule. Regulations by the Massachusetts Board of Registration in Medicine might clarify the law's language for the medical community. But a better course is to amend the cancer provision to include *all* cancer patients and use more traditional informed consent language: "Patients diagnosed as having breast cancer must be informed of all alternative treatments, including surgical, radiological, chemotherapeutic, and combinations thereof, and the likely risks, survival rates, and problems of recuperation of each." Cancer is singled out for special treatment, because it is the disease most feared by the public, and the one physicians have traditionally failed to discuss openly with patients. The provision also underlines the legislature's intention that the cancer patient not be seen as an exception to the general applicability of the doctrine of informed consent.*

In his study of surgical residency training, *Forgive and Remember,*[24] Charles Bosk notes that the art of obtaining informed consent was never taught to residents, and that they viewed it as a housekeeping chore of relative insignificance. The problem seems endemic with surgeons, and may require continued legislative therapy until they themselves come to see candid communication with patients about alternative treatment modalities as part of "the right stuff."

*Since this essay was first published, more than a dozen states have enacted similar breast cancer informed consent laws. A 1981 survey of breast cancer treatment by the American College of Surgeons indicates that the laws directly affect treatment decisions. Figures for lumpectomies or wedge excisions of tumors under 1 cm were: National average, 4.8%; Mass. (2 year old law) 17.9%; California (one year old law) 10%; New York (no law) 2%. New York passed its statute in 1985. Dabice, R.L. and Cordes, S., "Informed Consent Heralds Change in Breast Treatment", *Medical News,* Nov. 11, 1985 at 1,4.

Beyond the
Good Samaritan
Should Doctors Be Required to
Provide Essential Service?

Traditionally in American medicine the doctor–patient relation-ship has been viewed as purely voluntary. The courts have generally upheld the tradition that a physician has no obligation to accept every-one who seeks his or her services, and the American Medical Asso-ciation has stated it as an ethical precept: "A physician may choose whom he will serve."

While it may have been well suited to a frontier society, the volun-tary concept has serious drawbacks today. Recognizing this problem, some courts have expanded the concept of "acceptance" of a patient, both in the doctor's office and in the hospital. Likewise, there is some recent recognition that physicians are more than freelance artists who can rightfully sell their services to whomever they please and with-hold them from whomever they please for any reason or no reason at all. These moves can be seen as part of a trend aimed at eroding the purely voluntary nature of the doctor–patient relationship.

This erosion began in the institutional setting. Although hospitals are still often permitted to refuse to admit most patients for elective procedures, it is now the general rule that hospitals with emergency facilities have a *legal* obligation to accept and care for any and all individuals who are suffering from *emergency conditions* and who present themselves at the emergency ward. Though courts have used varying rationales to support this conclusion including reasonable reliance by the patient,[25] state statutes and regulations,[26] and federal tax laws applying to "charitable" institutions,[27] the results have been the same: admission must be granted.

A more perplexing question has been the role of the individual physician confronted with an emergency outside the hospital, either in his office or while traveling in the community. It has generally been held that physicians, like other private citizens, have no legal obligation to come to the aid of their fellow citizens in distress, even life-threatening distress. In an effort to encourage physicians to render such emergency services (for example, on the highway), most states passed so-called "Good Samaritan" statutes during the 1960s, which provided that if a physician or nurse did not stop and render aid (and did not request a fee), they could not be sued for malpractice if they acted with "due care."

These statutes were passed more than a decade before the term "crisis" was being used to describe the malpractice situation, but on the assumption that fear of lawsuits was preventing physicians from attempting to save lives. However, the real reasons most physicians did not stop probably had little to do with lawsuits (since no successful lawsuits have ever been filed against a "good samaritan" physician or nurse), but involved either not being able to give emergency care, or not wanting to get involved.[28] The statutes, accordingly, have not solved the problem.

In early 1977 a Massachusetts ophthalmologist was accused of refusing to see a one-week-old infant with a serious eye infection because the mother was a Medicaid recipient. The mother and child had to be driven into Boston from Cape Cod for treatment. Though the accusation could not be proven at a hearing, it did cause the Massachusetts Board of Registration in Medicine to consider how it could require physicians in the state to render "essential" services to those in need of them.

As part of this debate, the Massachusetts Medical Society proposed that the physician's "ethical obligation" of treating all emergency patients be made a legal one. Subsequently, on October 20, 1977, the Massachusetts Board became the first licensing board in the country legally to require all physicians under its jurisdiction to render emergency services to the best of their ability. The Regulation reads in part:

> A licensee shall render medical services to a person experiencing a medical emergency. A medical emergency is a set of circumstances which immediately threatens a person's life or is likely to cause serious injury absent the provision of immediate professional assistance. A licensee shall assume that a person who is referred to him by another licensee for the purpose of securing medical services of an emergency nature is experiencing a medical emergency.

Likewise, the AMA's Judicial Council has recently recommended rephrasing the Ethical Canon on acceptance for treatment to read: "Physicians may choose whom they will serve *except in emergencies*" [emphasis supplied].

The problem of defining the term "essential" services remains, and approaches are still being debated in Massachusetts. A recent case from Virginia, though not directly on this point, illustrates how courts can broaden the notion of "acceptance" for treatment to bring individual nonemergency cases requiring treatment within the traditional doctor–patient framework.

A blind woman brought an action against a physician for "breach of his duty to treat." She alleged that she had called his office earlier and made an appointment to be seen for "treatment of a vaginal infection." She arrived at the defendant's office on a Saturday morning accompanied by her four-year-old son and her guide dog. She was told that she would not be seen unless the guide dog was removed from the waiting room. She insisted that the dog remain because she "was not informed of any steps which would be taken to assure the safety of the guide dog, its care, or availability to her after treatment."

As a result of her stance, the physician evicted her from his office, refused to treat her, and did not assist her in any way to find other medical attention. She alleged that she was humiliated in the presence of her son and other patients, and that her condition became aggravated during the next two days while she sought other medical care. The defendant alleged that even if these charges could be proven, he had a legal right to refuse to treat her under these circumstances. A lower court agreed with the physician, but the Supreme Court of Virginia reversed, and remanded the case for trial.

The court stated the general rule about physicians not having to accept everyone who applies for services, and expressed the view that merely having an appointment with a physician was itself insufficient to establish a doctor–patient relationship because "it connotes nothing more than that the defendant had agreed to see her." However, the allegation that the appointment was for the "treatment of a vaginal infection" *was* found to be sufficient to create a doctor– patient relationship. In the court's words:

> The unmistakable implication is that plaintiff had sought and defendant had granted an appointment at a designated time and place for the performance of a specific medical service, one within the defendant's professional competence, *viz.*, treatment of a particular ailment.[29]

Of secondary importance to the discussion here is the question of how, once established, a doctor–patient relationship can be lawfully terminated. In this case the court held that if termination was based on the plaintiff's insistence on keeping her guide dog in the waiting room it was probably unlawful because the state's "White Cane Act" guaranteed her right to bring her guide dog with her into any place "to which the public is invited." This statute was found to apply to physician's offices where physicians permit certain members of the public to make appointments to see them. While this case was pending, the act was amended to explicitly cover "medical and dental offices."

How would the court have dealt with the case if the woman had just shown up at the office without an appointment, but in need of treatment? One can only speculate, but it is likely that the physician would not have been found to have a *legal* obligation to treat the patient. Does the physician have an ethical obligation to treat such patients? There may be no "right" answer to this question, but the relevant issue seems to be whether or not the condition that the patient is suffering from is such that treatment for it is "essential" for the patient's well-being, and that the physician is capable of rendering the needed service.

It would be desirable to define a set of circumstances that is broader than "emergency" conditions, and much narrower than elective conditions, under which physicians will be required to treat patients.

For the purposes of provoking discussion, let me suggest that physicians have an ethical (and should have a legal) obligation to treat all patients who present themselves to them if such treatment would be deemed "essential" by a reasonably prudent medical practitioner. The term "essential" means any condition which, while not immediately threatening the patient's life, health, or mental well-being (such conditions are "emergencies" and must be treated immediately), is, if treatment is delayed, likely to require more extensive or more expensive treatment or is likely to cause substantial pain and discomfort to the patient. The requirement to treat such conditions applies only where the alternatives available in the community are such that, if treatment is not rendered by the medical practitioner to whom the patient has made application, treatment is not likely to be rendered in a timely manner, and to cases in which the physician is qualified to render the essential service.*

*This essay provoked no discussion, and no other state licensing authority even followed Massachusetts to the extent of requiring physicians to render *emergency* care.

Adam Smith in the Emergency Room

In his irreverent *House of God*,[30] Samuel Shem describes "gomers" (get out of my emergency room) as "human beings who have lost what goes into being human beings." They are elderly, demented patients, usually transferred from nursing homes, with multiple illnesses that medicine cannot cure. Law Number I of the *House of God* was: "Gomers don't die." The problem of medical residents in the 1970s, according to Shem, was to keep such patients out of the emergency room (and therefore out of the hospital) because they would make your life miserable and there was nothing you could do for them.

In the mid-1980s, the new cry is to keep the uninsured and the poor (the new gomers) out of the emergency rooms based on Economic Law Number I of the House of Adam Smith: "Poor people can't pay."

The New Context

This new and disturbing development would have been completely unacceptable as recently as a decade ago. Although we remain the only industrial nation not to have a system of national health insurance, we have always considered the responsibility of our hospitals, as the purveyors of a social good, to provide emergency care to those in need, regardless of ability to pay. As one court put it in 1973, "It would shock the public conscience if a person in need of medical emergency aid would be turned down at the door of a hospital having emergency service because that person could not at that moment assure payment for the service."[31] Current efforts to transform medical care from a social good to an economic good threaten to erode this

46

community ethic, and it is in this context that the right to emergency room treatment has become such an important symbolic issue.

Public policy currently places the emphasis squarely on cost-containment: quality of care and equity of access take distant second and third places. The poor and uninsured suffer. Specifically, thirty-five million people (15 percent of the population) lack health insurance of any kind today, an increase of ten million people since 1977. Although many are poor and unemployed, most are not. Gail R. Wilensky of Project HOPE has noted, "In fact, the most likely to be uninsured are young adults between the ages of 18 and 24; and a third of the uninsured are children under 18." Any policy of refusing emergency care or transferring emergency patients who are uninsured will thus fall hardest on those in our society least able to protect themselves: children and young adults.

The Reagan Administration had set out to put a price on individual lives that can be used in cost-benefit analyses of proposed health and safety regulations. Federal administrative agencies have been using $400,000 to $7 million for one human life. On the other hand, even the Administration recognized that the price we were willing to pay for the hijacked US airline passengers in Lebanon was probably at least an order of magnitude higher than what we would be willing to pay to save "statistical lives" by, for example, tightening up airport security.

Emergency room patients fall in the middle. They are "statistical lives" in the sense that we do not know who they will be. But they are not like the statistical lives that are lost in automobile crashes, industrial accidents, or as a result of exposure to toxic chemicals. These people, *before* they die, will come to a hospital emergency room seeking medical assistance. Turning them down requires the physician, nurse, or clerk to decide to risk the life of an identified human being. More than that, if we permit the market to rule, it requires that the physician risk the patient's life at a specified price—the cost of emergency care for the condition that brought the patient to the emergency ward. Permitting hospitals and physicians to make such cost/benefit analyses from the perspective of the hospital's budget should "shock the public's conscience," and be ethically and legally unacceptable.

Arnold Relman, editor of the *New England Journal of Medicine*, cogently notes that "steps necessary to ensure adequate emergency care of the indigent and uninsured are, unfortunately, at odds with the currently fashionable philosophy in Washington... that hospitals are basically businesses." He argues persuasively that hospitals should continue to be viewed as "community resources" with an obligation "to treat all emergency patients brought to their doors..." Hospitals are turning away or transferring more and more indigent and uninsured from their emergency rooms.[32]

This is the inevitable consequence of trying to make emergency care an economic good, subject to the market. As Uwe Reinhardt of Princeton has noted, "Unfortunately, as every freshman in economics quickly learns, the one feat the Invisible Hand usually cannot achieve is the distribution of commodities on a basis other than ability to pay." As with other basic community services such as police protection, the fire department, and water purification, emergency health care can be converted from a public good into an economic good only by excluding a large segment of the community from protection. What emergency treatment obligations do hospitals and their physicians have?

The Hospital's Duty

A recent case from Arizona indicates that the courts are not likely to back down from their view that public policy demands emergency treatment to be provided to *all* persons experiencing a medical emergency when they arrive at a hospital emergency ward.[33] The case involved Michael Thompson, a thirteen-year-old child who was pinned against a wall by an automobile that had fallen off a jack and rushed by ambulance from the scene directly to the Boswell Memorial Hospital in Sun City. He arrived at 8:22 PM and was examined and treated by the emergency room physician who found that Michael's left thigh was severely lacerated; there was no pulse in the leg; the left foot and toes were dusky, cool and clammy; and the bone was visible at the lower end of the laceration near the knee. The physician administered fluids, ordered blood, and called in an orthopedic surgeon.

The surgeon examined Michael, consulted by phone with a vascular surgeon, and determined that Michael's condition was "stabilized" and that he was "medically transferable." At 10:13 PM, Michael was placed in an ambulance and transferred to the County Hospital where his condition worsened. His condition later stabilized, and surgery was performed at about 1:00 AM. He survived, but has serious residual impairment of the left leg, caused by the delay in restoring blood flow, which had stopped when the femoral artery was transected.

The hospital stipulated that the surgery could have been performed at Boswell, and that the transfer was "for financial reasons." A Boswell administrator testified that emergency "charity" patients are transferred from Boswell to County whenever a physician, in his professional judgment, determines that "a transfer could occur." The emergency room physician did so determine, and a witness for the plaintiff testified that the physician told Michael's mother, "I have the shitty detail of telling you that Mike will be transferred to County..." His mother "begged" the doctor not to send her son there.

This case can be seen as a "premature discharge" case rather than an emergency refusal, because the patient was seen and some treatment begun. On either view, it highlights the issues involved in emergency room transfers. The primary question before the court was whether the hospital violated the law in transferring this child solely because he lacked the proper insurance. The court reaffirmed Arizona law that it is public policy that a general hospital may not deny emergency care to any patient without a cause.

In interpreting this policy, the court noted that the state director of health services had adopted regulations based in part on the emergency section of the standards of the Joint Commission on Accreditation of Hospitals (JCAH), which state: "No patient should arbitrarily be transferred if the hospital where he was initially seen has means for adequate care of his problem." The JCAH makes clear that services that are "available and medically indicated" should be provided "regardless of the source of payment."

Arizona has a statutory scheme that pays private hospitals for care rendered in an emergency. But this statute does not permit private hospitals to determine whether or not to render emergency services. In this regard, they have the same duty as public hospitals:

> ...as a matter of public policy, licensed hospitals in this state are required to accept and render care to all patients who present themselves in need of such care. The patient may not be transferred until all medically indicated emergency care has been completed... without consideration of economic circumstances.

The court concluded that the child was suffering an emergency condition, that emergency surgery was the indicated treatment, and accordingly that the hospital had a legal duty to provide Michael with that emergency surgery. All this is straightforward and consistent with almost all past cases on this subject. But where does this leave the physician in charge of the hospital's emergency room?

The Physician's Role

The physician's role is pivotal. First, it is the physician's duty to determine whether or not an emergency exists. If one does exist, both law and medical ethics require the physician to treat the patient if he or she can. But the definition of a medical emergency is sometimes unclear. The broadest definition is supplied by the American College of Emergency Physicians, whose view is that a patient has made an appropriate visit to an emergency department when "an unforeseen condition of a pathophysiological or psychological nature develops which a prudent lay person, possessing an average knowledge of health and medicine, would judge to require urgent and unscheduled medical attention most likely available, after consideration of possible alternatives, in a hospital emergency department." The group gives examples of such conditions, including relief of acute or severe pain, hemorrhage, or threat of hemorrhage, and obstetrical crises and labor. All are sufficient to require a physician's attention. Once the physician sees the patient, according to the American Hospital Association, "a true emergency is any condition clinically determined to require immediate medical care."

The real questions, it seems, are: What does "immediate" mean? And what reasons are sufficient to justify a decision to transfer a patient to another facility? All the physicians in the Arizona case agreed that the child was suffering from an emergency condition that required emergency surgery. They seem to have believed, nonetheless, that the child was "medically transferable." This belief seems to have been primarily founded on the fear that the hospital would not be properly paid for the surgery. It *is* proper for a physician exercising *medical* judgment to determine, according to good and accepted medical standards, whether or not a patient is experiencing a medical emergency, if immediate treatment is indicated, and if such treatment *can* reasonably be provided at the hospital.

It is *not* proper for a physician who determines that a medical emergency exists to decide to attempt to transfer a patient based solely or primarily on financial considerations. Physicians cannot and should not permit themselves to be used as financial hatchet men by profit-maximizing hospital managers. They should act as conscientious objectors to hospital policies that put patients at risk, and their professional associations should vigorously support physician actions consistent with good patient care.

But what if "immediate" treatment is not necessary, and the patient can be safely transferred without any foreseeable risk to his or her health or any decrease in the chance for recovery, and nonetheless the county hospital (or tertiary care center) refuses to take the patient? The physician can either arrange for the patient to stay at the hospital or discharge the patient. In this regard, another 1984 Arizona case held that "since cessation of hospital care may not be medically indicated despite the cessation of the emergent condition...the private hospital may not simply release a seriously ill, indigent patient to perish on the streets."[34] The hospital's obligation to provide care *after* the emergency condition is stabilized *continues* until the patient is properly transferred or is medically fit for discharge from the hospital, and the physician should ensure that the hospital meets this patient care obligation.

Three Suggestions

For-profit hospitals and financially strapped public and nonprofit hospitals will undoubtedly attempt to redefine their legal duty toward emergency patients and to influence their medical staffs to adjust their professional ethics to narrower definitions of emergencies and broader criteria justifying transfer. The public and physicians should resist these efforts vigorously. Whatever "minimum" we owe all members of our community regarding medical treatment, it must include emergency treatment. Otherwise, the transformation of medicine from a profession dedicated to the alleviation of illness and suffering to a business unconcerned with suffering, disability, or even death will be complete.

Three actions seem reasonable. First, professional associations should reaffirm the ethical requirement of their members to assist all those needing emergency medical care, and only permit transfer for *better* care. Second, states, through statutes and regulations, should define "emergency" broadly (rather that narrowly) and add criminal penalties for hospitals and physicians in emergency departments that refuse such services. Third, uninsured individuals should be encouraged to carry cards that set forth their state's law regarding emergency treatment, and contain a form for the emergency room physician to sign if he or she refuses to treat them in what they consider an emergency situation; that form should certify that no emergency condition exists and that transfer can be accomplished without risk and will provide superior services should transfer be ordered.

Such a card has been used by the Legal Services of Middle Tennessee for years, and even when it does not help procure necessary services, it identifies the physician who determined that emergency services were unnecessary (*see* below).

The only nonstatutory right to medical care US citizens have is the right to be treated in an emergency room for an emergency condition. Of course hospitals should be paid for their services; but the fact that society has not yet worked out a payment mechanism that is universally acceptable is insufficient justification for physicians and hospitals to alter radically their traditional ethic of caring. The emergency rule should remain: Treat first and worry about payment later.

BILLFOLD CARD

I, Dr. _____, have examined _____

and, in my professional opinion, this person can be denied medical care without worsening of their injury or illness.

I hereby authorize _____ Hospital to refuse care to this person because they lack the money or insurance to pay the bill.

Physician: _____ Witness: _____

We hereby attest that _____ refused to sign this statement.

Witness: _____ Witness: _____

TO WHOM IT MAY CONCERN:
Under Sections 53-5201 and 5202 of the Tennessee Code Annotated, all Tennessee hospitals (even private, profit making hospitals) which operate general medical and surgical services must furnish emergency hospital services to any person who has been injured or suffers from an acute illness, if that injury or illness is liable to cause severe injury or illness if left untreated. State law makes it a crime to deny emergency care under T.C.A. §53-5203. Aside from the fact that denial of necessary hospital care is a crime, a hospital can be sued for money damages if it fails to provide necessary care to a person because he cannot make a deposit and/or pay for treatment. The Tennessee Court of Appeals has previously awarded damages in such a situation in a malpractice case called **Methodist Hospital v. Ball**, 362 S.W. 2d 475 (1961).

Health Law Project of Legal Services of Middle Tennessee, Inc.
(800) 342-3317 or in Nashville call (615) 244-6610

Sex in the Delivery Room
Is the Nurse a Boy or a Girl?

No male nurse is appropriate to assist in labor and delivery, but all male physicians are acceptable. If this statement appeared on a true-false examination, most responders would answer "false." But one responder, with considerably more authority than the average test-taker, recently marked this statement true. Since Elsijan R. Roy is a district court judge in Arkansas, only a higher court is empowered to grade her performance on this question.[35]

Background of the Case

Gregory Backus is a qualified male registered nurse who graduated from the Baptist Medical Center and trained in the obstetrics and gynecology (OB-GYN) department as a student nurse. Shortly after graduation, he requested placement as a full-time RN in the labor and delivery section of the OB-GYN department. His request was refused on the basis that the hospital "did not employ male RNs in the OB-GYN positions because of the concern of our female patients for privacy and personal dignity which makes it impossible for a male employee to perform the duties of this position effectively."

Backus filed a discrimination charge with the Equal Employment Opportunity Commission (EEOC). The EEOC issued a "right-to-sue" letter, after which he filed suit against the hospital under Title VII of the Civil Rights Act of 1964, alleging that sex was "not a bona fide occupational qualification" for the position he sought. The Act requires the employer to demonstrate that the challenged policy is reasonably related to the essence of its business and that there is a factual basis for believing that it is impossible or impractical to deal with persons on an individual basis.

The hospital responded that its policy was based on a recognition and respect for the "privacy rights of its patients": obstetrics is a "unique section of the hospital" because "an obstetrical patient constantly has her genitalia exposed." The duties of an OB nurse were described as "sensitive" and "intimate," including checking the cervix for dilation, shaving the perineum, giving an enema, assisting in the expulsion of the enema, and sterilizing the vaginal area. In the recovery room, the nurse checks the patient for bleeding, gives uterine massages, and changes the perineal pads. OB nurses are "randomly assigned" to patients upon admission, the hospital explained, and are "strangers" to the patients.

The court agreed that random assignment of male nurses would violate the "constitutional right to privacy" of patients, and heavily emphasized the proposition that "strangers" of the opposite sex should not be permitted to view the patient's naked body. This conclusion was bolstered by the testimony of some physicians, nurses, and patients at the hospital that they would object to a male nurse in OB. One female physician testified that half her patients would object to a male nurse and that an even greater percentage of their husbands would object.

The court determined that it was not possible for the hospital to "schedule around" or rotate a male OB nurse because (1) if a patient objected to a male nurse, a female nurse would have to be switched from her patient to the objecting patient; and (2) the hospital policy that no member of the opposite sex examine a patient's genital area without a chaperon ("to curb any risk of a molestation charge") would necessitate assigning a female nurse to chaperon the male nurse and result in duplication of staff and added expense. The court characterized this as a "business necessity" sufficient to justify the hospital's policy.

Both the hospital's arguments and the court's analysis merit comment. Persons interested in promoting patients rights have long known that most hospital rules and policies exist primarily for the convenience of the hospital staff, not for the patient. Thus, when a hospital announces that an employment policy is based on patients rights, one should be immediately skeptical. In the early 1970s, for example, many hospitals argued that their policies prohibiting fathers from entering the delivery room were really to protect the privacy of other OB patients, and not for the convenience of the OB physicians and nurses who opposed the practice. Likewise, some hospitals now argue against the twenty-four-hour-a-day visitation rights on the basis that the patient should be protected from oversolicitous friends and relatives. "Routine procedures" are justified as protecting patients, when they are really meant to protect the hospital staff from a negligence charge. Wheelchairs, for example, are required until one gets to the hospital door—after that, the former patient is usually on his or her own.

If the hospital really wanted to promote patients rights in the OB ward, it would not assign nurses randomly, but would permit patients to choose their primary nurse. Such choice could include the nurse's sex as well. But of course this would be "inconvenient" for the hospital staff. If the hospital really believed its sexist argument, it would have to reconsider the assignment of female nurses to male patients where tasks necessitating genital exposure were required.

The hospital's chaperon policy is at least more honest. Its purpose as stated is not to protect the patient from sexual assault by the doctor or nurse, but "to protect the hospital from charges of molestation." This may be a legitimate business purpose, but it could be accomplished simply by limiting a male nurse's OB duties to women accompanied in labor and delivery by their husbands or others.

The Court's Rationale

The court's rationale is overtly sexist. The judge believes that male physicians can specialize in obstetrics, but male nurses cannot. She gives two reasons for her opinion, which readers may find more persuasive than I do. At Baptist, there were eleven OBs on the staff, only two of whom were women. So not all women could have a female physician even if all wanted one. Therefore, the judge determined that it was acceptable for OB services to be performed by male physicians:

> ...the patient knows and accepts the fact of this necessity; if all female patients demanded female surgeons, obstetricians, gynecologists, and the like, a lot of them would go untreated. The same could not be said of their insistence on female attendants, who may or may not even be nurses, for such unskilled tasks as handling bedpans and giving baths.

Shortly following this statement the judge concludes with another nonsequitur:

> Due to the intimate touching required in labor and delivery, services of all male nurses are inappropriate...it is their very sex itself which makes all male nurses unacceptable.

As if unsure of her "necessity rationale," the judge adds another reason for distinguishing male doctors from male nurses: unlike the physician, the nurse is "an *unselected individual* who is intruding on the obstetrical patients' right to privacy" (emphasis in the original). This ignores the fact that the patient does not always choose the doctor, who may be a covering physician or resident; and if selection is the critical issue, patients should be permitted to select their primary nurse, who will spend a lot more time with them than the physician will.

The judge's real reason seems to be that she does not believe that nurses are true "health care professionals," at least in the sense that physicians are. Male doctors can perform any legitimate manipulation of the female genitalia, and the judge expects female patients to "accept" this as "necessity." But she also expects women to object to male nurses even to the point of agreeing that it makes sense for a hospital to "outlaw" male OB nurses altogether. Similar logic would permit hospitals to bar women to object to them because of their sex. Since most practitioners are male, there is no "necessity" for females in these specialties. One is also left wondering how the judge would have decided the case if the patients had objected to black nurses or physicians.

Sex Stereotyping

The clear thrust of both the hospital policy and the judge's opinion is to reinforce sexual stereotypes in health care. Keep it simple, they say: little boys should grow up to be doctors, not nurses—that's women's work. But perhaps this is taking the opinion too seriously. Neither the hospital nor the judge can turn back the clock. Women are here to stay in medicine, and men are likely to become more and more involved in nursing. The opinion reminds us of the way things used to be, and demonstrates that former arrangements and sex roles were based not on logic, but on biased stereotyping. Recognizing the basis for some of these outmoded hospital "rules" can help us change them for the benefit of patients and staff alike.[*]

[*]Before the appeal could be heard, Mr. Backus left his nursing job at Baptist Medical Center. The appeals court refused to rule on the merits of the case, concluding that his departure rendered the dispute moot. *Backus v. Baptist Medical Center*, 671 F.2d 1100 (8th Cir. 1982).

Conception

Introduction

The most fascinating issues in the field of health law are concentrated at the beginning and end of life. The essays in this section are concerned with the beginning, and most with what have become known as "new reproductive technologies" or simply, "new conceptions." New methods of conception have traditionally been at the core of parables of the future. In George Orwell's world of *1984*, for example, artificial insemination by a donor of superior genetic stock was mandatory, and sexual pleasure and family units were destroyed to help maintain the tension necessary in totalitarian society devoted to perpetual warfare. In a more likely scenario, Aldous Huxley's *Brave New World*, the family was also destroyed; but he portrayed a society controlled not by force and fear, but by gratification and drugs. Abolition of the family was followed by complete sexual freedom. Reproduction, however, was handled exclusively by the state, in "hatcheries" in which embryos were produced and monitored in an artificially controlled laboratory environment:

Of course, they didn't content themselves with merely hatching out embryos: any cow could do that. "We also predestine and condition. We decant our babies as socialized human beings, as Alphas or Epsilons, as future sewage workers or..." He was going to say "future World controllers," but correcting himself, said "future Directors of Hatcheries," instead.

It is oversimplistic to equate the separation of sexual intercourse from human reproduction as necessarily involving the gradual elimination of the family. Nonetheless, these new reproductive technologies have the potential to profoundly affect our view and definition of motherhood, of women's role in society, and of the very meaning of human reproduction itself. In addition, now that we can capture, produce and manipulate the extracorporeal embryo, we must deal with its status and decide whether to afford it legal protection. These essays primarily explore the legal issues that the new reproductive technologies raise, as well as the legal issues posed by one reproductive strategy that involves no new technology: surrogate motherhood. As this book goes to press, a number of cases around the country are in court that involve surrogate mothers who, although previously agreeing to have a child for a childless couple, have changed their minds after the birth, and want their child back. The future of the "new reproductive techniques" remains somewhat in doubt, although general support for infertile couples and a "right to procreate" remains strong.

All of the essays in this section originally appeared in the *Hastings Center Report*. "Redefining Parenthood" was published in October, 1984; "Artificial Insemination" in August, 1979; "Contracts to Bear a Child" in April, 1981; "The Baby Broker Boom" in June, 1986, and "Surrogate Embryo Transfer" in June, 1984.

Redefining Parenthood
and Protecting Embryos
*Why We Need New Laws**

Dependable birth control made sex without reproduction possible. Some saw the separation of procreation from sex as an affirmation of pleasure and love; others saw it as a sin against nature. One consequence was a relaxation of the inhibition against sex with multiple partners; venereal disease replaced pregnancy as the worst physical consequence. Nonetheless, venereal disease in women has its own potential side effect—sterility. Other factors, such as postponing pregnancy until the late 30s, also increased the incidence of infertility.

Now medicine is closing the circle opened with the advent of sex without reproduction by offering methods of reproduction without sex; including artificial insemination by donor (AID), in vitro fertilization (IVF), and surrogate embryo transfer (SET). As with birth control, artificial reproduction is defended as life-affirming and loving by its proponents, and denounced as unnatural by its detractors. How concerned should society be about artificial reproduction?

In England, working groups have produced three reports on artificial insemination over the past three decades. In 1984, the Warnock Commission issued Britain's latest and most comprehensive report on "human assisted reproduction."[1] In Australia, working groups have reported on IVF and are looking at frozen embryos. In the United States, the Ethics Advisory Board of the then-Department of Health, Education and Welfare issued our most recent national report in 1979. Lately it has been suggested that a new National Commission be established to study and monitor developments in this area. The

*Portions of this essay are adapted from testimony presented before the US House of Representatives' Subcommittee on Investigations and Oversight of the Committee on Science and Technology, Aug. 8, 1984.

reason for all these panels and commissions is that artificial reproduction raises profound social and ethical issues touching the nature of family relationships, and the nature and value of the human embryo.

"To allay public fear" concerning the consequences of new developments, the Warnock Commission developed sixty-three recommendations: thirty-three involving a proposed licensing board to regulate clinical services and research, seven involving the National Health Service's infertility program, and twenty-three for new British laws, including proposals to create seven new crimes involving human reproduction and embryo research.

The Warnock approach is legal overkill, since it seems premature to outlaw as criminal so many aspects of artificial reproduction. Nonetheless, just because we cannot answer all the questions raised by these new techniques does not mean we should remain silent about those areas in which action is imperative to protect important societal interests. Protecting the child and family, and protecting the real and symbolic value of the human embryo are cases in point. Two areas merit quick and decisive legal action: defining maternity and paternity at the moment of birth, and protecting the human embryo from commercial exploitation.

Identification of the Mother and Father

Artificial insemination by donor for the first time separated the genetic father of a child from the act of sexual intercourse in its conception. The procedure has become recognized as a standard "treatment" for infertility of a married couple when the husband is sterile. Customarily the sperm comes from an anonymous vendor who (as long as the husband of the woman consents) is not considered the legal father of the child. Thus we have developed a technique in which the mother's husband replaces the genetic father as the legal father of the child.

This model places personal agreements or contracts among the parties ahead of genetic or "blood line" considerations, and has been suggested as the model to apply to embryo transfer. The question is: Does it fit? The Warnock Commission, for example, recommended that:

> In order to achieve some certainty in this situation [egg donation]...legislation should provide that when a child is born to a woman following donation of another's egg the woman giving birth should, for all purposes, be regarded in law as the mother of that child,

and that the egg donor should have no rights or obligations in respect to the child.

This treats the donation of sperm and egg on an equal footing and seems reasonable as far as it goes; but it does not go far enough. We need a rule that applies equally to all births. The reason we require legislation to guarantee the obligations of the social or rearing father and extinguish the rights and obligations of the genetic father is that society has always assumed that the genetic father is the father of the child for all purposes.

Nonetheless, to protect children and foster families, children born during the course of a marriage are legally presumed the legitimate offspring of the couple. To challenge this legal presumption successfully, state laws commonly require the husband to disown the child affirmatively and present evidence in court proving beyond a reasonable doubt that he is not the genetic father. This is a reasonable rule because it protects the child and helps ensure that it has another parent, in addition to its mother, who is identifiable as financially responsible for its well-being, and who is likely to help rear the child.

In dealing with the mother, however, an additional biological consideration makes the identification issue more complicated. In males, one need only distinguish the genetic from the social or rearing father. But, as John Robertson has noted, in females we might have to distinguish among the genetic mother, the gestational mother, and the rearing mother. When these three are different women, which one should the law presume is the legal mother with the obligation to support and nurture the child?

This is a completely novel question. Previously, at birth there was never any question who the mother was since she was always both the genetic and the gestational mother. Only the identity of the father was uncertain, and this uncertainty was clarified by a social decision to presume paternity in the mother's husband. Now, in cases of surrogate embryo transfer, and embryo transfer to a woman other than the egg donor, the identity of the genetic and gestational mother will be different. Which of these two women has a greater claim to be identified in law as the child's "mother," and is her claim superior to any claim a rearing mother might have who is neither genetically nor gestationally connected to the child?*

---

* This question has now actually been addressed by a lower court in Detroit. See "The Baby Broker Boom" at p. 77.

The current legal presumption that the gestational (birth) mother is the legal mother should remain. This gives the child and society certainty of identification at the time of birth (a protection for both mother and child), and also recognizes the biological fact that the gestational mother has contributed more of herself to the child than the genetic mother, and therefore has a greater biological investment and interest in it. If any agreements regarding transfer, relinquishing of parental rights, or adoption are to be made, they should be made only by the gestational mother, and only after she has had a reasonable time after the birth to consider all her and her child's options.

To protect both the integrity of the family and the interests of the children involved, the current legal presumptions should remain and be codified so they cannot be modified by private contract: the child's father should be presumed to be the husband of the child's mother, and the child's mother should be presumed to be the woman who gave birth to the child. We can permit the husband to overcome this presumption by presenting proof of nonpaternity beyond a reasonable doubt (as long as he did not consent to the procedure that resulted in the birth), but the maternal presumption should be conclusive and irrefutable.

Protection of the Extracorporeal Embryo

Whether and how we should protect the extracorporeal embryo depends upon how we view it. We need not, of course, consider it a person to afford it legal protection, any more than we need consider a dog a person to protect it against cruelty, or a dolphin a person to protect it against destruction, or a national park a person to protect it against loggers. Nor need we grant the embryo any legal rights of its own to afford it legal recognition.

With in vitro fertilization in which all fertilized eggs are replaced in the uterus of the ovum donor (embryo replacement) the issue of embryo protection concerns mainly its care for the short time it is in vitro. Protection becomes a much more important issue if some of the embryos are not replaced, but are frozen for future implantation or research.

The human embryo is equal to more than the sum of its constituent parts. It not only has the complete genetic complement of a human being, but is also a powerful symbol of human regeneration. We can thus value it and afford it legal recognition even though we do not so value or legally recognize either the egg or the sperm. On the other hand, we do recognize and accept a significant loss of embryos in nature, and make no efforts to recover and freeze them. We also permit destruction of and research on fetuses under certain circumstances. Thus what we seem to be dealing with are the added obligations that we incur by actively intervening in the natural process of human reproduction by using freezing techniques as a method of embryo storage.

To protect the gamete donors, and the interests of society in protecting the integrity of assisted reproduction techniques, embryos should only be frozen with the informed consent of both gamete donors, and only for a specific and specified purpose, usually used in subsequent cycles when a pregnancy is not obtained. When the purpose is fulfilled, the frozen embryo should be destroyed.

Likewise, when something other than the original use of the embryo is contemplated, the informed consent of both gamete contributors should be required. If only one is alive, that one should have the decisionmaking authority, since that person has the most interest in and is thus likely to be most protective of the embryo. When they both die, the embryo should be destroyed. Recommendations like those of the Warnock Commission, to permit the frozen embryo to pass to the storage facility, which may dispose of the embryo as it sees fit subject only to certain licensing laws, treat the embryo too much like unclaimed luggage, and give insufficient weight to its origins and symbolic value.

Sales of Frozen Embryos

In the United States there is an almost universal consensus that human kidneys should not be bought and sold,* and the arguments against the sale of human embryos are even more compelling. In this case the specter of a coercive offer that exploits the bodies of the poor is replaced by the potential of a commercial market in prefabricated, selected embryos, which encourages us to view embryos as things or commodities that are simply means to whatever ends we design, rather than as human entities without a market price. Ian Kennedy has argued that we know intuitively that a human embryo is more valuable than a hamster or other experimental animal, and that is why we have

*See "Life, Liberty and the Pursuit of Organ Sales" at p. 378.

trouble permitting experiments on human embryos. Likewise, we know intuitively that a human embryo is more "valuable" than a kidney and of much more symbolic importance regarding human life: that is why we feel that embryos should not be the subject of commerce. The reason is not so much the embryo itself (although many will find its intrinsic value sufficient justification to outlaw sales), but the implications for the children that will result following the sale and implantation of a frozen embryo.

Embryos will be bought and sold, if at all, on the belief that they will produce a healthy child, and possibly one of a certain physical type, IQ, stature, and so on. When the child is not born as warrantied or guaranteed, what remedies will the buyer have against the seller? Even a brief glance at the sales provisions in the Uniform Commercial Code (UCC) illustrates that this is not an area in which we can permit sales or if we do permit them, it is an area in which we need a new set of sales statutes.

The UCC provides, for example, that "if the goods or the tender of delivery fail in any respect to conform to the contract, the buyer may (a) reject the whole; or (b) accept the whole; or (c) accept any commercial unit or units and reject the rest" (sec. 2-601). Section (c) might be read to apply to twins or triplets, and section (a) leaves us wondering who is responsible for the child. Likewise, "if the seller gives no instructions within a reasonable time after notification of rejection the buyer may store the rejected goods for the seller's account or reship them to him or resell them for the seller's account..." (sec. 2-604) This could be read as applying more directly to the frozen embryo itself, but its potential application to the child produced as a result of the embryo transfer process simply illustrates the inappropriateness of sales in this area at all, and the ease with which sale of human embryos can become confused with sale of human children. In this regard the Warnock Commission is correct in recommending legislation to ensure that "there is no right of ownership in a human embryo," but the commission is simplistic in lumping "gametes and embryos" together in its discussion of a policy on sales.

The lesson of Aladdin is not only that we must be cautious in letting genies out of bottles, but also that we should not exchange "new lamps for old" until we know the value of each. Ideally, the state and federal government should stay out of the arena of human reproduction. Unfortunately, if the children resulting from new techniques such as surrogate embryo transfer and the use of frozen embryos are to be adequately protected, this is no longer possible. Private contractual agreements tend to favor the interests of the infertile couple over those of the potential child.

Action on three levels is warranted: (1) a model state law designed to clearly define the identity of the legal mother and father of all children, including those born to other than their genetic parents, should be drafted and enacted; (2) professional organizations, with public participation, should develop and promulgate guidelines for sound clinical practice*; and (3) a national body of experts in law, public policy, science, medicine, and ethics should be established to monitor developments in this area and report annually to Congress and the individual states on the desirability of specific regulation and legislation.

At all levels, the primary focus should be on protecting the interests of the children, even if their protection sometimes comes at the expense of some infertile couples. This general policy will also protect the integrity of artificial reproduction itself.

*One notable early effort has been undertaken by the American Fertility Society, "Ethical Considerations of the New Reproductive Technologies," **46** *Fertility and Sterility* 1S (Supp. Sept. 1986).

Artificial Insemination

Beyond the Best Interests of the Sperm Donor

Recent ethical and legal discussion concerning novel ways of reproduction has focused on in vitro fertilization, while assuming that most of the issues surrounding another technique that has been successfully used an estimated 250,000 times have been more or less resolved. Artificial insemination by donor (AID) is a cottage industry on the verge of mass marketing. The May 14, 1979, issue of *Advertising Age* noted that two commercial sperm banks each fill over 100 orders a month, and one is preparing to market sperm directly to consumers. In the near future, a "home insemination kit," complete with sperm and instructions on use, is possible. The following ad, now directed to physicians, could appear soon in popular magazines and newspapers: "From our panel of excellent donors, you select one for yourself based on blood type, ethnic origin, race, height, weight, and coloration of skin, hair and eyes."

In *Island*, his vision of the ideal society, Aldous Huxley sees a time when everyone will use AID voluntarily (at least for a third child) with donors picked from a "central bank of superior stocks" to increase the general IQ of the population. In George Orwell's more sinister society, described in *1984*, artificial insemination is mandatory, and all marriages must be approved in advance by the Party. Although neither of these futures seems an immediate threat, the perceived legal difficulties, the general desire for secrecy, and the cottage-industry nature of AID have all conspired to prevent any meaningful standards from developing, and make any future for AID the product of chance rather than policy.

We are faced with a technology that has the potential to make major changes in our reproductive habits, that has been poorly thought out, and if it is "gaining public acceptance," is probably doing so under false pretenses. Until very recently the best one could do was conjecture about the practices in doctors' offices and infertility clinics. However, three researchers from the University of Wisconsin have published the results of their survey of 379 practitioners of AID who accounted for approximately 3500 births in 1977 (of an estimated total of 6000 to 10,000 annually in the United States).[2]

The results are disturbing. Besides pointing to a general lack of standards and the growing use of AID for husbands with genetic defects and for single women, the findings tend to indicate that current practices are based primarily on protecting the best interests of the sperm donor rather than those of the recipient or resulting child. Two areas merit immediate attention: donor selection and record keeping.

Donor Selection

The term "donor" is a misnomer. Virtually all respondents in the Wisconsin study purchased ejaculates, 90 percent paying from twenty to thirty-five dollars per ejaculate. A more accurate term would be "sperm vendors." While this distinction may seem trivial, it has legal consequences. For example, it makes no sense to designate the form signed by the vendor as a "consent form" since he is not a patient and is not really consenting to anything. It is a contract in which the vendor agrees to do certain things for pay. Moreover, continued use of the term "donor" gives the impression that the sperm vendor is doing some service for the good of humanity and deserves some special protection, rather than simply performing a service for pay. The problems with paid "donors" have been amply explored in Richard M. Titmuss's classic study of the blood market, *The Gift Relationship;* similar problems arise in the sperm business.

The actual selection and screening of sperm vendors, however, is more important than the term employed to describe them. The Wisconsin study found that 80 percent of all physicians use medical students and hospital residents all or most of the time. In this regard sociobiologists have found that animals will employ that reproductive strategy that maximizes the spread of their genes. In the words of Richard Dawkins in *The Selfish Gene*: "Ideally what an individual would like (I don't mean physically enjoy, although he might) would be to copulate with as many members of the opposite sex as possible, leaving the partner in each case to bring up the children." Artificial insemination, of course, adds an entirely new technology to use in pursing this strategy.

There can be little debate that physicians in all of these situations are making eugenic decisions—selecting what they consider "superior" genes for AID. In general they have chosen to reproduce themselves (or those in their profession), and this is what sociobiologists would probably have predicted. While this should not be surprising, it should be a cause for concern. Physicians may believe that society needs more individuals with the attributes of physicians, but it is unlikely that society as a whole does. Lawyers would be likely to select law students; geneticists, graduate students in genetics; military personnel, students at the military academies, and so on. The point is not trivial. Courts have found in other contexts that physicians have neither the training nor the social warrant to make "quality of life" decisions. Selecting donors in this manner, rather than matching for characteristics of the husband, for example, seems to be primarily in the best interests of the physician rather than the child, and probably cannot be justified.

More than this, the Wisconsin survey revealed that even on the basis of simple genetics, physicians administering AID "were not trained for the task" and made many erroneous and inconsistent decisions. Specifically 80 to 95 percent of all respondents said they would reject a donor if he had one of the following traits, and more than 50 percent of all respondents would reject that same donor if one of these traits appeared in his immediate family: Tay-Sachs, hemophilia, cystic fibrosis, mental retardation, Huntington's chorea, translocation or trisomy, diabetes, sickle-cell trait, and alkaptonuria. This list includes autosomal recessive diseases in which carriers can be identified, and those in which they cannot, dominant, X-linked, and multigenic diseases.

The troubling findings are that the severity and genetic risk of the condition was not reflected in rejection criteria, and that genetic knowledge appears deficient. For example, 71 percent would reject a donor who had hemophilia in his family, even though this X-linked gene could not be transmitted unless the donor himself was affected. Additionally, although 92 percent said they would reject a donor with a translocation or trisomy, only 12.5 percent actually examined the donor's karyotype. Similarly, while 95 percent would reject a carrier of Tay-Sachs, fewer than 1 percent actually tested donors for this carrier state.

Physicians might be giving medical students far more credit than they deserve for knowledge of their own genetic and family history, honesty, and freedom from venereal disease. Even so, the conclusion must be that if prevention of genetic disease is a goal, it cannot be accomplished by the means currently in use. The findings also raise serious questions about the ability of these physicians to act as genetic counselors, and suggest that other nonmedical professionals may be able to do a better job in delivering AID services in a manner best calculated to maximize the interests of the child and not just those of the sperm donor.

Record-Keeping

While the Wisconsin survey found that 93 percent of physicians kept permanent records on recipients, only 37 percent kept permanent records on children born after AID and only 30 percent kept any permanent records on donors. The fear of record-keeping seems to be based primarily on the idea, common in the legal literature, that if identifiable, the donor might be sued for parental obligations (child support, inheritance, and so on) by one of his "biological children" sired by the AID process, and that this suit might be successful. The underlying rationale is that unless anonymity is assured, there would be no donors. There are a number of responses to this argument:

1. It is important to maintain careful records to see how the sperm "works" in terms of outcome of the pregnancy. If a donor is used more than once, a defective child should be grounds for immediately discontinuing use of the sperm for the protection of potential future children. Since the survey disclosed that most physicians have no policy on how many times they use a donor and 6 percent had used one for more than 15—with one using a donor for 50 pregnancies—this issue is more likely to affect the life of a real child than the highly speculative lawsuit is to affect the life of a donor.

2. No meaningful study of the characteristics and success of donors can ever be made if there are no records kept concerning them.

3. In those cases where family history is important (and it is important enough to ask every donor about his) the AID child will never be able to respond accurately.

4. Finally, and most important, if no records are kept the child will never, under any circumstances, be able to determine its genetic father. Since we do not know what the consequences of this will be, it cannot be said that destroying this information is in the best interests of the child. The most that can be said for such a policy is that it is in

the best interests of the donor. But this is simply not good enough. The donor has a choice in the matter, the child has none. The donor and physician can take steps to guard their own interests, the child cannot.

Given the recent history of adopted children and their efforts to identify their biological parents, it is likely that if AID children learn they were conceived by AID, they would want to be able to identify their genetic father. It is now accepted practice to tell adopted children that they are adopted as soon as possible, and make sure they understand it. It is thought that they will inevitably find out some day, and the blow will be severe if they have been deceived. In AID the consensus seems to be not to tell on the grounds that no one is likely to find out the truth since to all the world the AID pregnancy appears to have occurred normally.

Moralists would probably agree with Joseph Fletcher, who has argued that the physician should not accept the suggestion that a husband's brother be used as a donor without the wife's knowledge (the husband's intent is to keep the "blood line" in his children) because this would be a violation of "marital confidence." A similar argument can be made concerning the strategy of consistently lying to the child; that it is a violation of "parental confidence." There is some evidence that AID children do learn the truth, and the only thing that the fifteen state legislatures that have passed laws pertaining to AID agree on is that the resulting child should be considered legitimate, an issue that will never arise unless the child's AID status is discovered by someone.

Not keeping records can also lead to bizarre practices. For example, some physicians use multiple donors in a single cycle to obscure the identity of the genetic father. The Wisconsin survey found that 32 percent of all physicians utilize this technique, which could be to the physical detriment of the child (and potential future children of a donor with defective sperm) and cannot be justified on any genetic grounds whatsoever.

A number of policies would have to be changed to permit open disclosure of genetic parenthood to children. The first is relatively easy: a statute could be enacted requiring the registration of all AID children in a court in a sealed record that would only be available to the child; the remainder of the statute would provide that the genetic father had no legal or financial rights or responsibilities to the child. Variations would be to keep the record sealed until the death of the donor, or until he waived his right to privacy in this matter, or to only disclose genetic and health information. In the long term, a more practical solution may lie in only using the frozen sperm of deceased donors, in which case full disclosure could be made without any possibility of personal or financial demands on the genetic father by the child.

Conclusions

Current AID practices on donor screening and record-keeping are based primarily on protecting the interests of practitioners and donors rather than recipients and children. The most likely reason for this is found in exaggerated fears of legal pitfalls. Policy in this area should be dictated by maximizing the best interests of the resulting children. The evidence is that current practices are dangerous to children and must be modified. Specifically, consideration should be given to:

1. Removal of AID from the practice of medicine and placing it in the hands of genetic counselors or other nonmedical personnel (alternatively, a routine genetic consultation could be added to each couple who request AID);

2. Development of uniform standards for donor selection, including national screening criteria;

3. A requirement that practitioners of AID keep permanent records on all donors that they can match with recipients (I would prefer this to become common practice in the profession, but legislation requiring court filing may be necessary);

5. Establishment of national standards regarding AID by professional organizations, with public consultation; and

6. Research on the psychological development of children who have been conceived by AID, and their families.

S. J. Behrman concludes his editorial on the Wisconsin survey by questioning the "uneven and evasive" attitude of the law in regard to AID, and recommending immediate legislative action: "The time has come—in fact, is long overdue—when legislatures must set standards for artificial insemination by donors, declare the legitimacy of the children, and protect the liability of all directly involved with this procedure. A better public policy on this question is clearly needed."[3]

Agreement with the need for "a better public policy" is not synonymous with immediate legislation. The problem with AID is that there are many unresolved problems with it, and few of them are legal. It is time to stop thinking about uniform legislation and start thinking about the development of professional standards. Obsessive concern with self-protection must give way to concern for the child.

Contracts to Bear a Child

Compassion or Commercialism?

Many medical students supplement their income by selling their blood and sperm. Although this practice seems reasonably well accepted, society does not permit individuals to sell their vital organs or their children. These policies are unlikely to change. Where on this spectrum do contracts to bear a child fall? Are they fundamentally the sale of an ovum with a nine-month womb rental thrown in, or are they really agreements to sell a baby? Though this formulation may seem a strange way to phrase the issue, it is the way courts are likely to frame it when such contracts are challenged on the grounds that they violate public policy.

Surrogate Motherhood

In a typical surrogate mother arrangement, a woman agrees to be artificially inseminated with the sperm of the husband of an infertile woman. She also agrees that after the child is born she will either give it up for adoption to the couple or relinquish her parental rights, leaving the biological father as the sole legal parent. The current controversy centers on whether or not the surrogate can be paid for these services. Is she being compensated for inconvenience and out-of-pocket expenses, or is she being paid for her baby?

Two personal stories have received much media attention. The first involves Patricia Dickey, an unmarried twenty-year-old woman from Maryland who had never borne a child, and who agreed to be artificially inseminated and give up the child to a Delaware couple without any compensation. She was recruited by attorney Noel Keane of Michigan, known for his television appearances in which he has said that for a $5000 fee he will put "host mothers" in touch with childless couples. Ms. Dickey explained her motivation in an interview: "I had a close friend who couldn't have a baby, and I know how badly she wanted one...It's just something I wanted to do."[4] The outcome of Dickey's pregnancy has not been reported.

More famous is a woman who has borne a child and relinquished her parental rights. Elizabeth Kane (a pseudonym), married and the mother of three children, reportedly agreed to bear a child for $10,000. The arrangement was negotiated by physician Richard Levin of Kentucky who is believed to have about 100 surrogates willing to perform the same services for compensation. Levin says, "I clearly do not have any moral or ethical problems with what we are doing."[5] Mrs. Kane describes her relationship to the baby by saying, "It's the father's child. I'm simply growing it for him."[6]

Even this brief sketch raises fundamental questions about the two approaches. Should the surrogate be married or single; have other children or have no children? Should the couple meet the surrogate (they were in the delivery room when Mrs. Kane gave birth to a boy)? Should the child know about the arrangement when he grows up (the couple plans to tell the child when he is eighteen)? Is monetary compensation the real issue (the sperm donor has agreed to give Ms. Dickey more sperm if she wants to have another child for her own— could this cause more problems for both him and her)? What kind of counseling should be done with all parties, and what records should be kept? And isn't this a strange thing to be doing in a country that records more than a million and a half abortions a year? Why not attempt to get women who are already pregnant to give birth instead of inducing those who are not to go through the "experience"?

These questions, and many others, merit serious consideration. So far legal debate has focused primarily on just one: can surrogate parenting properly be labeled "baby selling"? Some have argued that it can be distinguished from baby selling because one of the parents, the father, is biologically related to the child, and the mother is not pregnant at the time the deal is struck and so is not under any compulsion to provide for her child. But the only two legal opinions rendered to date disagree. Both a lower court judge in Michigan and the Attorney General of Kentucky view contracts to bear a child as baby selling.

Court Challenge in Michigan

In the mid-1970s most states passed statutes making it criminal to offer, give, or receive anything of value for placing a child for adoption. These statutes were aimed at curtailing a major black market in babies that had grown up in the United States, with children selling for as much as $20,000. Anticipating that Michigan's version of this statute might prohibit him from paying a surrogate for carrying a child and giving it up for adoption, Noel Keane sought a declaratory judgment. He argued that the statute was unconstitutional since it infringed upon the right to reproductive privacy of the parties involved. The court was not impressed, concluding that "the right to adopt a child based upon the payment of $5000 is not a fundamental personal right and reasonable regulations controlling adoption proceedings that prohibit the exchange of money (other than charges and fees approved by the court) are not constitutionally infirm." The court characterized the state's interest as one "to prevent commercialism from affecting a mother's decision to execute a consent to the adoption of her child," and went on to argue that: "Mercenary considerations used to create a parent–child relationship and its impact upon the family unit strike at the very foundation of human society and is patently and necessarily injurious to the community."[7]

The case is on appeal, but is unlikely to be reversed. The judge's decision meant that Ms. Dickey, and others like her, could not charge a fee for carrying a child. It did not, however, forbid her from carrying it as a personal favor or for her own psychological reasons.

The Kentucky Statutes

One of the prime elements of surrogate mother folklore held that contracts to bear a child were "legal" in Kentucky. On January 26, 1981, Steven Beshear, the Attorney General of the Commonwealth of Kentucky, announced at a Louisville news conference that contracts to bear a child were in fact illegal and unenforceable in the state. He based his advisory opinion on Kentucky statutes and "a strong public policy against baby buying."

Specifically, Kentucky law invalidates consent for adoption or the filing of a voluntary petition for termination of parental rights prior to the fifth day after the birth of a child. The purpose of these statutes, according to the attorney general, is to give the mother time to "think it over." Thus, any agreement or contract she entered into before the fifth day after the birth would be unenforceable. Moreover, Kentucky, like Michigan, prohibits the charging of a "fee" or "remuneration for the procurement of any child for adoption purposes." The Attorney General argued that even though there is no similar statute prohibiting the payment of money for the termination of parental rights, "there is the same public policy issue" regarding monetary consideration for the procurement of a child: "The Commonwealth of Kentucky does not condone the purchase and sale of children."[8] The Attorney General has since brought an action to enjoin Richard Levin and his corporation from making any further surrogate mother arrangements in the state.*

Should We Care?

Surrogate parenting, open or behind a wall of secrecy, is unlikely ever to involve large numbers of people. Should we care about it, or should we simply declare our disapproval and let it go at that? I don't know, but it does seem to me that the answer to that question must be found in the answer to another: what is in the best interests of the children? Certainly they are more prone to psychological problems when they learn that their biological mother not only gave them up for adoption, but never had any intention of mothering them herself. On the other hand, one might argue that the child would never have existed had it not been for the surrogate arrangement, and so whatever existence the child has is better than nothing.

One of the major problems with speculating on the potential benefits of such an arrangement to the parties involved is that we have very little data. There is only anecdotal information available on artificial insemination by donor, for example. It does not seem to harm family life. But the role of the mother is far greater biologically than that of the father, and family disruption might be proportionally higher if the mother is the one who gives up the child. The sperm donor in the Patricia Dickey case is quoted as having said:

> It may sound selfish, but I want to father a child on my own behalf, leave my own legacy. And I want a healthy baby. And there just aren't any available. They're either retarded or they're minorities, black, Hispanic...that may be fine for some people, but we just don't think we could handle it.

Is this man really ready for parenthood? What if the child is born with a physical or mental defect—could he handle that? Or would the

---

* This case is dealt with in the following essay, "The Baby Broker Boom," at p. 77.

child be left abandoned, wanted neither by the surrogate nor by the adoptive couple? The sperm donor has made no biological commitment to the child, and cannot be expected to support it financially or psychologically if it is not what he expected and contracted for.

Perhaps the only major question in the entire surrogate mother debate that does have a clear legal answer is: Whose baby is it? On the maternal side, it is the biological mother's baby. And if she wants to keep it, she almost certainly can. Indeed, under the proper circumstances, she may even be able to keep the child and sue the sperm donor for child support. On the paternal side, it is also the biological child of the sperm donor. But in all states, children born in wedlock are presumed to be the legitimate children of the married couple. So if the surrogate is married, the child will be presumed (usually rebuttable only by proof beyond a reasonable doubt) to be the offspring of the couple and not of the sperm donor. The donor could bring a custody suit, if he could prove beyond a reasonable doubt that he was the real father, and then the court would have to decide which parent would serve the child's "best interests."

It is an interesting legal twist that in many states with laws relating to artificial insemination, the sperm donor would have no rights even to bring such a suit. For example, to protect donors the Uniform Parentage Act provides that "The donor of semen provided...for use in artificial insemination of a woman other than the donor's wife is treated in law as if he were not the natural father of a child thereby conceived."[9] The old adage, "Mama's baby, papa's maybe" aptly describes the current legal reaction to a surrogate who changes her mind and decides to keep the child.

DHEW's Ethics Advisory Board's final recommendation on in vitro fertilization and embryo transfer was that a "uniform or model law" be developed to "clarify the legal status of children born as a result of in vitro fertilization and embryo transfer." This makes some sense, although it seems premature. We need a set of agreed-on principles regarding artificial insemination by donor and surrogate mothers if legislation on in vitro fertilization and embryo transfer is to have a reasonable chance of doing more good than harm.*

*Elizabeth Kane, an original promoter of surrogate motherhood, has more recently helped organize opposition to it. She tells her story in *Birth Mother: The Story of America's First Legal Surrogate Mother*, Harcourt Brace Jovanovich, NY, 1988.

The Baby Broker Boom

Should babies be treated as commodities? Should reproduction be commercialized? Should motherhood be determined by contract? A few years ago these questions seemed absurd. But the hope that surrogate motherhood would wither of its own weirdness is now beginning to seem quaint. Indeed, two recent court decisions strongly support commercial surrogate mother agreements. If surrogate mother companies were listed on the New York Stock Exchange, these cases would have sent their stock soaring.

The Kentucky Case

Almost since its inception, Surrogate Parenting Associates, Inc. (SPA) was in trouble in its home state of Kentucky. In 1981, the Attorney General instituted proceedings against the corporation to revoke its charter. He charged that by entering into commercial surrogate arrangements in which a woman would be paid to be artificially inseminated, and then to bear a child for whom she would relinquish parental rights (for later step-parent adoption by the father-sperm donor's infertile wife), the corporation violated the state's prohibition against the "purchase of any child for the purpose of adoption." The statute was amended in 1984 to add the words, "or any other purpose, including termination of parental rights."[10]

A trial court ruled against the Attorney General, an appeals court in his favor, and the Supreme Court of Kentucky has now sided with the corporation.[11] The court declared that the intention of the legislature in prohibiting baby selling was solely "to keep baby brokers from overwhelming an expectant mother or the parents of a child with financial inducements to part with the child." It therefore approved of baby sales if the price was agreed to *before* conception, and the surrogate mother retained the right to cancel the contract up to the point of relinquishing her parental rights.

Surrogate motherhood is a nontechnical application of artificial insemination that requires no sophisticated medical or scientific knowledge or medical intervention. But the court saw surrogate motherhood as modern science, and did not want to interfere with "a new era of genetics," "solutions offered by science," and "new medical services."

The majority's opinion thus misses the focus of the Attorney General's argument: surrogacy's essence is not science, but commerce. The only "new" development in surrogacy is the introduction of physicians and lawyers as baby brokers who, for a fee, locate women willing to bear children by AID and hand them over to the payer–sperm donor after birth. The novelty lies in treating children like commodities.

Justice Vance, one of two dissenting justices, understood this. He noted that the corporation's "primary purpose is to locate women who will readily, for a price, allow themselves to be used as human incubators and who are willing to sell, for a price, all of their parental rights in a child thus born." His rationale was that payment is made to the surrogate in two parts. The first part "of the fee is paid in advance for the use of her body as an incubator." But the second portion of the fee is not paid unless and until "her living child is delivered to the purchaser, along with the equivalent of a bill of sale, or quitclaim deed, to wit—the judgment terminating her parental rights." As the judge persuasively argues, the last payment must be made for the child, since if the child is not delivered, the last payment need not be made.

The majority probably thought it was approving very limited baby selling: permitting a father–sperm donor to purchase the gestational mother's interest in his genetic child if the gestational mother contracted to make such a sale prior to conception and still desired to sell her child after its birth. But limiting baby buying to fathers does not make baby buying any more tolerable than permitting a father to kidnap his biological child from its mother would make kidnapping tolerable. If mothers are to give up their parental rights to fathers, it should be voluntarily, and without a monetary price that converts the child into a commodity. That is what the Kentucky legislature undoubtedly had in mind when it outlawed baby selling.

The Michigan Case

The Kentucky court did not address baby selling in the case of full surrogacy: a surrogate who "gestates" an embryo to which she has

made no genetic contribution. But a Michigan trial court has. Twenty
-three year old Shannon Boff was pregnant with a child genetically
unrelated to her at the time the question of her motherhood came up.[12]
For the first reported time in the US, in vitro fertilization (IVF) had
been used to fertilize an ovum from an infertile woman (who lacked
a uterus), and the resulting embryo was implanted into another
woman, who agreed to act as a surrogate mother by gestating the fetus.

This raised an undecided legal question: Should the genetic or the
gestational mother be considered the "legal" mother? That is, which
woman should have legal rearing rights and responsibilities? The
genetic parents, who had paid $40,000 for this "project" ($10,000 of
which went to Ms. Boff) wanted to have their own names listed on the
child's birth certificate, not the names of Ms. Boff and her husband.

Unfortunately, the case was a setup. Even though both "competing" sets of parents were represented by legal counsel, they all wanted
the judge to rule the same way. Since she did, there will be no appeal
and no further judicial analysis of the question. Nor did the judge
appoint anyone to represent the interests of the potential child. Like
the Kentucky court, the Michigan judge decided to let contracts and
commerce rule the day, rather than deal with any wider social issues,
or consider the best interests of any child.

In so doing, the judge consistently put form over substance. For
example, in determining that Ms. Boff's husband should not be presumed to be the father of his wife's child, the judge accepted the argument of their attorney that he could not be presumed the father under
the AID statute because he signed a "nonconsent to any type of artificial insemination of his wife." But given his active participation in
the entire project (he said he rubbed and drew faces on his wife's
enlarged stomach and treated the pregnancy as if his wife was carrying
their own child), his signature is hardly the "clear and convincing
evidence" the statute requires. Moreover, the entire Paternity Act
under which the case was brought covers only children "born out of
wedlock," so the court may have had no jurisdiction at all over this
case.[13]

The discussion of maternity is taken even less seriously. Like the
Kentucky court, the Detroit judge saw her primary task as trying to
make the law conform with and comfort modern science. Promoting
private contract and personal profit were also seen as appropriate
judicial strategies. To get to this point, the judge found it necessary
to rule that the state's paternity statute must be applicable to women
as well as men, to afford women "equal protection of laws."

This is, of course, true only if there are no significant differences between maternity and paternity. But if there are no significant differences, then the female gamete donor should logically be treated "equally" to a male gamete donor: the child would then have two genetic "fathers," and would have a [gestational] mother as well. Not to so recognize the gestational mother's status dehumanizes her (and all mothers?), turning her into a mere incubator. Of course, had Ms. Boff asserted her rights and identity as the child's mother, the judge would almost certainly have upheld her claim.

In applying the paternity statute to maternity, the court concluded that the gestational mother (whom the court referred to as the "birthing mother"), is acting as a "human incubator for this embryo to develop." Where the incubator "contracted to do this" via IVF, and where subsequent tissue typing confirms the genetic links of the child to the gamete donors, then "the donor of the ovum, the biological mother, is to be deemed, in fact, the natural mother of this infant, as is the biological father to be deemed the natural father of this child."

Besides putting contract above biology, this conclusion begs the question of who the child's mother is during pregnancy, and also makes identification of the child's mother at birth impossible. It thus fails to protect either the child or its mother where decisions regarding the newborn infant's care need to be made quickly. The judge dealt with this by saying that her decree would depend upon HLA tissue-typing confirming the identity of the genetic parents, a procedure that would not resolve the issue until at least a few days after the birth.

Although commerce won out in court, Ms. Boff said she would leave the baby business herself: "I'm going into retirement; any more babies coming from me are going to be keepers."

Identifying the Mother

The contrary conclusion—that the woman who gestates a child should be considered the child's legal mother for all purposes—is not based on antiscientific, anachronistic, or sentimental views of motherhood. Rather, it is a recognition of the gestational mother's greater biological contribution to the child, including risks and physical contributions of the nine months of pregnancy, and the need to protect the newborn by always providing the child with at least one immediately identifiable parent.

The gestational mother, for example, contributes more to the child than the ovum donor does in the same way she contributes more to the child than a sperm donor does. Other considerations also argue for this traditional view of motherhood. What if there are three "competing" mothers, as happens if the ovum donor is anonymous (as most sperm donors are), the gestational mother a surrogate, and the contracting rearing mother simply someone who wants to raise the child? In this scenario the only relationship the rearing mother has is monetary: she paid the surrogate a fee to gestate the embryo and give up the child. If we *really* believe money and contracts should rule, then the identity of the child's mother will depend upon contract and payment only, and both genetics and gestation (and therefore all biological ties) will be irrelevant.

Neither of these results seems reasonable, and the traditional presumption would always provide the child with an identifiable mother who would be the same woman who biologically contributed the most to the child. Accordingly, the traditional assumption should continue to be utilized, even in this "brave new world," and whether or not any contracts are signed or any money changes hands. The Kentucky court's ruling, of course, is consistent with this view. The gestational mother could honor her prior contract, but could also change her mind and retain *her* child anytime before formally relinquishing her parental rights.

Commercial surrogacy promotes the exploitation of women and infertile couples, and the dehumanization of babies. If the courts think this is a small price to pay to promote the "baby business," then it's time for state legislatures to define motherhood by statute.*

---

*In February 1988, the New Jersey Supreme Court ruled in the case of *Baby M* that the surrogacy contract entered into between William Stern and Mary Beth Whitehead was void as against public policy, and illegal because it violated the state's laws against baby selling, as well as laws involving adoption, custody, and termination of parental rights. This opinion was unanimous and well-reasoned, and is likely to be widely influential. *See* Annas, G. J., "Death Without Dignity for Surrogate Motherhood: The Case of Baby M," *Hastings Center Report,* April 1988.

Surrogate Embryo Transfer
The Perils of Patenting

The world's first surrogate embryo transfer (SET) baby has arrived on the scene with an army of legal and ethical issues. The birth followed by six years the birth of in vitro fertilization (IVF) baby Louise Brown, and the transfer procedure employed was touted as a potential alternative to IVF because it could be done as an office procedure without anesthesia or surgery. An IVF pioneer cautions, however, that the use of a "surrogate" mother, even for a short period of time, the inconvenience of synchronizing the menstrual cycle of the donor and recipient, and the medical risks to the donor, will make this method less appealing to many.[14]

The procedure, as developed by a research team led by John Buster, under a grant from Fertility and Genetics Research, Inc. (FGR), a Chicago-based for-profit company, involves five steps:

1. Synchronization of ovulation times between the embryo donor and the recipient woman.
2. Insemination of the donor woman with sperm from the infertile recipient's husband.
3. Lavage, or washing out, of the donor's uterus after about five days following fertilization.
4. Recovery of the embryo from the lavage fluid.
5. Transfer of the embryo to the recipient's uterus.

In a report of its preliminary experience, the UCLA team (the only group in the world currently experimenting with this technique) reported that in twenty-nine attempts, twelve embryos were recovered and transferred. Two pregnancies developed and there were two complications: one ectopic pregnancy in a recipient and one retained pregnancy in a donor.[15] The procedure raises all the legal and ethical issues presented by surrogate motherhood and IVF, although since the embryo donor never really knows if she's pregnant or not prior to the lavage procedure, the bonding of mother and child inherent in surrogate motherhood may not present a major problem in this process.[16] But surrogate embryo transfer (a more accurate term than "ovum transfer," since what is transferred is not an ovum but an embryo or a "conceptus") brings an issue to third-party-assisted reproduction not previously encountered: process patenting.

The company that funded the research with $500,000 of venture capital has applied for patents on both the instruments developed to recover and transfer the embryo, and the SET process itself. This latter application has already raised cries from a number of quarters. John Fletcher has argued that "patenting knowledge seems self-defeating... [and] flies in the face of openness in science and applications from science."[17] And Ervin Nichols, Director of Practice Activities at the American College of Obstetrics and Gynecology, says he was "astounded" at the patent application: "It's an almost unheard of precedent in medicine. It would mean that any time anybody develops a new and different technique, that it would be patented, and then nobody else could do it unless he had a license to."[18]

Patenting Medical Processes

On March 16, 1984, the company, sensitive to the reaction, released a memorandum summarizing its patent counsel's advice. The company argues that process patents are "legal, used and enforceable," that medical devices and medical drugs are often patented, and that "medical methods and processes can be patented and often are so." In support of this contention an attachment to the memorandum cites twenty-eight examples of US patents issued for medical methods and processes. Four sample methods patents are also attached. The memorandum concludes: "The principle behind all patents is one of protection to the 'inventor genius' and the 'investor' who has invested considerable time and dollars to develop a new and unique device, drug or process."

No one seriously questions the propriety of patenting a machine, drug, or medical device, and the company is surely correct in asserting that this is often done. On the other hand, the company exaggerates when claiming that medical processes are often patented. The twenty-eight examples it has assembled are an assortment of primarily surgical procedures; most of which, it seems safe to conclude, are rarely used. They include, for example, a "method and apparatus for direct electrical injection of gold ions into tissue such as bone"; and "surgical method of fixation of artificial eye lenses."

Of the four actual patents attached to the memorandum, the most recent and most extensive one involves a "method of suturing the organs of the gastrointestinal tract" submitted by Ernest Akopov of the USSR. My guess is that no one has ever been licensed to carry out this procedure in the United States and that neither individual patients nor scientific progress has been hurt by this patent since no one is likely to want to perform the operation described in precisely the manner in which it is patented. The other three attachments also describe surgical procedures (incision closing, use of an electrically heated surgical instrument, and vascular repair).

Arguments Favoring Process Patents

The patenting of medical processes is not unprecedented, but it is exceptional and usually involves procedures on the fringes of surgery, often in connection with specific instruments. None of this provides a precedent for SET, and a review of the policy arguments is required to arrive at a reasoned view.

John Buster himself argues most persuasively in favor of patenting the SET process. Without private investor financing, he maintains, he would not have been able to do the research that led to SET, since NIH would not fund the research, and he could not charge private patients. He would not have gotten this funding if the investors stood no chance to profit from their investment via patenting and licensing the products and processes resulting from his research.

Second, Buster argues that by enforcing the patent, and permitting only those individuals licensed by the company to engage in SET, control over the diffusion of this technology is retained. The company can thereby control the quality of the product delivered and protect the public both from quacks using the technology and physicians using it for questionable purposes. In the words of Buster's coworker, Maria Bustillo: "Hopefully, we will be able to select the physicians who want to set this up in their area. We're hoping to control the medical personnel who will be directing the program in each city."[19] The profit motive, it is asserted, will give the company an incentive to make SET available to all qualified persons who want to use it, so its diffusion will not be stifled.

Arguments Against Process Patenting

The primary generic argument against patenting medical processes is the potential conflict of interest it poses for physicians regarding reporting their research in an unbiased manner, both in the professional literature and in the popular press. It could also restrict independent, unbiased evaluation by other investigators who might be denied a license to confirm or refute the observations of the group having the patent. This is critical for processes since, unlike drugs and devices that require premarket testing and approval by the Food and Drug Administration, no independent agency has authority over the safety and efficacy of medical procedures.

Although Buster is not the person who applied for the patent, he and his team have recently been offered stock options in FGR, the company that is applying for the patent. The fact that they waited until the research under the original grant was completed before purchasing the stock options was a laudable attempt to deal with the potential conflict of interest issue.

On the other hand, their dealings with both the professional and popular media may have been tainted by their enthusiasm for their process and its economic potential. For example, in response to the concern of prospective FGR business partners about the success rates of SET, Buster, in a lay magazine, reportedly estimated annual United States demand for SET at 50,000 women, and said that with a large enough donor pool he could give these women a 1 in 20 to a 1 in 3 chance of getting pregnant the first time: "Basically what we're saying is that in a best case situation, it's possible for an infertile woman to enter our pool, ovulate, and end up pregnant 35 percent of the time. A little more than one-third of the time she enters the pool, she'll walk out pregnant."[18] The lay reader will focus on the 35 percent success rate (the most favorable possible) rather than the median in the estimate of 1 in 11, or the reported success rate (almost a year later) of less than 7 percent.[19]

One expects the professional literature to be somewhat more objective. In the same March 1984 issue of the *Journal of the American Medical Association* (JAMA) in which Buster's initial results are reported, there appears an article entitled "Legal Issues in Nonsurgical Human Ovum Transfer." The article will probably seem balanced to most of JAMA's nonlegal readers, but it is biased in favor of SET. In fact, the author at one point goes so far as to suggest that we should worry less about the rights of SET children than about the desires of infertile couples.

Neither the author nor the editors of JAMA tell the readers that the author was paid by John Buster and his research team to write a document, under contract from them, that helped get their initial SET protocol through UCLA Harbor's Institutional Review Board in 1981. Her JAMA article is based on that work. Thus what should have been presented in the professional literature as an editorial in favor of SET, is instead presented as an opinion of a neutral observer.

The generic argument for scientific accuracy and fidelity to the truth is powerful. Nonetheless, it is probably true that the quest for promotion, tenure, and prizes has an effect on full disclosure at least as potentially negative as the profit motive. Although the IVF process has not been patented, its record of quality control and public and professional disclosure is no better or worse than one would anticipate with SET under patent. We are really left with only the specific argument relating to human reproduction: the government should not be involved in controlling or supervising, directly through police powers, or indirectly through patenting powers, the process of human reproduction.

Conclusion

One can only imagine how the SET patent might be enforced, but if private investigators or paid informants are used to monitor the SET process itself, this is an invasion of privacy every bit as intolerable as the vision of police searching the marital bedroom for "telltale signs of the use of contraceptives," one rationale that was used to strike down Connecticut's anticontraceptive status as an unconstitutional invasion of a couple's privacy. The Supreme Court noted that the anticontraceptive law, "in forbidding the use of contraceptives rather than regulating their manufacture or sale, seeks to achieve its goals by means having a maximum destructive impact upon [privacy]."[20] We can permit commerce in drugs and devices, stomach surgery, and even SET process patenting for cattle and other animals, without permitting commerce in methods of human reproduction. The subject matter does not lend itself to patent infringement enforcement without potentially unbearable privacy violation.

SET may well prove an effective and useful method to enhance the fertility of infertile couples. The generic ethical and legal issues it raises in common with IVF and surrogate motherhood, however, must be directly addressed. Its unique issue, process patenting, should be rejected unless the patent holder can persuade the medical profession and the public that effective quality controls will be utilized and enforcement that could compromise reproductive privacy will not be attempted.*

---

*By 1987, the UCLA group had broken up, and FGR was no longer pursuing its plans for SET or the patent.

Pregnancy and Birth

Introduction

The essays in this section deal with the rights and interests of pregnant women during their pregnancy, and the rights of newborn children. Throughout this section the question of the pregnant woman's moral and legal obligations to her fetus is addressed in varying contexts. A pervasive theme is whether it is ever reasonable to convert a pregnant woman's moral obligation to protect her fetus into a legal obligation, enforced by the state. These essays consistently argue that such a legal "solution" to fetal neglect would be misdirected and likely counterproductive.

Two other areas are also explored. The first is the growing use of genetic screening directed at detecting abnormalities in the fetus. The development of amniocentesis and maternal serum alpha-fetoprotein (MSAFP) testing especially have raised new legal issues regarding the physician's obligation to inform pregnant women of the existence of these tests. The failure of physicians to so inform has resulted both in suits by parents of handicapped infants (whose births, the parents allege, would have been avoided by abortion had the indicated screening test been performed), and by the infants themselves, who allege that they should be compensated for their life in a handicapped state—a life they would not be alive to live had proper screening tests,

Pregnancy and Birth

followed by their abortion, been performed. We will also see how such lawsuits led the American College of Obstetricians and Gynecologists to recommend routine use of MSAFP screening for all pregnant women.

The concluding three essays in this section chronicle the "Baby Doe" debate, which centers on the fate of handicapped newborns who require sophisticated medical interventions to survive. What is the proper role of the government after birth, and how can handicapped infants be protected from those who would prefer to see them die? This is an extremely difficult question. We may be tempted, like the drunken minister in Sam Shepard's play Buried Child, to simply throw up our hands and exclaim, "I don't know what to do. I don't know what my role is." But we must do better than this. Even if we conclude that federal "hot lines" are not constructive, the law certainly can and should help protect handicapped newborns, and help make sure they receive the treatment and support due them as citizens.

All of the essays in this section are from the Hastings Center Report, and the first, "Fetal Neglect," is the most recent, having been published in December, 1986. "Medical Paternity" is from June, 1979; "Righting the Wrong" from February, 1981; "Genetic Screening" from December, 1985; "Homebirth" from August, 1978; "Forced Cesareans" from June, 1982; "Disconnecting the Baby Doe Hotline" from June, 1983; "Baby Doe Redux" from October, 1983; and "Checkmating the Baby Doe Regulations" from August, 1986.

Fetal Neglect

Pregnant Women as Ambulatory Chalices

In Margaret Atwood's *The Handmaid's Tale*, most women are sterile, and the few who retain the capacity to bear children have reproduction as their exclusive function. As one handmaid describes her station: "We are two-legged wombs, that's all; sacred vessels, ambulatory chalices." This future scenario strikes many as unlikely, but we need not speculate. A recent criminal indictment directly raises the issue of when it is legally acceptable to treat a pregnant woman as a container, while treating the welfare of the fetus she contains as more significant.

Pamela Stewart Monson is the subject of what may be the first criminal charge ever brought against a mother for acts and omissions during pregnancy. Criminal charges were filed against her in California in October, 1986. Although we will have to await trial to learn the facts, the reports that are available suggest that sometime very late in her pregnancy Mrs. Monson was advised by her physician not to take amphetamines, to stay off her feet, to avoid sexual intercourse, and, because of a placenta previa, to seek immediate medical treatment if she began to hemorrhage. According to the police, Mrs. Monson noticed some bleeding the morning of November 23, 1985. Nevertheless, she remained at home, took some amphetamines, and had intercourse with her husband. She began bleeding more heavily, and contractions began sometime during the afternoon. It was only later, perhaps "many hours" later, that she went to the hospital. Her son was born that evening. He had massive brain damage, and died about six weeks thereafter.

The Statute

The police and the District Attorney have both told the press their theory of the criminal complaint. Police lieutenant Randy Narramore has said, "We contend that she willfully disobeyed instructions [of the doctor] and as a direct result the child was born brain dead (sic) and later died."[1] District Attorney Harry Elias says simply, she "did not follow through on the medical advice she was given."[2] Police officials wanted Monson prosecuted for murder, but the District Attorney has decided to prosecute under a California child support statute.

The support statute itself has been amended many times since it was first passed in 1872, and during much of the subsequent time applied only to fathers. The current version reads in relevant part:

> If a parent of a minor child *willfully omits*, without lawful excuse, *to furnish* necessary clothing, food, shelter or *medical attendance, or other remedial care* for his or her child, he or she is guilty of a misdemeanor punishable by a fine not exceeding two thousand dollars, or by imprisonment [for one year].[3] (emphasis added)

A later provision decrees that "a child conceived but not yet born is to be deemed an existing person insofar as this section is concerned."

Use of this statute, instead of a homicide charge, avoids the issue of causation, since violation of the duty itself violates the statute regardless of the consequences to the fetus or child. Thus, for example, failure of a father to provide food for his children would violate the statute, even if the mother was able to provide her children with food from another source. Why the father hasn't been charged in this case is puzzling.

There are very few cases that have attempted to define the scope of the statute, and the only cases that deal with its application to fetuses were decided during the Depression. These cases hold that a father can fail to provide food, clothing and shelter to the "unborn child" and that because these needs are "common to all mankind," proof of their "necessity" is not required. In regard to "medical attendance and other remedial care," however, "in a prosecution based on failure to furnish them, it would be incumbent on the prosecution to make affirmative proof of their necessity."[4] Such proof would, of course, require medical testimony. Monson did provide "medical attendance" by both seeking prenatal care and by seeking assistance in childbirth at a hospital.*

---

*See "Homebirth: Autonomy v. Safety" at p. 114.

Fetal Neglect

Use of this provision by the District Attorney is an attempt to extend child support statutes to create a crime of "fetal neglect." Does it make any sense to decree that the pregnant woman must, in effect, live for her fetus? That she must legally "stay off her feet" if walking or working might induce contractions? That she commits a crime if she does not eat only healthy foods; smokes cigarettes or drinks alcohol; takes any drugs (legal or illegal); has intercourse with her husband? etc. Should all these "do's" and "don'ts" be catalogued in a statute, or should they be the subject of her physician's advice? And how does such a criminal law change the nature of the doctor-patient relationship? It seems evident that in the minds of the police the doctor's patient was not Mrs. Monson at all, but her fetus. It also seems evident that although called "advice" and "instructions," the police believe that the physician was actually giving the fetal container, Mrs. Monson, orders—orders that she *must* follow or face criminal penalties, including jail. In this case, for example, the doctor had instructed Monson not to have intercourse with her husband during the remainder of her pregnancy, because it might induce contractions and a premature delivery. She nonetheless engaged in intercourse. The prosecution alleges that such "disobeying instructions" or "failure to follow through on medical advice" is grounds for criminal action. This strikes me as both silly and dangerous. Silly, because medical *advice* should remain *advice*: physicians are neither law makers nor seers. Dangerous because medical advice is a vague term that can cover almost anything.

After-the-fact prosecutions would not help individual fetuses. Effectively monitoring compliance would require actual confinement of pregnant women to an environment in which eating, exercise, drug use, and sexual intercourse could be controlled. This could, of course, be a maximum security country club, but such massive invasions of privacy can only be justified by treating pregnant women as nonpersons during their pregnancy; by seeing women as Margaret Atwood's handmaids who have only one function: to have the healthiest children we can make them have.

Other quandaries must be faced to apply child neglect statutes to fetuses. The first is that, unlike a child, the fetus is absolutely dependent upon its mother and cannot itself be "treated" without in some way invading the mother. The "fetal protection" policy enunciated by the prosecution seems to assume that like mother and child, mother and fetus are two separate individuals, with separate rights. But treating them separately before birth, and "balancing" their respective rights, can only result in favoring one over the other in disputes. Favoring the fetus radically devalues the pregnant woman, and treats her like an inert incubator, or as a culture medium for the fetus. This view makes women unequal citizens, since only they can have children, and relegates them to performing one main function: childbearing. It is one thing for the state to view the fetus as a patient; it is another thing to assume that the fetus' interests are in opposition to its mother, and to require the mother to be the servant of the fetus. The most extreme remedy in child neglect cases, for example, is for the state to take control of the child. The logical extension of child neglect remedies is for the state to take control of the pregnant woman—since it must do so to gain control of the fetus.*

Child neglect covers a wide variety of activities, but generally involves failure to provide necessities, like clothing, food, housing or medical attention, to the child. Such laws *do not*, however, require parents to provide optimal clothing, food, housing or medical attention to their children; and do not even forbid taking risks with their children (e.g., permitting them to engage in dangerous sports) or affirmatively injuring their children (e.g., corporal punishment to teach them a lesson). None forbid mothers to smoke, take dangerous drugs, or to consume excessive amounts of alcohol, even though these activities may have a negative effect on their children.

Even if we could define fetal neglect we are faced with the inherently sexist application of such a law. Can it be that the primary reason to attempt to make fetal abuse laws stricter than child abuse laws is that such laws almost always only apply to women? While this type of sex discrimination could survive current equal protection analysis, NYU's Sylvia Law seems correct in proposing a framework for equal protection analysis that would, in fact, protect women, and which fetal neglect laws could not survive:

> Laws governing reproductive biology should be scrutinized by courts to ensure that (1) the law has no significant impact in perpetuating either the oppression of women or culturally imposed sex-role constraints on individual freedom or (2) if the law has this impact, it is justified as the best means of serving a compelling state interest.[5]

*One logical action is to force women to undergo cesarean sections for the sake of their fetuses. See "The Most Unkindest Cut" at p. 119.

On the other hand, while it seems draconian to apply the child neglect standards to the mother's life style during pregnancy, the California statute could be interpreted to apply to fetuses in a way never envisioned by the legislature. Certainly the legislature never intended that women need provide any more "clothing, food or shelter" to their fetuses than they provide for themselves. But what about the terms "medical attendance" and "remedial care"? Historically, medical attendance for the fetus meant medical attendance for the mother. If fetal surgery becomes a therapeutic reality sometime in the future, however, "remedial care" may be applicable. Suppose, for example, that instead of the instructions the doctor gave Mrs. Monson, he had diagnosed her fetus as suffering from blocked ureters, and "recommended" surgery on the fetus to attempt to correct the problem. Would her failure to agree to this surgery be tantamount to "fetal neglect" for failure to provide her fetus with "necessary remedial care"?

This carries us into even more problematic waters. If women *must* "consent" to such "care" of their fetuses, they are not only relegated to the role of containers, but their own rights are also made so subordinate to those of a fetus that the container may be opened to gain access to the fetus without its consent, even when opening the container risks damage to it.

Waiving One's Right to Abortion

Some have argued that it is nonetheless fair to subject pregnant women to fetal neglect statutes because after they waive their right to abortion and decide to have a child, they take on added obligations to the future child, including providing it with such things as "necessary medical attendance." This argument seems misplaced for at least two reasons. First, such a "waiver" never in fact takes place. Women do not appear before judges or even notaries to waive their rights at any time during the pregnancy. Indeed, the vast majority of pregnancies in marriage are planned and delighted in, and viewing all pregnant women as potential aborters seems bizarre. Moreover, insofar as the right to terminate one's pregnancy is constitutionally-protected, it remains a woman's legal right to the time of birth, at least when her life or health is at stake. Secondly, and more importantly, women have a constitutional right to bear children if they are physically able to. To have a legal rule that there are no restrictions on a woman's decision to have an abortion, but if she elects childbirth the state will require her to surrender her basic rights of bodily integrity and privacy, creates a state-erected penalty on her exercise of her right to bear a child.[6] Such a penalty would (or at least should) be unconstitutional.

Conclusion

Attempts to define fetal neglect, and to set up a prenatal police force to protect fetuses from their mothers, are steps backward in terms of both women's rights and fetal protection. Women's rights will only be fostered when we treat them equally, and not subject them to oppressive laws that can never apply to men. The best chance the state has to protect fetuses is through actions to enhance the status of all women through fostering reasonable pay for the work they do, and equal employment opportunities, and providing a reasonable social safety net, quality prenatal services, day care programs, etc. It is probably not coincidental that government is trying to blame the pregnant victims of poverty for their problems at the same time government is cutting funds for maternal and child health care and nutrition. If the state really wants to protect fetuses it should do so by improving the welfare of pregnant women—not by oppressing them.

Converting pregnant women into vessels, even "sacred vessels," may seem a simple solution to a complex problem; but using the criminal law to transform a woman's and society's moral responsibility to fetuses into solely a woman's legal responsibility is not helpful. Indeed, even from a strictly utilitarian perspective, it is likely to lead to more harm to fetuses, since many women will quite reasonably avoid physicians during pregnancy if failure to follow the "advice" these physicians give them can result in their being thrown into a "country club" before birth, and into a jail afterwards.*

---

*On Feb. 26, 1987 Municipal Court Judge E. Mac Amos effectively dismissed all charges against Monson on the grounds that her actions did "not constitute a violation" of California's child support statutes.

Medical Paternity and Wrongful Life

When medical innovations make it possible to prevent the births of children with significant physical or mental handicaps, children who will require very expensive care during their lifetimes, litigation is a predictable aftermath. Thus, both genetic counseling and amniocentesis are the subject of an increasing number of courtroom confrontations between parents and physicians. While making physicians pay for the defects of their pregnant patients seems to some to smack of "medical paternity," it is the measure of damages, not the application of traditional medical malpractice principles, that makes these cases appear novel.

Because amniocentesis has been recognized as "routine" in certain cases, and because obstetricians are expected to provide accurate genetic counseling (or refer patients to someone who is qualified for this task), it is essential to understand what law now exists. In this regard a pair of recent New York cases provide a good starting point. While these cases follow precedents in other states very closely, they have been the subject of many irresponsible comments in the press: "A lot of doctors will interpret the decision as almost requiring amniocentesis"; "This decision could force a lot of doctors to hand out mimeographed sheets of paper so that they're in the clear or you'll have doctors randomly sticking needles into the amniotic sac;" "It would be a pity if a legal case forced doctors to perform a test which should be a medical decision"; and, from an insurance spokesman, who had not read the opinion, "The ruling is scarifying because it seems to expand liability beyond an already critical point ... it [the cost of malpractice insurance] could be astronomical."

The two cases were consolidated by the Court of Appeals, New York's highest court.[7] The first case involved Dolores Becker who in 1974, at the age of thirty-seven, became pregnant and was under the exclusive care of her obstetrician from the tenth week of her pregnancy until the child's birth. Her child was born with Down syndrome. Mrs. Becker alleged that she was not advised of the availability of amniocentesis or its ability to detect Down syndrome *in utero*, and that if she had been so advised she would have had an abortion if the test indicated her fetus was affected.

The plaintiffs in the companion case, Hetty and Steven Park, in 1969 had a child afflicted with polycystic kidney disease who died five hours after birth. Following this tragedy, they asked Mrs. Park's obstetricians what the chances were of future children being affected. The physicians allegedly replied (incorrectly) that the disease was not hereditary and that their chances of having a similarly affected child were "practically nil." Based on this information Mrs. Park again became pregnant, and again gave birth to a child suffering from polycystic kidney disease. Their second child, however, lived for two-and-a-half years. On written motions and without any of the issues having as yet been tried, the lower courts had agreed that the parents and their children had standing to sue for both wrongful birth and monetary loss, but could not sue for psychiatric and emotional distress. The question before the Court of Appeals was not who would win the lawsuits, if and when they were heard, but whether, *assuming everything the plaintiffs alleged was true*, the plaintiffs could recover damages.

Even given the lack of factual ambiguity, the court found that the questions presented cast "an almost Orwellian shadow" of "genetic predictability" and had to be resolved in a way that would necessarily transcend "the mechanical application of legal principles." The court faced two issues: (1) should a defective child be permitted to sue a physician (or anyone) for "wrongful life" on the theory that the child would have been better off never having been born?; and (2) should the parents of a defective child whose birth would have been prevented, but for the negligence of a physician, be able to recover from the negligent physician the costs of treating the defect?

Wrongful Life

At common law, courts routinely denied actions for "wrongful death" on the theory that this was a criminal and not a civil matter, and that such wrongs died with the victim. This rule caused great financial hardship to surviving spouses and children; and with the rise of the industrial age and more dangerous modes of transportation, England and all of the states in the United States eventually passed statutes permitting legal actions to be brought against the person who negligently or deliberately caused the death.

Neither courts nor legislatures, on the other hand, had ever recognized an action for "wrongful life." Negligence actions for "wrongful conception" or "wrongful pregnancy" resulting from failure of sterilization procedures have been recognized by some courts, but the plaintiff in these cases is the parent, not the child. Likewise, some courts have permitted the newborn to sue when the physician himself caused the child's defect, such as by negligently transfusing the child's mother with incompatible blood.

To prevail in a malpractice action the plaintiff, in this case the child, must prove four elements: duty, breach, damages, and causation. There is no problem with three of these. However, the damages issue is perplexing, and the New York court refused to permit children to sue because they have suffered no "legally cognizable injury." In the court's words: "Whether it is better never to have been born at all than to have been born with even gross deficiencies is a mystery more properly left to the philosophers and the theologians."

Recognition of such an action would also raise the possibility of an action brought for less than a "perfect" child. Finally, the court found it was impossible to place the party in the position he would have occupied but for the negligent act, since the plaintiff would not have existed at all:

> Simply put, a cause of action brought on behalf of an infant seeking recovery for wrongful life demands a calculation of damages dependent upon a comparison between the Hobson's choice of life in an impaired state and nonexistence. This comparison the law is not equipped to make.

Therefore, following the "wrongful death" history and previous decisions, the court concluded that the recognition of any such action, and the measurement of damages for it, "is best reserved for legislative, rather than judicial, attention." The cases by the children were, accordingly, dismissed.

Negligent Prenatal Advice

The actions by the parents, however, were upheld. Accepting their version of the facts, the court found that the parents had stated a proper malpractice claim (that is, the doctor had breached a duty to the patient that had caused damage) against the physician involved and that the alleged damages (monetary expenses for the care and treatment of the children) could be accurately measured. The court, however, refused to permit recovery for emotional or pyschiatric damages because permitting such an action would "inevitably lead to the drawing of artificial and arbitrary boundaries." One such consideration the court mentioned is balancing the emotional trauma with the parental "love that even an abnormality cannot fully dampen."

More specifically, the court ruled that if the parents could prove all of the allegations in a trial, the Beckers could recover the sums expended for long-term institutional care of their retarded child, and the Parks could recover the sums expended for the care and treatment of their child until her death. While the attorney for the Beckers predicted a multimillion-dollar verdict, it has since been learned that the couple gave up their retarded child for adoption before the decision. Their potential recovery will therefore be limited to actual expenses while they had custody of the child.*

The adoption points to a major problem in not allowing the child to recover damages: although the award is measured in terms of how much long-term care will cost, the parents are not obligated to spend any of it for the child's own care. Accordingly, if a multimillion-dollar verdict actually had been obtained, and the child thereafter put up for adoption, the parents would have become rich while the adoptive parents and the child would have no claim to the funds awarded on the basis of the child's defects. The legislature may wish to modify the law in this regard to permit the child to sue and require that any award, to either the parents or child, be placed in a trust fund for the exclusive benefit of the child regardless of his or her status as their child, an adopted child, or a ward of the state.

When the *Park* case went to trial, both sides agreed that damages were $67,358. The jury, however, believed that the doctors had in fact warned the parents, and accordingly found in favor of the doctors.

Conclusions

1. The law requires physicians to give accurate information to their patients, and forbids the withholding of vital information from them.

---

* This case was eventually settled out of court for $2,500, the amount the couple had previously spent on foster care.

These principles are consistent with the doctrine of "informed consent" and the reasonable expectations of pregnant women under a physician's care. "The physician does not guarantee a healthy child, but the reasonable expectation of the patient is that she will be apprised of any information the physician has that the child might be defective and of the alternative ways to proceed, so that the patient can determine what action to take."[8]

2. There is no requirement that obstetricians actually perform amniocentesis on any pregnant woman (indeed, many are not qualified to perform this procedure and for them to do so may itself be malpractice). Rather, the requirement is to inform women who are at risk of its availability and to refer them to an appropriate specialist or facility *if* they decide they want this diagnostic test.

3. The damages recoverable are much more limited than they could have been. Especially in the *Park* case, a strong argument can be made (as a dissenting judge notes) that damages for emotional harm directly suffered as a result of the physician's negligent assertion that polycystic kidney disease was not hereditary should be allowable. Such erroneous advice is grossly negligent, and allowing recovery for damages suffered as a result might have a salutary deterrent effect. Legislatures may wish to consider adopting statutes in this regard.

4. The issue of "wrongful life" is dead in the courts.* Only legislation will revive it. While I am in the minority, it does not seem to me that the concept of "wrongful life," no matter how denoted, is so strange. It is, for example, perfectly consistent with amniocentesis followed by abortion: both actions argue that no life *is* preferable to life with certain physical or mental defects. Further, since many defective newborns will never have the mental or physical ability to commit suicide, and may not have parents or others who can provide for their well-being, permitting them to sue for damages suffered on their own behalf is both rational and humane.

In short, the New York opinions are not novel, and can be interpreted as traditional judicial responses to the availability of new prenatal diagnostic techniques and genetic counseling responsibilities.

---

*As will be seen in the following essay, "Righting the Wrong," this prediction was premature.

Righting the Wrong
of Wrongful Life

A lower court in California has taken what might turn out to be a major step in remedying what Alexander Capron has termed "the wrong of 'wrongful life'"—denying compensation to a genetically defective child whose birth would have been prevented had proper information been made available to the parents. Previous courts had ruled against such children because they believed that there was no way to comprehend nonexistence, thus making it impossible to calculate damages based on a comparison of nonexistence to a defective existence, and that it seemed illogical to permit a child to recover for something he or she would not even be alive to sue for if he or she had not been born.

Capron has argued eloquently that neither of these arguments is persuasive. As to the first, we permit judges and juries to make similar distinctions and measurements, for example, in wrongful death cases, all the time. And as to the second, there is nothing illogical in the plaintiff saying, "I'd rather not be here suffering as I am, but since your wrongful conduct preserved my life, I am going to take advantage of my regrettable existence to sue you."[9]

The California case has been interpreted as against the weight of judicial authority and as creating a new cause of action against genetic counselors and obstetricians. It can, however, be seen as simply rectifying a legal anomaly that has up to now prevented a person who was severely injured, due to the negligence of another, from recovering damages from the negligent party. Why did the California court permit recovery on the part of the child while the highest court in New York would not? My conclusion, in a previous column about the New York cases, that "the issue of 'wrongful life' is dead in the courts," now seems premature.

The Curlender Case[10]

One can only guess why the California court did what it did. The courts were moving toward permitting recovery by the child (both parents and the child were initially denied recovery by the courts; and now parental claims are almost universally acknowledged). And the facts made it especially difficult to deny the child an opportunity to present her case to a jury. Shauna Tamar Curlender was born with Tay-Sachs disease. Her parents had previously retained the laboratories named as defendants to administer tests to determine whether or not they were carriers of recessive Tay-Sachs genes. The tests were reported to have been negative. The parents relied upon these tests. Because of the alleged negligence in the laboratory's performance, they had a daughter with Tay-Sachs disease, subject to severe suffering, who had a life expectancy of approximately four years.

The child's lawsuit sought damages for emotional distress and the deprivation of 72.6 years of life. She sought an additional $3 million in punitive damages on the grounds that the defendants knew their testing procedures were likely to produce a substantial number of false negatives and yet proceeded to use them "in conscious disregard of the health, safety, and well-being of the plaintiff ..." Since the complaint did not allege the date of the plaintiff's birth, the court could not determine whether the parents relied upon the test to conceive a child or to forego amniocentesis; nor does the court seem to care.

The court begins its inquiry with a worthwhile excursion through the history of so-called "wrongful life" cases. As Capron has noted, this term is unfortunate and misleading: the plaintiff is not attempting to recover because it has been born. "The wrong actually being complained of is the failure to give accurate advice on which a child's parents can make a decision whether not being born would be preferable to being born deformed." The court understands this, and quotes with favor the dissenting opinion in a 1979 New Jersey case in which the majority granted the right to recover to the parents of a Down syndrome child, but denied recovery to the child itself: "To be denied the opportunity—indeed the right— to apply one's own moral values in reaching that decision [concerning the child's future], is a serious, irreversible wrong."[11]

It also agrees with a Pennsylvania court that held, in a case involving the negligent performance of amniocentesis that failed to detect a fetus affected with Tay-Sachs disease, that "Society has an interest in insuring that genetic testing is properly performed and interpreted."[12]

On the basis of this history, and the court's own view of what the law should be, it makes three important observations concerning the handling of "the 'wrongful life' problem" by previous courts:

1. There is a major difference between a child who is unwanted or illegitimate, but healthy (the type of children involved in the original cases in which the courts first coined the term "wrongful life"), and a child who is born with a severe deformity or disease. Unlike a severe deformity, illegitimacy is simply not an *injury*.

2. There is a trend in the law to recognize that there should be recovery when an infant is born defective and its "painful existence is a direct and proximate result of negligence by others." In this regard the court notes that abortion is currently a woman's legal right, and that there has recently been a "dramatic increase in ... the medical knowledge and skill needed to avoid genetic disaster."

3. The injured parents and children have continued to sue for "wrongful life," in spite of past decisions against them, because of the seriousness of the wrong, an increasing understanding of its causes, and "the understanding that the law reflects, perhaps later than sooner, basic changes in the way society views such matters."

Given these observations, the court is determined to recognize the right of the child to sue for the negligent acts of others that led to her birth:

> The reality of the "wrongful life" concept is that such a plaintiff both *exists* and *suffers*, due to the negligence of others. It is neither necessary nor just to retreat into meditation on the mysteries of life. We need not be concerned with the fact that had defendants not been negligent, the plaintiff might not have come into existence at all. The certainty of genetic impairment is no longer a mystery ... a reverent appreciation of life compels recognition that the plaintiff ... has come into existence as a living person with certain rights.

This remarkable statement deals with the major imponderable in wrongful life cases by dismissing it with the phrase "We need not be concerned" about it. And this *is* the key to wrongful life recovery *by the child*. All the court's major concerns can be addressed by permitting the parents to recover damages. Certainly they have been injured: expecting a normal child, they must deal with an abnormal one. There are both emotional and monetary costs. But what about the child? He or she expected nothing, not even birth. He or she *never* had the possibility of being born healthy—only the chance to be either aborted or not conceived at all. From one way of looking at it, the child *could not* have been damaged by the testing laboratory's negligence, because without the negligence the child would not have existed at all. The argument is *not* that any life is better than none, no matter what the suffering; but rather that to be damaged one needs to be worse off after the negligent act complained of than before it. It cannot be said, in this sense, that the child is worse off existing than not existing if one assumes that non-existence is a state in which there are no rights and no rightful expectations.

On the other hand, we could determine that had the child, as fetus, had a voice in the matter, she would *herself* have opted to be aborted. Given the severity of her affliction such a decision would also have been in her best interests. Denying her parents the opportunity to make this decision *on her behalf*, thus injures her directly. For policy reasons we may also want to ignore the troublesome issue of nonexistence, and deal with the "reality" of the existence of a handicapped child who needs care and attention. It is correct to say that the testing laboratories did not cause the child's disease; but it is also correct that the child would not be in existence, suffering from the disease, but for the negligence of the testing laboratories. The court's position seems no less difficult to sustain than the more traditional one.

Concerning damages, the court denies recovery based on a seventy-year life expectancy, and instead requires damages to be based on the plaintiff's actual life expectancy of four years. Though one can quarrel a bit with this, it seems a fair compromise. After all, the child *never* would have had any opportunity for a seventy-year life span: it was no life or a four-year life. Under these circumstances, limiting damages for medical expenses and pain and suffering for actual life span seems reasonable. This may be especially so in light of the court's decision to permit the plaintiffs to proceed with their $3 million claim for punitive damages. It should be stressed, however, that the court did not find that the plaintiff did or should win this case, only that the plaintiff was entitled to try to prove in court what she alleged in the pleadings.*

---

*This case was eventually settled out of court for a reported $1.6 million.

Marching into Muddier Waters

The court should have stopped here. Instead it marches on into much muddier waters. It correctly notes a concern that recognizing the child's right to sue physicians and laboratories for negligence might logically lead to some courts recognizing the right of children to sue their own parents for allowing them to be born. While calling this fear "groundless," the court goes as far as any court ever has in approving the notion in principle. It specifically talks about a hypothetical case in which the parents have been warned of the probability of having a handicapped child, and yet decide to go ahead with the pregnancy. In such a case the court suggests no action could be brought against the physician or testing laboratory (of course), but remarkably it sees "no sound public policy which would protect those parents from being answerable for the pain, suffering, and misery which they have wrought upon their offspring."

In a similar vein, Margery Shaw, a geneticist and lawyer, has suggested that women who "abandon their right to abort" upon being informed that their fetus is defective, should incur a "conditional prospective liability" for negligent acts toward their fetus should it be born alive. She would permit such a defective child to sue its mother for negligence, and would also permit children harmed by fetal alcohol syndrome or drug addiction to sue their mothers. She might even go further:

> Withholding of necessary prenatal care, improper nutrition, expo-
> sure to mutagens and teratogens, or even exposure to the mother's
> defective intrauterine environment caused by her genotype, as in
> maternal PKU, could all result in an injured infant who might claim
> that his right to be born physically and mentally sound had been
> invaded.[13]

The court and Shaw are not alone, and their position is a *logical* extension of permitting the child to sue on its own behalf. But there are policy objections to this notion, the laudable purpose of which is to protect the unborn child. The most fundamental objection is that there is no "right to be born physically and mentally sound," and should not be. Such a "right" could almost immediately turn into a duty on the part of potential parents and their care-takers to make sure no "defective," different, or "abnormal" children are born. As Capron has argued, "The enforcement of such a rule by the state, through the courts and other agencies of social control, might even lead to unprecedented totalitarianism."[14] It could also lead to severe deprivations of liberty of pregnant women during the third trimester. One can envision confining certain women to ensure "proper nutrition" or withdrawal from alcohol or drug addiction, or to ensure a "healthful environment" for their fetuses. This seems absurd now but could seem reasonable in a world that viewed "normal" birth as an ultimate value.

Parents should be permitted, as they now are, to make good-faith judgments.* Some children will suffer, but this seems a less onerous result that the massive curtailment of liberty implicit in the "right to normalcy" notion. Caretakers have the obligation to provide parents with accurate information on which to base decisions; they do not have the right to make decisions for them. Only by keeping this distinction firmly in place can we help to assure that increases in genetic knowledge will lead to increases in human autonomy rather than to its destruction.

*California passed a statute soon after this case to prohibit children from suing their parents for wrongful life.

Is a Genetic Screening Test Ready When the Lawyers Say It Is?

When the President's Commission issued its final report on *Screening and Counseling for Genetic Conditions* in February 1983, there were no screening tests that were recommended for all pregnant women in the United States. Of the one possible candidate for this status, alphafetoprotein (AFP) screening, the Commission wrote somewhat cryptically:

> The controversy over AFP testing points up many of the difficult, important, and as-yet-unresolved questions about implementing large-scale screening programs. These questions take on added importance in light of the likely development of other genetic screening tests involving large segments of the population.

Neural tube defects (NTDs) are genetic disorders that occur with an overall incidence of one to two every 1000 live births in the United States. The two major types are anencephaly (lack of a brain) and spina bifida (a lesion on an incompletely closed spinal cord). Couples who have had a child with one of these conditions have a two to three percent risk of recurrence, and account for five to ten percent of all NTDs. Unfortunately, the remaining 90 to 95 percent of NTDs occur in families without any prior medical history. This means that to be effective, a screening test for NTDs must be used routinely on all pregnant women.

Alphafetoprotein is a major fetal serum protein secreted by the fetal kidneys, and normally present in amniotic fluid in measurable amounts. It may also, however, enter the amniotic fluid directly from exposed membrane surfaces on the fetus, as in anencephaly or open spina bifida. There is also an association between elevated levels of maternal serum alphafetoprotein (MSAFP) and NTDs; measuring second trimester concentration of MSAFP has been shown to be an effective means of identifying pregnant women who are at risk of having a fetus with a NTD.

Here is the process used in screening as described by geneticist-obstetrician Sherman Elias of the University of Tennessee. Following counseling and obtaining of informed consent, a blood sample is taken from the pregnant woman some time between sixteen and eighteen weeks of gestation as calculated from the first day of the last menses. In patients who show an elevated MSAFP (approximately five percent of those screened), the test is repeated. If the second MSAFP level is again elevated, an ultrasonographic evaluation is advised to detect conditions other than NTDs, such as gestational age, multiple pregnancies, fetal death or threatened abortion, Rh disease, and certain rare congenital abnormalities that are also associated with elevated MSAFP.

In less than half of this group (one to two percent of the original number screened) no explanation for the elevated level is found, and an amniocentesis for amniotic fluid AFP is recommended. Approximately five to ten percent of those undergoing amniocentesis show elevated AFP levels, the majority of which will be associated with NTDs. Properly administered, this series of tests detects 80 to 90 percent of all anencephalies and 63 to 90 percent of all open spina bifidas. More recently it has been shown that such testing may also detect 20 to 40 percent of all Down syndrome fetuses.

Why Not Advise the Test for Everyone?

The reasons for not rushing into routine MSAFP screening are evident from this description. To be useful to all pregnant women, high-quality laboratory, counseling, ultrasound, and amniocentesis services must be available and accessible across the country. In the absence of such services, MSAFP could simply increase cost and parental anxiety (since the vast majority of elevated MSAFP samples will turn out *not* to be associated with NTDs or any other abnormality), and possibly lead to unnecessary abortions.

These and other reasons led the American College of Obstetricians and Gynecologists (ACOG) to oppose FDA approval of AFP test kits, and to tell its members in an October 1982 Technical Bulletin (No. 67) entitled "Prenatal Detection of Neural Tube Defects" that although the test was useful for those patients who had had a previously affected child, "routine maternal serum AFP screening of all gravida is of uncertain value":

> Maternal serum AFP screening should be implemented only when it can be performed within a coordinated system of care that contains all the requisite resources and facilities to provide safeguards essential for ensuring prompt, accurate diagnoses and appropriate follow-through services. *When such coordination of resources and services is not possible*, the risks and costs appear to outweigh the advantages and *the program should not be implemented.* (emphasis added)

Enter the Lawyers

Nonetheless, with no stated justification other than the 1983 FDA approval of commercial kits for the radioimmunoassay of AFP, ACOG members received a May 1985 "Alert" from the College's Department of Professional Liability entitled "Professional Liability Implications of AFP Tests." The Alert stated in part:

> It is now imperative that you investigate the availability of these tests in your area and familiarize yourself with the procedure, location and mechanism of the follow-up tests to screen for neural tube defects. It is equally imperative that *every prenatal patient* be advised of the availability of this test and that your discussion about the test and the patient's decision with respect to the test be documented in the patient's chart. (emphasis added)

The rationale for this advice was not medical, but legal: to give the physician "the best possible defense" in a medical malpractice suit premised on the birth of a baby with a neural tube defect. While in the long run this alert may help encourage physicians to provide accurate information about MSAFP testing to their patients, in the short run it simply creates confusion. For example, two physicians wrote incorrectly in the letters section of the *New England Journal of Medicine* that the Alert meant that ACOG had recognized "the value of maternal alpha-fetoprotein screening."

Such confusion is understandable. It is unprecedented and counter-productive for a professional medical association, dedicated to articulating and promoting sound medical practice, to permit its *legal* department to promulgate *medical* standards.   Certainly MSAFP screening should be offered to pregnant women where appropriate counseling and follow-up care are available.   But, as ACOG correctly noted in 1982, where such care and counseling are not available, "the risks and costs appear to outweigh the advantages."

Setting Standards of Practice

ACOG's Department of Professional Liability has been very agitated about what it sees as a crisis of medical malpractice insurance unavailability and unaffordability, a crisis that has reportedly led about ten percent of ACOG's Fellows to give up obstetrics altogether and almost 20 percent to drop high-risk obstetrics.   But if malpractice litigation is a problem, letting lawyers set medical standards of care is hardly the solution.

The general rule has always been that "good medicine is good law." Physicians who follow good and accepted medical standards in their practice will be found in compliance with the general standard of "reasonable prudence" for physicians, and thus cannot be found negligent in a malpractice suit. It may also be said of the ACOG Alert that "bad law is bad medicine." If routine MSAFP screening has come of age in the United States, ACOG's scientific committees, not its legal liability department, should so advise its members.

Perhaps ACOG's lawyers were thinking about the famous case of *Helling v. Carey*[15] in which two ophthalmologists were found negligent for not having performed a glaucoma test on a young woman whose glaucoma, when it was finally discovered, had progressed to a point where her peripheral vision was lost and her central vision severely reduced.   In their defense, the physicians argued that it was the standard of care in the ophthalmology community *not* to routinely screen patients under the age of forty for glaucoma because the incidence of the disease in this population (about one in 25,000) was so low.   On appeal from a verdict for the physicians, the Supreme Court of Washington noted that the standard of care is judged on the basis of "reasonable prudence."   Even though ophthalmologists did not routinely offer this test to patients under 40 years of age, the court decided that not to so offer it was a failure of reasonable prudence. The glaucoma test, it found, was simple, inexpensive, safe, accurate, and could detect an arrestable and "grave and devastating" disease.

The court quoted with approval the past statements of Justices Oliver Wendell Holmes and Learned Hand, both to the effect that while reasonable prudence is usually measured by custom or what is in fact done, custom can never be its sole measure. As Justice Hand put it in the famous *T.J. Hooper* case:

> [A] whole calling may have unduly lagged in the adoption of new and available devices. It never may set its own tests, however persuasive be its usages. *Courts must in the end say what is required; there are precautions so imperative that even their universal disregard will not excuse their omission.*[16]

But what would this same court have said about MSAFP testing? The initial screening test is relatively inexpensive and safe, but accuracy requires highly trained professionals to do a complex series of follow-up tests, including ultrasound and amniocentesis. And even with this, there will be false negatives. Nor can this series of tests be fairly termed "simple." It does detect a serious condition, one that can even be described as "grave and devastating," but the condition cannot be "arrested" or "treated" except by aborting the affected fetus.

Whether the *Helling* court (or any other court) would require a physician to inform a pregnant patient about MSAFP testing would probably hinge on the existence of those factors ACOG identified in its 1982 Technical Bulletin: "a coordinated system of care that contains all the requisite resources and facilities to provide safeguards essential for ensuring prompt, accurate diagnoses and appropriate follow-through services."

In ACOG's September 1985 *Newsletter*, its Director of Fellowship Activities, Keith White, tried to retreat from the Alert without retracting it. The attempt was important, since the Alert could be introduced in court as evidence of a standard of care adopted by ACOG. White states flatly: "The College has not, and does not recommend routine screening of maternal serum for AFP ..." He nonetheless concludes with a strangely equivocal statement: "The Department of Professional Liability *does not* set standards; we are trying to 'tell it like it is.'"

But this does not help either the physician or the patient. Surely no one can take seriously the argument that it is a good defense in a malpractice suit to have documented in your patient's chart that you advised her of the "availability" of MSAFP testing, and noted her "decision" about it; even though ACOG does *not* recommend the test for routine screening, and no facilities for testing and followup were reasonably available in your geographic area.

ACOG's lawyers overreacted. It is refreshing to see physicians rushing to defend lawyers, but to "tell it like it is," it won't work. Energy should be directed toward setting up a MSAFP screening system that is available to all patients regardless of economic status or source of payment, a system based on high professional standards of testing, with solid and helpful counseling at each step along the way. That seems the only reasonably prudent course.*

---

*ACOG will probably soon recommend that MSAFP screening be made available to all pregnant women as part of routine prenatal care. Elias, S. & Annas, G. J., "Routine Prenatal Genetic Screening," *New Engl J Med* **317**: 1407–1409, 1987.

Homebirth
Autonomy vs Safety

The level of hostility between consumer groups and physicians has probably been more intense on homebirth than on any other issue. Many consumers, frustrated in their demands for more humane birthing conditions in hospitals, have abandoned their hopes for institutional change and embraced the homebirth alternative. Physicians have reacted in a variety of ways. The American College of Obstetricians and Gynecologists (ACOG), for example, issued a statement in May 1975:

> Labor and delivery, while a physiologic process, clearly presents potential hazards to both mother and fetus before and after birth. These hazards require standards of safety which are provided in the hospital setting and cannot be matched in the home situation.

A later ACOG survey, based on highly questionable data from eleven of fifty state health departments, found mortality risks to infants born at home to be two to five times greater than to those born in hospitals. Obstetricians have accordingly been highly discouraged from taking part in homebirths, and some who have done homebirths have faced loss of hospital privileges. Pediatricians and neonatologists recently entered the arena, arguing that homebirth is a lunatic fringe movement and that it could legally constitute child neglect. That consumers and providers are still far apart was underscored at an April 1978 meeting of the National Association of Parents and Professionals for Safe Alternatives in Childbirth (the organization's third national meeting) in Atlanta.

Those attending the conference heard stories of police and social workers being summoned during or shortly after homebirths by pediatricians alleging child neglect, of physicians being threatened with loss of hospital privileges and even licensure revocation for taking part in homebirth, of nurses who faced license revocation for homebirth participation, and of midwives with various backgrounds under extreme pressure to abandon their vocation.

Since pregnant women are not sick, have a good deal of notice before hospitalization, and have a number of alternatives, they are perhaps the most vocal and effective group of "patients" confronting today's health care system. The issues that they raise will not soon be resolved, and it is important that the legal context in which the current debate is being waged be understood. As a baseline, for example, it is worth noting that there is no law requiring women to deliver their babies in hospitals; and even if there were, it would be of dubious constitutional validity and unenforceable.

The foundations of the current law are based on a long tradition of protecting the public against medical quacks and a more recent desire to protect children from parental abuse and neglect. Since both licensure laws and child abuse and neglect statutes have ample rationales independent of childbirth, neither is likely to be abandoned. The challenge for those desiring change is to work for modifications within the framework of these laws.

Homebirths Attended by Physicians

An initial decision involves the birth attendant. The state has an interest in the outcome of the birth, and in the qualifications of the attendant. Therefore, only persons with certain qualification may legally attend births. In *Roe v. Wade*, the 1973 abortion decision,* for example, the United States Supreme Court noted that, though a woman had a right to have an abortion before viability, the state had such a strong interest in protecting the life of a viable fetus that, after viability, it could constitutionally forbid abortion altogether except in cases where the woman's life or health was at stake, and a state could constitutionally require that abortions only be performed by licensed physicians.*

Although no state forbids physicians to attend homebirths, it is often very difficult for a woman to find a willing physician. One deterrent seems to be fear of potential malpractice liability. This fear is based primarily on ignorance of the law, which has led to an increase

*See the Section on Reproductive Liberty for a detailed discussion of the US Supreme Court's abortion decisions from 1975 through 1986, and pp. 142–143 for a summary of *Roe v. Wade*.

in so-called "defensive medicine." One type of "defensive" posture would be to refuse to do home deliveries on the grounds that hospital births, while unnecessary, are standard medical practice; and failure to hospitalize could look bad to a jury and could be taken as de facto negligence. Indeed, at least one physician—a specialist in medical malpractice—has argued that a physician-attended homebirth in New York City is de facto evidence of physician negligence.[17]

This assertion is simply wrong. So long as the decision to have a homebirth is made by the woman after she has been fully informed of the potential risks and complications, and so long as all generally accepted medical steps have been taken to screen the woman for medical complications and provide emergency backup facilities, it is highly unlikely that any malpractice action against a physician would be successful. Indeed, because of the absence of interventions like anesthesia, and the presence of a well-informed and participating woman, a physician is less likely to be sued for malpractice in the homebirth setting.

Midwives and Licensing

If, for whatever reason, the couple does not want or cannot obtain the services of a physician, the next choice would be a licensed midwife. Statutes and qualifications vary, but in states that have specific legislation providing for the licensing of midwives, it is likely that courts would find that any nonphysician or nonlicensed midwife, who claimed to be an expert in childbirth and attended a childbirth, would be guilty of the crime of practicing either medicine or midwifery without a license.

The court's reasoning would probably parallel that of a California Supreme Court decision which found that the legislature had an "interest in regulating the qualifications of those who hold themselves out as childbirth attendants...for many women must necessarily rely on those with qualifications which they cannot personally verify."[18]

Parental Obligations

The parents' primary duty is to the child, but so far only after it is actually born.* No state has yet attempted to require pregnant women

---

*But see, "Forced Cesareans: The Most Unkindest Cut of All" at p. 119; and "Fetal Neglect" at p. 91.

to take any affirmative action to safeguard their fetuses. However, all states have passed statutes that forbid parents from abusing their children, and that require them to provide necessary medical attention for their children. Parents may thus in general make a decision to have a homebirth with impunity. If they have reason to know that complications are likely to develop that will require hospital care to save the child from death or permanent injury, however, and the child dies or is permanently injured because of the homebirth, it is possible that a zealous prosecutor would bring a criminal complaint against the parents. In both cases the charge would be child neglect, and if the child died, possibly manslaughter (depending on the cause of death and its predictability). This possibility is intended to discourage parents from attempting to manage homebirths by themselves, and to encourage them to seek a licensed attendant at the homebirth and to seek hospitalization when it is indicated.

The Medical Profession and Homebirths

The medical profession's reaction to homebirth has not been enthusiastic. In Massachusetts, the Board of Registration in Medicine seriously considered a regulation prohibiting physicians from participating in homebirths. The regulation was rejected, not because the Board supported homebirth, but because it decided that it was inappropriate to attempt to interfere with the way a physician desires to practice and the way women want to have their babies. The Board also recognized that such a regulation would not prevent homebirths, but would only ensure that they would not be attended by a physician. It is remarkable that this issue even reached the Board, since only about 200 to 500 homebirths occurred in Massachusetts that year out of about 70,000 births.

Nevertheless, the issue has become highly politicized. A past president of the Massachusetts section of the ACOG was quoted as saying that home-birthers are "kooks, the lunatic fringe, people who have emotional problems they're acting out." In the legislature, a statute to permit the licensing of nurse-midwives specified that they could practice only in the hospital or clinic setting, and then only under the supervision of a physician.

In November 1976, the Obstetrics-Gynecology Department of Yale–New Haven Hospital adopted the following "department policy":

Hereafter (December 1, 1976) any physician with OB privileges at YNHH who intentionally participates in a non-emergency "home delivery" will be viewed as no longer fulfilling the professional expectations of the OB staff of the hospital, and will immediately have OB admitting privileges revoked.

This policy is legally dubious because it has nothing directly to do with in-hospital care and therefore can make no pretext about being for the safety of hospital patients. As to nonhospital patients, the hospital has no legitimate concern with their free exercise of the legal option of homebirth. The only interest it seems to have is economic; and if the attempt is to join in a conspiracy to restrain trade in homebirths, the activity might also be challenged under the antitrust laws. Finally, the promulgating authority seems ambiguous in this case, and summary revocation of privileges is inconsistent with the doctrine that staff qualifications must not only be "reasonably related to the operation of the hospital," but also "fairly administered."[19]

A more offensive tactic has been employed by a number of pediatricians who have reported couples having homebirths to the state authorities under relevant child abuse and neglect statutes. In at least one case in Idaho, a child was forcibly removed from a home for a period of time. Although the reporting of child abuse is generally protected by immunity statutes, the report must be made in "good faith." The mere hatred of homebirths is an insufficient reason for filing such a report. A physician who filed such a report on the sole basis that he or she disagreed with homebirths could be successfully sued for libel. It is unfortunate that some members of the medical profession have resorted to this dubious and traumatic tactic, and equally unfortunate that the only thing likely to stop this offensive is a successful suit against a reporting physician. Physicians who want to encourage women to have their babies in hospitals would be most effective in this quest by modifying hospital procedures to give women more control over the manner of the birth and bonding experience. Some physicians still think that the physician–patient relationship is the most sacred in society. For them it may be. But to their patients many other relationships, like mother–child and husband–wife, transcend it completely. Many consumers will rightly refuse to compromise these relationships and the values they represent for an unpredictable and often marginal additional level of safety.

Forced Cesareans
The Most Unkindest Cut of All

Recently, four Israeli obstetricians suggested that when women in labor refuse surgical intervention recommended to save the life of their fetuses, "It is probably that the patient hopes to be freed in this way of an undesired pregnancy...because it is an unplanned pregnancy, the woman is divorced or widowed, the pregnancy is an extramarital one, there are inheritance problems, etc."[20]

The view that women who refuse cesarean sections are in some way willfully abusing their fetuses seems prevalent and deeply held, at least by some male obstetricians and judges. It is reflected in two recent cases in which Georgia and Colorado judges ordered women who were refusing surgical interventions during labor to undergo them. This essay reviews the Georgia and Colorado cases and argues that the law is and should remain that a pregnant woman has a right to refuse surgery recommended for the sake of her fetus.

The Georgia Case[21]

Jessie Mae Jefferson was due to deliver her child in about four days when the hospital in which she would be attended sought a court order authorizing it to perform a cesarean section and any necessary blood transfusions should she enter the hospital and refuse. She had previously notified the hospital that it was her religious belief that the Lord had healed her body and whatever happened to her child was the Lord's will. At an emergency hearing conducted at the hospital, her examining physician testified that she had complete placenta previa with 99 percent certainty that her child would not survive vaginal delivery and a 50 percent chance that she herself would not survive it. On this basis, the court decided that the "unborn child" merited legal protection and authorized the administration of "all medical procedures deemed necessary by the attending physician to preserve the life of the defendant's unborn child."

The next day a public agency petitioned for temporary custody in the same court, alleging that the unborn child was "a deprived child without proper parental care" and seeking an order requiring the mother to submit to a cesarean section. The odds that the unborn child would die if a vaginal birth was attempted were put at 99 to 100 percent by the physician. The court granted the petition, on the basis that the "State has an interest in the life of this unborn, living human being [and] the intrusion involved...is outweighed by the duty of the state to protect a living, unborn human being from meeting his or her death before being given the opportunity to live."

The parents immediately petitioned the Georgia Supreme Court to stay the order; and on the evening of the same day as the hearing, that court denied their motion, with a two-sentence conclusory opinion, citing *Roe v Wade, Raleigh Fitkin*, and *Strunk v Strunk* as authority. A few days later, Mrs. Jefferson uneventfully delivered a healthy baby without surgical intervention.

The Colorado Case

The pregnant woman in the Colorado case was unmarried, and had previously given birth to twins. She was described as obese, angry, and uncooperative. An internal fetal heart monitor suggested fetal hypoxia, and a cesarean section was recommended. Because of the patient's fear of surgery, she refused. Her mother, sister, and the father of her child attempted unsuccessfully to change her mind. A psychiatric consultant concluded she was neither delusional nor mentally incompetent.

The hospital administration was notified, and a decision was made to request court intervention. The hospital staff petitioned the juvenile court to find the unborn baby a dependent and neglected child and order a cesarean section to safeguard its life. An emergency hearing was convened in the patient's room, following which the court granted the petition and ordered the surgery. The cesarean section was performed, resulting in a healthy child without complication for the mother. Since more than nine hours elapsed between the external fetal heart monitor tracings that indicated distress (and six hours from internal tracings) and the delivery, the physician was surprised that the outcome was not poor. He indicated that the case "simply underscores the limitations of continuous fetal heart monitoring as a means of predicting neonatal outcome."[22]

Both cases hold that a woman can be forced to undergo a cesarean section if her physicians determine it necessary to safeguard the life of her fetus. These are remarkable rulings from geographically and politically diverse areas of the country. But both were decided hours after they were argued, without time for thoughtful judicial consideration, and neither court made any attempt to analyze the rights of the pregnant woman. At least three questions arise: What is the state of law? What should the role of the judiciary be in labor room disputes? What position should physicians and the hospital take when confronted with a woman who refuses a cesarean section against medical advice?

The State of the Law

Both cases lack an analysis of the precedents, and place heavy and primary reliance upon *Raleigh Fitkin*,[23] and *Roe v. Wade.** The courts should have at least considered the severe limitations of these two cases. *Raleigh Fitkin* involved a woman who was approximately eight months pregnant. Physicians believed that at some time before giving birth she would hemorrhage severely and both she and her unborn child would die if she did not submit to blood transfusions that she had refused because she was a Jehovah's Witness. The trial court upheld her refusal, and the hospital appealed to the New Jersey Supreme Court. In the meantime the woman had left the hospital against medical advice, and the case was moot. Nevertheless, the court proceeded to determine that the unborn child was "entitled to the law's protection" and that blood transfusions could be administered to the woman "if necessary to save her life or the life of her child, as the physician in charge at the time may determine."

This opinion is of limited value. First, no one was forced to do anything as a result, i.e. no transfusion was actually performed, and no police were dispatched to apprehend the woman and return her to the hospital. Second, it was a one-page opinion, with little policy discussion. Third, the extent of bodily invasion involved in a blood transfusion is much less than that involved in a cesarean, major abdominal

*Discussed at pp. 142–143.

surgery. And fourth, the case was decided eight years before the US Supreme Court decision in *Roe v Wade* and more than a decade before the same New Jersey court decided the case of Karen Ann Quinlan. One question posed (and not yet resolved), for example, is: would the parents of Karen Ann Quinlan have been permitted to remove her from the respirator if she had been pregnant?

The second case, *Roe v. Wade*, does stand for the proposition that the state has a compelling interest in preserving the life of viable fetuses. But it does not have such an interest if "the life or health of the mother" is endangered by carrying the child to term. The question that needs to be discussed is the relevance of the additional danger (physical or mental) to the mother of undergoing a cesarean section where its purpose is to protect the health of the fetus. In the Colorado case, for example, it was noted that excessively obese patients are "generally considered at increased risk of anesthetic and surgical complications." When do such increased risk factors outweigh the child's right to be born via cesarean? And what would happen if, despite a court order, the patient refused to submit to the cesarean? The physician in the Colorado case cautions that "had the patient steadfastly refused it might not have been either safe or possible to administer anesthesia to a struggling, resistant woman who weighed in excess of 157.5 kg." Surely nothing in *Roe v Wade* gives either judges or physicians the right to favor the life or health of the fetus over that of the pregnant woman.

No mother has ever been legally required to undergo surgery or general anesthesia (e.g., bone marrow aspiration) to save the life of her dying child. It would be ironic and unfair if she could be forced to submit to more invasive surgical procedures for the sake of her fetus than for her child. It is premature to label the conclusions of two quickly decided cases that lack any meaningful analysis "the law."

Judges in the Hospital

Judges are not terribly good at making emergency decisions. Perhaps the most famous example is the opinion of Judge Skelly Wright in the *Georgetown College* case.[24] That case involved an emergency petition to permit blood transfusions to a Jehovah's Witness to save her life. A lower court judge refused to issue such an order, but Judge Wright did, less than an hour-and-a-half after he was approached by counsel for the hospital. He went to the hospital and interviewed the woman and her husband. The woman, a twenty-five-year-old with a seven-month-old child, was "not in a mental condition to make a decision." Her husband refused; but said if the judge ordered it, it would not be his responsibility. Because the judge believed that the woman's reasoning would be similar, he ordered the transfusions.

The full bench of the US Circuit Court of Appeals refused to review the case, but some of the members dissented from this refusal, and stated their concerns.[25] Judge Miller, for example, noted that Judge Wright was

> ...impelled, I am sure, by humanitarian impulses and doubtless was himself under considerable strain...In the interval of about an hour and twenty minutes between the appearance of the attorney at his chambers and the signing of the order at the hospital, the judge had no opportunity for research as to the substantive legal problems and procedural questions involved. He should not have been asked to act in these circumstances.

Judge Warren Burger, later Chief Justice of the US Supreme Court, quoted Justice Benjamin Cardozo on judicial restraint:

> The judge, even when he is free, is still not wholly free. He is not to innovate at pleasure. He is not a knight-errant, roaming at will in pursuit of his own ideal of beauty or of goodness. He is to draw his inspiration from consecrated principles. He is not to yield to spasmodic sentiment, to vague and unregulated benevolence.

It is inappropriate for judges to act impulsively, without benefit of reflection on past precedent and the likely future impact of their opinions. Both the cesarean section cases discussed in this essay suffer from lack of reflection. Obviously the delivery room is not conducive to such reflection, and judges do not belong there at all in such circumstances.

What Should the Law Be?

The law can take one of two paths, neither completely satisfactory. The first is to follow the lead of the Georgia and Colorado cases, and require women to submit to cesarean sections when their physicians deem them necessary to protect their fetuses. The problems with this approach are illustrated by these two cases. First, physician prediction of fetal harm is not very accurate. Indeed, in both of these cases serious errors were made. In the first, a 99 percent certainty turned out to be wrong; the supposed 1 percent reality occurred. And in the Colorado case, the fetal heart monitor significantly overstated the amount of damage to the fetus from delayed delivery. So permitting physicians to judge when fetuses are in danger may simply be giving them a license to perform cesarean sections whenever they want to, without regard to the pregnant woman's desires.

But suppose 100 percent accuracy. We may still want to permit women to refuse surgery, to protect their liberty as well as that of all competent adults. Practical considerations also support the woman over the fetus. Women may take matters into their own hands and not deliver in hospitals. Other interventions they might consent to will be unavailable at home, and an opportunity to try to change their minds will be lost. The question of what to do with a woman who continues to refuse in the face of a court order remains. Do we really want to restrain, forcibly medicate, and operate on a competent, refusing adult? Such a procedure may be "legal," especially when viewed from the judicial prospective that the woman is irrational, hysterical, and evilminded; but it is certainly brutish and not what one generally associates with medical care. It also encourages an adversarial relationship between the obstetrician and the patient, and gives the obstetrician a weapon to bully women he views as irrational into submission. Attempts at vaginal deliveries after one birth by cesarean section, for example, may fall victim to such a rule.

Could the case be distinguished from other medical interventions, including fetal surgery, if and when it becomes accepted medical procedure, or would women be forced to consent to these as well? And if one can lawfully force surgery, one should certainly be able to restrain the liberty of a woman for the sake of her fetus, e.g., by confining her during all or part of her pregnancy. It seems wrong to say that patients have the right to be wrong in all cases except pregnancy—in that case, why should only doctors have the right to be wrong?

The second alternative is to honor the rare case of a woman's refusal. I assume this is general practice at the vast majority of hospitals in the country, and I believe it is the proper practice ethically and legally. This may seem callous to the rights of fetuses since some fetuses that might be salvaged may die or be born defective. This will be tragic, but it is likely to be rare. It is the price society pays for protecting the rights of all competent adults, and preventing forcible, physical violations of women by coercive obstetricians and judges. The choice between fetal health and maternal liberty is laced with moral and ethical dilemmas. The force of law will not make them go away.*

*In Aug. 1987, ACOG's Committee on Ethics urged obstetricians not to seek court intervention to force treatment, but rather to concentrate on "education and counseling." See also Annas, G. J., "Protecting the Liberty of Pregnant Patients" *New Engl J Med* **316**: 1213–1214; 1987.

NOTICE

DEPARTMENT OF HEALTH AND HUMAN SERVICES
Office for Civil Rights

DISCRIMINATORY FAILURE TO FEED AND CARE FOR HANDICAPPED INFANTS IN THIS FACILITY IS PROHIBITED BY FEDERAL LAW. SECTION 504 OF THE REHABILITATION ACT OF 1973 STATES THAT

"NO OTHERWISE QUALIFIED HANDICAPPED INDIVIDUAL SHALL, SOLELY BY REASON OF HANDICAP, BE EXCLUDED FROM PARTICIPATION IN, BE DENIED THE BENEFITS OF, OR BE SUBJECTED TO DISCRIMINATION UNDER ANY PROGRAM OR ACTIVITY RECEIVING FEDERAL FINANCIAL ASSISTANCE."

<u>Any person having knowledge that a handicapped infant is being discriminatorily denied food or customary medical care should immediately contact:</u>

Handicapped Infant Hotline
U.S. Department of Health and Human Services
Washington, D.C. 20201
Phone 800-<u>368-1019</u> (Available 24 hours a day) - TTY Capability

In Washington, D.C. call <u>863-0100</u>

OR

Your State Child Protective Agency

Federal Law prohibits retaliation or intimidation against any person who provides information about possible violations of the Rehabilitation Act of 1973.

Identity of callers will be held confidential.

Failure to feed and care for infants may also violate the criminal and civil laws of your state.

Disconnecting the
Baby Doe Hotline

Three Down syndrome children—Phillip Becker,* an unnamed Johns Hopkins baby, and Baby Doe—have become the focus of public controversy over the issue of withholding treatment and nutrition from handicapped children. A month after Baby Doe died in Bloomington, Indiana, at the age of six days, the Secretary of Health and Human Services (HHS) put hospitals on notice that it was "unlawful" for a recipient of Federal financial assistance to withhold from a handicapped infant nutritional sustenance or medical or surgical treatment required to correct a life-threatening condition if: (1) the withholding is based on the fact that the infant is handicapped; and (2) the handicap does not render treatment or nutritional sustenance contraindicated.[26] In announcing the letter, the Secretary said: "The President has instructed me to make absolutely clear to health care providers in this nation that federal law does not allow medical discrimination against handicapped infants."

Approximately ten months later, shortly after the tenth anniversary of the US Supreme Court's abortion decision, and the week a Boston television station aired a sensationalized television series on withholding treatment that featured the film "Who Should Survive?" about the Johns Hopkins baby (a series viewed with approval by the President), the White House instructed HHS to issue a followup regulation with more force. Specifically, the substance of the May 1982 notice was to be included in a poster displayed conspicuously in each delivery ward, maternity ward, and intensive care nursery (*see* p. 125) Included in the notice was a toll-free, 24-hour-a-day "hotline" number. Individuals with knowledge of any handicapped infant being discriminatorily denied food or customary medical care were encouraged to call. HHS officials were given authority to take "immediate

*See pp. 219–231 for two essays on Phillip Becker.

remedial action" to protect the infant, and hospitals were required to provide access to the hospital and its records to agency investigators.

Normally, publication of a proposed rule must be followed by at least 30 days in which affected parties can comment, but HHS argued that the requirement was unnecessary in this case. The access and medical records requirements, HHS maintained, were "minor technical changes" in the law "necessary to meet emergency situations," and all other aspects of the interim final rule were "necessary to protect life from imminent harm." The agency argued in its notice that "any delay would leave lives at risk" and that "for even a single infant to die due to lack of adequate notice and complaint procedure is unacceptable."[27]

Legal Challenge

The American Academy of Pediatrics, the National Association of Children's Hospitals and Related Institutions, and Children's Hospital National Medical Center brought suit against HHS and its new Secretary, Margaret Heckler, to enjoin the interim final rule on March 18, four days before it was to become effective. US District Court Judge Gerhard Gesell denied a temporary restraining order, but granted expedited review and agreed to hear the case upon submission of written documents and arguments on April 8. Six days later he ruled that the HHS regulation was invalid because the agency had failed to follow proper procedures in promulgating it.[28]

Judge Gesell is on strong grounds when he criticizes the procedural aspects of promulgating the regulation. He correctly notes that the Administrative Procedure Act (APA) was "designed to curb bureaucratic actions taken without consultation and notice to persons affected." The right to notice and comment was established to make sure that the rule-making agency at least considered all factors that the affected individuals believed were relevant so that rational decisionmaking could be fostered and arbitrary and capricious decisionmaking avoided. In this case, the interim final rule was issued without notice or an opportunity to comment before it became effective. According to the record, the Secretary relied primarily on the Boston television videotape of its investigative series "Death in the Nursery," a series of news accounts about it, a McNeil-Lehrer broadcast, and survey articles from "medical and academic journals."

The judge noted that he could not substitute his own judgment for the Secretary's, but rather had to determine if the agency had considered the "relevant factors" in making its rule. If it did not, "the test of rationality cannot be sustained because it is arbitrary and capricious." On this note the judge found adequate consideration that there might be a problem, but no consideration as to the effect of a toll-free hotline, which could be triggered by an "anonymous tipster": "... the sudden descent of Baby Doe squads on the scene, monopolizing physician and nurse time and making hospital charts and records unavailable during treatment, can hardly be presumed to produce higher quality care for the infant."

The judge concluded that the primary purpose of the rule was to "require physicians treating newborns to take into account wholly medical risk–benefit considerations and to prevent parents from having any influence upon decisions as to whether further medical treatment is desirable." No alternative means of preventing discriminatory denial of food and treatment were explored, and no definition of "customary medical care" was given. Without such a definition the judge concluded that the regulation is "virtually without meaning beyond its intrinsic *in terrorem* effect." He declared it invalid due to the Secretary's failure to follow the 30-day notice and comment requirements of the Administrative Procedure Act. This is the holding of the case, and the judge could have stopped here; but he went on to discuss the statutory authority under which the rule was promulgated and the rule's constitutionality.

Statutory and Constitutional Issues

The rule was promulgated under authority of section 504 of the Rehabilitation Act of 1973, which forbids discrimination on the basis of handicap by any program receiving federal financial assistance. Though there was little debate as to the applicability of the statute to infants, some *amici* briefs suggested that the rule required doctors to take all heroic measures needed to prolong life as long as possible, despite expense and prognosis. Noting that previous opinions suggested that 504 was never meant to be applied so "blindly and without any consideration of burdens," the judge recommended congressional clarification. But he refused to rule on the adequacy of statutory authority for the regulation until a specific application of the regulation raised this issue.

The judge likewise refused to deal with the constitutional issues of vagueness and due process. However, he did note that "to the extent the regulation is read to eliminate the role of the infant's parents in choosing an appropriate course of medical treatment, its application may in some cases infringe upon [right to privacy interests]."

In his conclusion, Judge Gesell characterized the regulation as "arbitrary and capricious" and the hotline as "hasty" and "ill-considered." He opined that any federal intervention in the delivery rooms and newborn intensive care units should "obviously reflect caution and sensitivity" and that "at a minimum, wide public comment prior to rule-making is essential."

A Strategy of Fear

The first reasonable question to ask is: Can HHS promulgate this regulation if it follows proper notice and comment procedures? The answer is a qualified yes. HHS has the statutory authority to promulgate appropriate regulations to prevent discrimination against the handicapped in federally financed facilities, and infants with handicaps qualify. But there are potential problems of vagueness and due process, which need more adequate attention than they have received by HHS. The term "customary medical care" needs clarification, as does the method used to determine if a complaint justifies investigation, and the procedures used to investigate. The more important question is: Should HHS adopt this or a similar rule? The answer seems to be no.

The negative response is not a knee-jerk rejection of White House right-to-life tactics. Indeed, in coupling their attack on "infanticide" with their attack on abortion, the right-to-lifers are not so far off the mark as many would like to assume. Judge Gesell, for example, resorts to language reminiscent of that used by Justice Harry Blackmun in *Roe v Wade* in describing the potential impact on the family of a severely handicapped child:

> The Secretary did not appear to give the slightest consideration to the advantages and disadvantages of relying on the wishes of the parents who, *knowing the setting in which the child may be raised,* in many ways *are in the best position to evaluate the infant's best interests.* Ignoring parental preferences again may increase the risk that parents will withdraw the infant from hospital care entirely, and *the long-term interests of physically disabled newborns may be affected by thrusting the child into situations where economic, emotional and*

marital effects on the family as a whole are so adverse that the effort
to preserve an unwanted child may require concurrent attention to
procedures from adoption or other placement. (emphasis supplied)

This language is alarmingly ambiguous. If the judge is talking
about maintaining the child in the hospital so the doctors can continue
to attempt to get parental approval for treatment, or barring that, can
make arrangements for adoption or other placement of the infant, this
seems reasonable. If, on the other hand, he is suggesting that parents
should be able to refuse treatment for a handicapped infant because
survival may adversely affect the quality of life of the parents
themselves and other family members, this goes far beyond even the
arguments advanced in the first Phillip Becker case. Treatment
should be denied to a handicapped infant for only one reason: the
treatment is not in the infant's best interests. Withholding treatment
in the interests of others would be a monumental and unjustified
change in the law, and runs contrary to the rationale of child abuse and
neglect statutes.

Nonetheless, the hotline strategy should be withdrawn because it
produces fear rather than reflection, it has been ineffective during its
short life, and it discourages the development of reasonable substantive rules and the procedures for fairly applying them. Judge Gesell
seems right in characterizing the regulation as having mainly an *in
terrorem* effect. The result is not more compassionate decisionmaking. Instead the incentive is to avoid decisionmaking by always treating everyone as heroically as possible without regard for prognosis or
the pain and suffering inflicted by the treatment itself.

Surgeon General C. Everett Koop has said: "The rule does not
require prolonging the act of dying, but, rather, protecting the act of
living through appropriate nourishment and care." The thought is
laudable, but the guidance in his expression is almost useless. Because of the fear instilled by the rule, the dying process is likely to be
prolonged. Instead of saving lives, the hotline will increase infant
suffering.

From March 17 to April 14, the hotline fielded about 600 calls.
Almost all were comments or requests for posters or information.
Twenty percent were wrong numbers or hung up. Only 16 callers
made a specific allegation. Of these, only five were deemed to merit
investigation and to date none has been found to warrant further
action. Either there are no instances of inappropriately withholding
care, or the hotline is an ineffective means of identifying them.

It should be clear by now that what is needed in neonatal decision-making is precise articulation of agreed-upon principles and improved procedures for applying them to individual cases. The notion that the situation is too complex for any rules is untenable, at least if we define "rules" broadly. For example, the rules "The primary basis for the decision should be the best interests of the infant," and "Care should not be withheld solely on the basis of mental retardation" seem reasonable. Since past decisions, like the one about Baby Doe, seem to have been made without adequate information, some form of mandatory consultation and possibly even internal hospital committee review may also be indicated. The White House and Secretary should abandon the simple "single bullet" approach to this complex problem and join in a dialog designed to develop substantive rules and reasonable procedures that have a chance to benefit handicapped newborns.

Finally, if the White House "504 strategy" is pursued, it should not be limited to infants. Handicapped individuals of all ages deserve the aggressive support of the federal government to help ensure that they are not denied needed medical services on the basis of handicap. Indeed, a more constructive federal strategy would entail legislation to ensure that no one, handicapped or not, is denied needed medical services because of inability to pay. Hospital notices to this effect would be welcomed by most citizens.

Baby Doe Redux
Doctors as Child Abusers

Drafted in haste, invalidated by a federal judge, and severely criticized by organized medicine, the "Baby Doe" regulations have been dressed up and reissued by the Reagan Administration. Like their prototype, these regulations turn the traditional notion of child abuse on its head. In the standard view, physicians are seen as protectors; the hospital as a safe haven. Physicians treat abused children and report incidents of child abuse to the appropriate authorities to protect them from further harm.

The Baby Doe regulations, however, identify the hospital itself as a site of child abuse, and at least some physicians as willing co-conspirators with parents in abusing and neglecting children by withholding food and needed medical care. Since the Administration can no longer trust physicians, a new reporting mechanism is needed to police the potential child abusers. The proposed "solution" is a hotline that anyone who suspects physician-condoned abuse can call.

No wonder pediatricians are upset. Seeing themselves, and having been seen by society, as the primary health care advocates for children, they are disheartened and insulted to be told now that the federal government no longer trusts them not to abuse their patients. Do past abuses justify a system of continual federal supervision of hospital activities, teams of federal investigators available upon anonymous hotline tips, and threats of loss of federal funds unless vague mandates are followed?

Baby Doe's New Clothes

The original Baby Doe regulations were invalidated by Judge Gerhard Gesell on the basis that they were "arbitrary and capricious" because the Department of Health and Human Services (HHS) did not

132

comply with the Administrative Procedure Act in giving at least 30 days' notice of the regulation and an opportunity for interested parties to comment on it, and did not consider the relevant factors involved sufficiently to satisfy a test of rationality.

The judge also noted a number of substantive objections to the regulation: the lack of a definition of "customary care," the invasive and disruptive police tactics used by federal "Baby Doe squads," the triggering mechanism of an "anonymous tipster," and the failure to explore any alternative methods of protecting handicapped newborns.

The White House, through HHS, reissued the Baby Doe regulations on July 5, 1983.[29] The administration decided to deal with the procedural issues raised by the rule and to ignore or gloss over the much more difficult substantive issues.

The new proposal provides a 60-day period for filing written comments, and more than 11,000 comments were filed by the closing date, September 6, the "overwhelming majority" of which were favorable. Should HHS review these and consider the relevant factors the comments discuss, it will remedy the procedural defects that caused its initial regulations to be invalidated. The proposal itself is *identical* with the original one, with four exceptions:

1. The hotline notice need now be posted only at "each nurse's station."

2. A minimum size requirement for the notices is set at 8-1/2 by 11 inches (the original regulation had no size requirement, but HHS had sent out very large posters when copies were requested).

3. The state child protective agency's phone number *shall* (instead of may) be added to the poster.

4. Most significantly, an entirely new section mandates that each state's child protective services agency establish procedures designed "to prevent medical neglect of handicapped infants" within 60 days of the effective date of the regulation, including provisions for mandated health care provider reporting, review of such reports, provision of services, and notification to HHS of reports.

Impact of Changes

The nursing station requirement, which removes the notice from areas where patients and their families are likely to see it, makes sense only in the light of HHS Secretary Margaret Heckler's comments at her confirmation hearing. Without citing any evidence, she testified that the Baby Doe regulation was needed because nurses were afraid to report cases of child neglect to the appropriate authorities. Her department seems to believe that nurses have been unwilling but passive participants in child abuse. Nothing short of supplying them with a hotline number that they can use anonymously and with immunity will induce them to take their role of child abuse reporters seriously. This view of modern nurses is extremely demeaning, and at odds with their role as team members in most specialized pediatric units.

The Administration never considered the size of the notice a major issue. Indeed, the initial decision to print such large posters was apparently made independently in HHS's print shop by a supervisor who routinely makes all notices regarding the handicapped large enough that the visually impaired can read them. This laudable goal, of course, was inapplicable in this instance; the change to a smaller size poster will be welcomed by those who saw the original size as a gratuitous insult.

The only substantive change is one not addressed in the previous litigation, and one that is also the subject of proposed federal legislation (which could require it even if these regulations are not adopted): mandated changes in state child abuse and enforcement procedures. Since child abuse has traditionally been an area left to the states, having such wide-range changes mandated by the federal government will cause some uneasiness. Nevertheless, this section seems to be an implicit recognition that the federal government simply does not have the expertise to investigate child abuse complaints. Although this reasoning is correct, diverting child abuse resources to hospital care or newborns may permit more common child abuse and neglect problems in other settings to go undetected or unremedied. HHS does not seem to have considered this possibility.

A Political Ploy

These regulations were promulgated in response to the "Baby Doe" case in Bloomington, Indiana, in which an infant with Down syndrome and a correctable esophageal atresia was not treated, with court approval. This case appropriately sparked much outrage. The judicial opinion was unprecedented: in every previous similar case brought to court, treatment had been ordered. The judicial process failed to protect the child. But regulations should not be based on singular, extraordinary events. HHS apparently agrees, and cites three other cases during the past decade in which Down syndrome infants with treatable esophageal or intestinal atresia were not treated.

Without supporting the hotline regulation, one can agree with HHS that these are all cases of child abuse, and that action should have been taken to obtain treatment for these infants. The law clearly calls for treatment, and ignorance of the law, rather than maliciousness, was probably responsible for failure to act in these cases. There were no prosecutions, and even the one case in which HHS was involved resulted only in an amendment to the hospital's internal procedures. The record, in short, does not indicate that existing legal mechanisms to protect infants have been tried and found wanting: it indicates one case of failure, and nonuse in three other cases. Proposing a new scheme based on federal regulations under these circumstances seems more a political ploy than a serious attempt at reform.

This conclusion is bolstered by the Administration's inability to come to grips with the major criticism of the regulation: the vagueness of the "customary medical care" standard physicians are required to follow. The problem, of course, is that there is no "customary medical care" in the difficult and problematic cases, and this is precisely what makes it difficult for physicians to deal with them.

HHS believes that it is being helpful in its expanded comments when it says that this standard does not require treatment that is "medically contraindicated" and permits withholding of treatment on the basis of "legitimate medical judgment." Since these terms add nothing, HHS explains that ultimately the physician's responsibility "corresponds with the responsibilities of parents or guardians under state child abuse and neglect laws to provide medically indicated treatment for their children." But, of course, this is where we came in. The suggested definition is circular; "customary medical care" is care that must be provided to avoid violation of the child abuse and neglect laws. The idea that standards are likely to be developed in any way but through individual cases being decided by courts of law seems farfetched. Certainly *ad hoc* "Baby Doe squads" descending on hospitals in the middle of the night are unlikely to be helpful to anyone. Nor does this "standard" provide any guidance to the nurses who are expected to use the hotline to call for help in appropriate circumstances.

As further evidence of the political nature of the proposed regulation, there is no analysis of alternative means of dealing with discrimination against handicapped infants. In other areas the Administration has attempted to avoid unnecessary regulations, and insists on subjecting even health-related regulations to cost-benefit and cost-effectiveness analysis. No attempt has been made to do any of this homework in this instance. The practical, substantive criticisms raised against the initial regulations have not been addressed because the Administration does not seem interested in drafting a practical regulation: it seems intent on making a political statement that will satisfy some supporters.

We certainly can do better. We need more reflection, more accurate information, consultation, and public involvement in decision making regarding handicapped newborns. But we pay a very high price by rushing to assume that every physician is a potential child abuser, and that every nurse must be a police informant to protect newborns from harm. Hospitals must continue to be seen as safe havens for children. If they are not safe, there are less destructive alternatives for reform.

Those guilty of abusing handicapped newborns should be prosecuted for this criminal activity. Questionable cases should be resolved in court, where all the relevant facts can be examined by a politically appointed and accountable decisionmaker, and a decision reached in a public forum on the basis of clearly articulated principles. Only in this case-by-case manner are socially acceptable substantive principles likely to be developed.

There is no simple solution to this complex problem. The proposed hotline, promoting only fear and distrust, fails even to address it directly. Nor do ethics committees, privately constituted groups with indeterminate membership, hazy mandates, undefined principles of decisionmaking, and nonexistent procedural requirements, bring us closer to our goal of avoiding arbitrary discrimination. In fact, no improvements in procedures will be more than marginally beneficial until we reach some societal accord on the proper substantive principles that should govern decisions in individual cases.

Checkmating the
Baby Doe Regulations

Describing the history of the Baby Doe regulations as "checkered," and using a chess analogy to dismiss the dissent of Justice Byron White, a plurality of the US Supreme Court has checkmated the Administration's attempt to use section 504 of the Rehabilitation Act of 1973 as a basis for its Baby Doe regulations.[30]

The decision on the validity of the final regulations split the court 5 to 3, with Chief Justice Burger concurring only in the judgment, and Justice Rehnquist not participating in the decision. Justice Stevens, writing for the four member plurality, stated that the case was limited to whether or not the four mandatory provisions of the regulations are authorized by section 504, and that the answer to this question rested "entirely on the reasoning of" the Second Circuit in deciding the case of Baby Jane Doe.[31]

The Case of Baby Jane Doe

Baby Jane Doe was born in October 1983, suffering from spina bifida, hydrocephaly, and microcephaly. Her parents refused to consent to surgery to close her meningomyelocele. An unrelated party who received a confidential tip brought suit in New York to have the surgery performed. A lower court agreed to authorize the surgery, but the Appellate Division reversed the following day, and a week later New York's highest court ruled that the lower court should not have heard the case at all because the party who brought it had no relationship with "the child, her parents, her family, or those treating her illness."

In the meantime, an independent investigation by the New York State Child Protective Services concluded there was no cause for state intervention. Nonetheless, HHS sought Baby Jane Doe's medical records to make its own determination as to whether or not a section

504 violation had occurred. The US District Court concluded that no such violation had occurred because the hospital failed to perform the surgery because of parental refusal, not because of the infant's handicap. The Court of Appeals focused solely on the question of whether section 504 authorized HHS to investigate a case like this. Section 504 provides:

> No otherwise qualified handicapped individual in the United States ...shall, solely by reason of his handicap, be excluded from participation in, or be denied the benefits of, or be subjected to discrimination under any program or activity receiving Federal financial assistance.

The Court of Appeals concluded that even though Baby Jane Doe fit the definition of a "handicapped individual," she was not "otherwise qualified" because this phrase referred to handicapped persons who could benefit from services in spite of their handicap, rather than cases in which the handicap itself is the subject of the services. The court bolstered its conclusion by reviewing the legislative history of the Rehabilitation Act of 1973 which had to do primarily with employment and vocational education, and never envisioned any federal role in individual medical treatment decision.

HHS did not appeal the Baby Jane Doe decision. The AHA, the AMA and others, however, used it as the basis for a separate suit in US District Court to enjoin the four mandatory sections of the Baby Doe regulations (notice posting, mandatory reporting, access to medical records, and expedited action to effect compliance). The lower court invalidated the regulations, and the Second Circuit summarily affirmed based on their Baby Jane Doe opinion. Accordingly, it is fair to say that the original Baby Jane Doe case is the subject of this appeal.

The Supreme Court's Opinion

Somewhat inexplicably, Justice Stevens did not examine the reasoning of the Second Circuit in any detail, although he seemed to agree with it. In a footnote, for example, he stated, "The legislative history of the Rehabilitation Act does not support the notion that Congress intended intervention by federal officials into treatment decisions traditionally left by state law to concerned parents and the attending physicians or, in exceptional cases, to state agencies charged with protecting the welfare of the infant."

Instead, the centerpiece of the Court's analysis was that it is parents not hospitals who are refusing to treat handicapped newborns. State laws regarding child neglect apply to parental refusals, but section 504 does not; since in the absence of parental consent the infant is neither "otherwise qualified" for treatment nor has the infant been denied care "solely by reason of his handicap." The plurality noted that HHS's original position, based on the Bloomington Indiana case, was that section 504 did apply to parental refusals and that the hospitals violated 504 by even allowing such infants to remain in their care. But HHS abandoned this position in its final rules, noting that section 504 did *not* authorize HHS to override parental refusals. Instead HHS relied on two possible categories of 504 violations to justify the rules: (1) a hospital's refusal to furnish a handicapped infant with medically beneficial treatment "solely by reason of handicap"; and (2) a hospital's failure to report cases of suspected medical neglect of handicapped newborns to a state child protective services agency.

Justice Stevens concentrated almost all of his analysis on the "axiom of administrative law" that an agency's basis for a regulation must include "a rational connection between the facts found and the choice made; and that an agency's action "must be upheld, if at all, on the basis articulated by the agency itself." Stevens argued that no such "rational connection" exists for the first rationale because HHS could find *no* case in which a hospital denied treatment to a handicapped infant "solely by reason of handicap." This was true in the original Baby Doe case from Bloomington, Indiana; and in the case of Baby Jane Doe. Moreover, in the 49 cases HHS investigated in 1983 and used as a basis for the final rule, none involved a finding that a hospital had refused to provide care for which parents had consented. Physician surveys that indicated attitudes supportive of nontreatment were found irrelevant because HHS had not used them as a basis for the regulation, and did not contend that physicians rather than parents were making the nontreatment decision.

Regarding the second rationale for the reporting regulations, Stevens says three things: (1) the record reveals *no* case where a hospital failed or was accused of failing to make such report; (2) much more evidence of Congressional intent is needed when the federal government commands a state; and (3) the way the states discharge their child abuse and neglect protection functions is "wholly outside the nondiscrimination mandate to section 504." Section 504 would apply if the hospital refused to treat a handicapped child because of the handicap where parents wanted the treatment; but mandatory reporting is not needed since it is assumed that the parents themselves would bring such a case to the attention of the authorities.

The plurality's decision not only invalidated the four mandatory rules, but also affirmed the lower court's much broader injunction which forbids "continuation or initiation of regulatory and investigative activity directed at instances in which parents have refused to consent to treatment...and efforts to seek compliance with affirmative requirements imposed on state child protective service agencies."

The opinion is a strong affirmation of federalism. In this area federal authority stems primarily from funding, and if the federal government wants to improve the lot of the handicapped, it can do so by making a much wider range of services available to them. On the other hand, areas of child abuse and neglect are traditionally areas of state authority, and the court sees no reason to suppose that Congress meant to change this when it enacted section 504.

Justice Byron White, joined by two others, dissented. In his view, HHS could properly take attitudinal surveys of physicians into account in promulgating regulations whose purpose was to "foster an awareness by health care professionals of their responsibility not to act in a discriminatory manner with respect to medical treatment decisions for handicapped infants." He reached this conclusion because he differed with the plurality on the centrality of parental consent to treatment decisions. He thought the survey literature made it clear that the "parental consent decision does not occur in a vacuum. In fact, the doctors (directly) and the hospital (indirectly) in most cases participate in the formulation of the final parental decision and in many cases substantially influence that decision."

Discussion

Like its progenitors, the case itself is "checkered." Justice Stevens objects to Justice White's dissent on the grounds that, "like bishops of opposite colors, the opinions of Justice White and the Court of Appeals do not even touch one another," because Justice White limited his discussion to cases in which treatment is completely unrelated to the handicap; whereas the Court of Appeals was concerned only with cases in which the handicap was related to the condition to be treated.

This oversimplifies the disagreement, but the bishop analogy seems appropriate to much of the past four years of discussion about the Baby Doe regulations.

From the start, for example, the Reagan Administration has been concerned primarily with playing "right to life" politics, and not with enforcing section 504. So close was the administration's political agenda in pursuing the Baby Doe regulations to its views on abortion, that when the President was asked about the U.S. Supreme Court's June, 1986 abortion decision* at his June 11, 1986 news conference, he responded by talking about the Baby Doe decision instead. Noting correctly that the Court had ruled that the federal government was "getting into something that properly was the province of the state," he went on to say, "...I feel very strongly that we're talking about human life. And the case that prompted this entire act was one in which the determination is made that this life is to be taken away and yet it isn't done as you would with an animal, it isn't done with a merciful putting to sleep or—they just let it starve to death."

Second, child neglect by parental treatment refusal has *always* been a matter of state law. It still is. The Court does *not* hold that parents have a right to refuse treatment for their handicapped children. The holding is that review of such decisions is a matter for the states. Since there is *no* evidence that the states are doing an inadequate job, and since Congress did not authorize HHS to get directly involved when it enacted 504, there is no reason why the status quo concerning the respective powers of the Federal and State governments should be changed in this area. The 1984 Child Abuse Amendments enacted by Congress explicitly recognize the respective roles of Federal (funding) and state (child neglect enforcement) governments. Third, section 504 was never explicitly or implicitly meant to apply to individual medical treatment decisions in which the handicap itself influences the possible benefit that can be derived from the treatment. Unfortunately, though this point is acknowledged by the plurality, it is not discussed, so what application, if any, HHS could make of section 504 *if* it developed a sufficient factual record for rule-making is not determined by this case.

The proper legacy of the Baby Doe regulations is twofold: it highlights the inherent powers of the states in the area of child neglect; and it demonstrates that the most effective way the federal government

---

*Discussed in "Reaffirming *Roe v. Wade* Again" at p. 174.

can help the handicapped is not only to enforce section 504, but primarily to fund service programs that directly benefit the handicapped. The infant care review committee recommendations in the Baby Doe regulations were not addressed by this decision because they are not mandatory. Perhaps the most useful result of the controversy and discussion engendered by the Baby Doe regulations would be voluntary continuation of these committees.[32]

There is no simple solution to this complex problem, but the combination of carefully crafted published professional standards, and public review of individual problematic cases in court provides the most likely method by which the best interests of handicapped children will remain central in treatment decisions.

Reproductive Liberty

Introduction

Unlike almost all of the other health law and bioethics issues dealt with in this book, the issue of abortion has primarily been defined by the United States Supreme Court. Thus law has taken the lead in delineating the contours of the debate over the limits of reproductive liberty that pregnant women should have, and most of that debate, in turn, has focused on the Supreme Court's landmark 1973 decision in *Roe v Wade*. This decision followed two other major opinions on "the right to privacy," and has itself been followed by more than a dozen other decisions on abortion over the past decade. The essays in this section chronicle and explain the past decade of decisions. To put them in perspective, a few words must be addressed to the two decisions that preceded *Roe*, and to that opinion itself.

The first case, *Griswold v Connecticut*, 381 US 479 (1965), involved the constitutionality of a Connecticut statute that made it a crime to use contraceptives. In striking down this statute, the Supreme Court enunciated a constitutional "right to privacy" which it said was suggested by "penumbras" that emanate from specific guarantees in the Bill of Rights. Defining a specific "zone of privacy," Justice William Douglas, writing for the Court, focused on sexual relations in marriage:

We deal with a right of privacy older than the Bill of Rights—older than our political parties, older than our school system. Marriage is the coming together for better or for worse, hopefully enduring, and intimate to the degree of being sacred. It is an association that promotes a way of life, not causes; a harmony in living, not political faiths; a bilateral loyalty, not commercial or social projects...an association for as noble a purpose as any involved in prior decisions.

To secure the argument the Court also noted that the statute could not be enforced without massive and unthinkable governmental intrusion into people's lives and homes: "Would we allow the police to search the sacred precincts of marital bedrooms for tell-tale signs of the use of contraceptives? The very idea is repulsive to the notions of privacy surrounding the marriage relationship."

Strictly speaking, *Griswold* applied only to married couples. But during the same term in which *Roe v Wade* was being discussed, the Court decided another case which involved a Massachusetts statute that made it a crime to sell, lend, give away or exhibit any contraceptive device. That case is *Eisenstadt v Baird*, 405 US 38 (1972). The constitutional issue, as the Court defined it, was whether the state could treat unmarried citizens differently than married ones in the area of contraception. The Court concluded that there was no constitutionally acceptable rationale for treating married and unmarried individuals differently, and that the basis for the right to privacy under *Griswold* was the sexual relationship itself, not marriage:

The marital couple is not an independent entity with a mind and heart of its own, but *an association of two individuals* each with a separate intellectual and emotional makeup. If the right to privacy means anything, it is the right of the *individual*, married or single, to be free from unwarranted governmental intrusion into matters so fundamentally affecting a person as the decision whether to bear or beget a child.

This extremely important decision thus bridged the gap between *Griswold* and *Roe* by going further than the Court had to go in broadening the right to privacy to include a decision not only to "beget" a child, but to "bear" one as well.

The 1973 opinion of *Roe v Wade*, 410 US 113, followed. It is the most important, most controversial, and most well-known US Supreme Court decision in recent history. At issue in *Roe* was a Texas statute that made it a crime to "procure an abortion" or to attempt one, except to save the life of the mother. Justice Harry Blackmun wrote the opinion of the Court. Building on a series of cases that had described a "right to personal privacy, or a guarantee of certain areas or zones of privacy," the Court determined that such a "right to privacy" existed "in the Fourteenth Amendment's concept of personal liberty ..." The Court concluded that this right "is broad enough to encompass a woman's decision whether or not to terminate her pregnancy":

> The detriment that the State would impose upon a pregnant woman by denying this choice altogether is apparent. Specific and direct harm medically diagnosable even in early pregnancy may be involved. Maternity, or additional offspring, may force upon a woman a distressful life and future...

Stopping short of concluding that the right to abortion is absolute, the Court instead recognized that the state has interests in the health of the mother and the life of the fetus that may at times be sufficiently "compelling" to permit the state to limit abortion. With regard to maternal health, the Court found this interest compelling only after the point in pregnancy at which abortion becomes more dangerous to the woman's health than carrying the fetus to term (in 1973, about the end of the first trimester). With regard to fetal life, the Court found the state could claim a compelling interest in its protection after viability, which the Court defined as the time the fetus was capable of surviving independent of the mother (in 1973, about the beginning of the third trimester). The Court thus concluded that the state could regulate abortion procedures to protect the pregnant woman's health after the time in pregnancy when abortion was no longer safer than childbirth; and could outlaw abortion altogether (except when the life or health of the pregnant woman was at stake) after the fetus was viable.

In dividing pregnancy into three stages, and weighing state interests in regulating abortion differently during each, the US Supreme Court was acting more like a superlegislature than a court. The decision has been severely criticized on this basis. As we shall see, however, it has survived more than a decade of intense attack, and remains the law of the land today. Even if the Court ultimately modifies it, both the Court and the legislatures of the individual states will be faced with the same question: What is the appropriate role for the government to play in regulating pregnancy terminations?

The essays in this section chronicle the Court decisions of the past decade. But the writing that captures the human dimension of the abortion issue most poignantly is not a court decision, but a 1985 novel, *The Cider House Rules*, by John Irving. The major character of the novel is Wilbur Larch, and near the beginning, before he decides to work in an orphanage in St. Cloud, Maine, he confronts himself and the history of his profession with regard to abortion. He reflects on the not-so-distant past when abortion was, for the most part, legal, and when procedures much more complex (like in utero decapitation and fetal pulverization) were routinely taught to medical students:

> By the time he got back to Portland, he had worked the matter out. He was an obstetrician; he delivered babies into the world. His colleagues called this "the Lord's work." And he was an abortionist; he delivered mothers, too. His colleagues called this "the Devil's work," but it was *all* the Lord's work to Wilbur Larch. As Mrs. Maxwell had observed, "The true physician's soul cannot be too broad and gentle."

The legal question addressed by the abortion decisions concerns the limitations the state may constitutionally place on a decision a woman makes to have an abortion performed by a licensed physician. All of the essays in this section were originally published in the *Hastings Center Report*. Unlike most other sections, they appear here in chronological order. "Round Two" was published in October, 1976; "Let Them Eat Cake" in August, 1977; "Parents, Children and the Supreme Court" in October, 1979; "The Irrelevance of Medical Judgment" in October, 1980; "*Roe v. Wade* Reaffirmed" in August, 1983; and "*Roe v. Wade* Reaffirmed, Again" in October, 1986.

Abortion and the
Supreme Court
Round Two

Although there is general societal agreement that under some circumstances (for example, to save her life), a woman has a constitutional right to an abortion, debate continues as to what those circumstances are and how many restrictions the state may place on this right. Two 1976 decisions by the United States Supreme Court on this issue will probably serve mainly to intensify rather than resolve many basic disagreements.

Early in 1973 the Court determined (seven of the nine Justices concurring in the result) that the "right to privacy" was "broad enough to encompass a woman's decision whether or not to terminate her pregnancy." This right was categorized as fundamental, and therefore could only be infringed upon if a state could demonstrate a "compelling interest" in regulating the activity. To analyze the state's interests, the Court fashioned a test based loosely on the trimesters of pregnancy. In the first trimester the Court determined that a state could not interfere with the decision of a woman and her physician concerning abortion (based upon the safety of an abortion compared with carrying a child to term). After the first trimester, when abortion procedures are more dangerous, the Court found regulations relating to the safety of the pregnant woman permissible. And after the fetus is "viable" (the third trimester), the Court found the state's interest in protecting fetal life sufficient to proscribe abortion altogether unless the life or health of the woman herself is at stake.[1]

Unresolved Issues

Though the Court's initial opinion swept very broadly, it left a number of issues unresolved. Many were the subject of a decision which,

147

like *Roe v Wade*, was written by Justice Harry Blackmun. In this case,[2] in which the constitutionality of a Missouri statute was under challenge, the Court decided that:

(1) "Viability" can be constitutionally defined as "that stage of fetal development when the life of the unborn child may be continued indefinitely outside the womb by natural or artificial life-supportive systems"; and the application of such a definition "must be a matter for the judgment of the responsible attending physician."

(2) A state may constitutionally require a woman to give her informed, voluntary, and written consent to the abortion.

(3) A state may not constitutionally require the consent of a spouse to an abortion during the first trimester.

(4) A state may not constitutionally require the consent of a minor's parent or parents to an abortion during the first trimester.

(5) A state may not proscribe a method of abortion (saline amniocentesis) that is employed by a majority of physicians, and is safer than continuing a pregnancy to term.

(6) A state may require certain records concerning all abortions be kept "for the advancement of medical knowledge" so long as they are kept "confidential."

(7) A state cannot constitutionally require a physician to attempt to preserve the life of a previable fetus delivered during an abortion.

All nine Justices agreed with points 1, 2, and 6—and these issues may be taken as being fairly well settled. Of the remaining points, 3, 5, and 7 were decided by a 6–3 margin, whereas issue 4, parental consent, was decided by only a 5–4 margin. The arguments concerning point 7 turned mainly on technical issues of statutory interpretation. The resolution of point 5, the attempted outlawing of saline abortions, was decided by the majority's perception of what was actually occurring in medicine. Unlike *Roe v Wade*, the Court's determination of current medical practice was not bolstered by any reference to the medical literature. Even though the majority declined to make use of the medical literature, three Justices characterized this part of the decision as turning the Court into the country's "ex officio medical board with powers to approve or disapprove medical and operative practices and standards throughout the United States." This reaction was not altogether unjustified, as previously such decisions had been left largely to the state legislatures and state regulatory agencies, both of which have more expertise and fact-funding ability than a court in matters of actual medical practice.

Almost all the concurring and dissenting opinions in this case dealt with the issues of spousal and parental consent. The competing arguments on these issues make it clear both that the questions are not yet settled, and that in dealing with them the Court is acting very much like a legislative body, often arguing what is best for society, rather than what is constitutionally required. The Court's opinion on these two issues is limited to the first trimester, and thus speculation as to how these issues would be decided if a state statute made these requirements applicable only to the second or third trimester is in order. The majority of the Court found the delegation to someone else of a "veto power" over the woman's decision constitutionally defective, three Justices stating that "since the State cannot regulate or proscribe abortion during the first [trimester]...the State cannot delegate authority to any particular person...to prevent abortion during that same period."

Spousal Consent

The majority of the Court noted that ideally both husband and wife would concur in the abortion decision. However, it is an "obvious fact that when the wife and the husband disagree on this decision, the view of only one of the two marriage partners can prevail." The Court found that, since the mutuality of the marriage relationship (the only interest the state had claimed for this requirement), cannot be strengthened by giving the husband a veto power, since the right to privacy is an individual one, and since "it is the woman who physically bears the child and who is more directly and immediately affected by the pregnancy," the balance must be resolved in her favor as the decision-maker.

It is likely that these same considerations would govern during the second trimester as well. However, in the third trimester, the state can constitutionally proscribe elective pregnancy terminations altogether. It can, accordingly, be argued that after viability the state has the authority to delegate veto power. Further, since the husband's interest in the fetus is substantially increased after it is viable, and since the medical risks become very close to the risks of childbirth, a state statute requiring consent of the husband in the third trimester might withstand constitutional attack.

Three dissenting Justices were prepared to uphold the spousal consent requirement even in the first trimester. The Chief Justice and Justice Rehnquist concurred in the views of Justice White that:

> A father's interest in having a child—perhaps his only child—may be unmatched by any other interest in his life. It is truly surprising that the majority finds in the United States Constitution, as it must in order to justify the result it reaches, a rule that the State must assign a greater value to a mother's decision to cut off a potential human life by abortion than to a father's decision to let it mature into a live child.

Parental Consent

The issue of a parental consent is more difficult. The state, under the parens patriae doctrine, has a duty to protect the welfare of minors. There was evidence before the Court that girls as young as ten and eleven years old had sought abortions. Moreover, the counseling at one abortion clinic was described as being done in groups of both minors and adults who were strangers to one another, not being performed by a physician, and being concerned mainly with abortion procedures, complications, and birth control techniques. The lower court had found a compelling state interest sufficient to justify the parental consent requirement as safeguarding "the authority of the family relationship." The majority of the Supreme Court disagreed. It noted that "[c]onstitutional rights do not mature and come into being magically only when one attains the state-defined age of majority. Minors, as well as adults, are protected by the Constitution and possess constitutional rights." The Court was unable to find a state interest of sufficient strength to compromise the minor's right to an abortion in the first trimester. Regarding parental authority, Justice Blackmun wrote:

> It is difficult to conclude that providing a parent with absolute power to overrule a determination made by the physician and his minor patient will serve to strengthen the family unit. Neither is it likely that such a veto power will enhance parental authority or control where the minor and the nonconsenting parent are so fundamentally in conflict and the very existence of pregnancy already has fractured the family structure.

In somewhat hedging his conclusion, however, Justice Blackmun noted that the Court was not holding that "every minor, regardless of age or maturity may give effective consent to an abortion." In fact, the Court's holding is that a minor, like an adult, can consent to an abortion if her consent is competent, voluntary, and informed. Thus "nonmature" or "noncompetent" minors are still required to have their parent's consent to an abortion, even in the first trimester, as they would for any medical procedure.

As for the second trimester, the arguments advanced by the state for parental consent would seem no stronger than during the first. However, the fact that the child has waited so long before deciding, that the pregnancy may be impossible to hide from the parents in any event at that stage, and that abortion procedures are considerably more dangerous during the second trimester, might enable a requirement of parental consent in the trimester to withstand constitutional challenge. A similar argument has been used against a general parental consent requirement by Planned Parenthood officials who argue "many teenagers will delay telling their parents, forcing them into second trimester abortions, which carry a much higher risk."[3] After viability, of course, arguments similar to those concerning spousal consent would apply, and the state could probably require parental consent for a minor's pregnancy termination.

A much more difficult and interesting question concerns the type of requirements a state may constitutionally impose on a minor who desires an abortion. For example, may a state require simple notification of parents, rather than their consent? Though notification of a husband may not be appropriate because the couple may be separated, with questions of divorce, adultery, alimony, and custody at issue,[4] a notification requirement may well be appropriate for a minor child living with her parents. As Justice Stewart, one of two judges joining in a concurring opinion to help constitute the 5–4 decision on parental consent, argued:

> There can be little doubt that the State furthers a constitutionally permissible end by encouraging an unmarried pregnant minor to seek the help and advice of her parents in making the very important decision whether or not to bear a child. That is a grave decision, and a girl of tender years under emotional stress, may be ill-equipped to make it without support. It seems unlikely that she will obtain adequate counsel and support from the attending physician at an abortion clinic, where abortions for pregnant minors frequently take place.

The majority of the Court actually seemed to be encouraging a notice requirement, even in the first trimester. In a companion case[5], the Court returned to the Massachusetts Supreme Judicial Court for interpretation a state statute that required the minor either to obtain her parent's consent or to appear before a Superior Court Judge prior to having an abortion. The Court noted that it would be an unconstitutional statute if it gave the parents a veto power over their child's decision. On the other hand, it could be constitutionally interpreted if it only encouraged the child to consult with her parents regarding this decision. The Court noted, "In this case, we are concerned with a statute directed towards minors, as to whom there are unquestionably greater risks of inability to give an informed consent."

Four of the Justices in the Missouri case would even have found the parental consent "veto" requirement constitutionally permissible. Justice White, writing for three of them, argued that "Missouri is entitled to protect the unmarried minor woman from making the decision in a way which is not in her own best interests, and it seeks this goal by requiring parental consultation and consent. This is the traditional way by which States have sought to protect children from their own immature and improvident decisions [such as work, travel, marry, view certain motion pictures]..." In a separate dissent, Justice Stevens questioned the psychological assumptions of the majority:

> It is unrealistic, in my judgment, to assume that every parent–child relationship is either (a) so perfect that communication and accord will take place routinely or (b) so imperfect that the absence of communication reflects the child's correct prediction that the parent will exercise his or her veto arbitrarily to further a selfish interest rather than the child's interest...In each individual case factors much more profound than a mere medical judgment may weigh heavily in scales. The overriding consideration is that the right to make the choice is exercised as wisely as possible.

The Court was seriously split on these important issues; we will have to await their ultimate resolution. What is apparent is that the issue of parental consent and consultation regarding abortion is very much alive. The issue of spousal consent, while not yet dead, is terminally ill.*

---

*In the fall of 1987, with only eight Justices, the Court split 4–4 on the constitutionality of a parental notice requirement.

Let Them Eat Cake

In one of his dissents to the Supreme Court's June 1977 trilogy of abortion decisions, Justice Harry Blackmun argues that the Court's refusal to find unconstitutional a state Medicaid regulation, which authorized payment for childbirth for indigent women but not for abortions, is "disingenuous and alarming, almost reminiscent of 'let them eat cake.'" And Justice William Brennan adds, "a distressing insensitivity to the plight of impoverished pregnant women is inherent in the Court's analysis." He quotes a remark of Anatole France as an appropriate parallel: "The law, in its majestic equality, forbids the rich as well as the poor to sleep under bridges, to beg in the streets, and to steal bread."

In the first case[6] the Court holds that Title XIX of the Social Security Act did not require a state to fund the cost of *all* abortions permitted under state law, but that a state could limit reimbursement to "necessary" or "therapeutic" abortions. In its second opinion[7] the Court holds that the Constitution did not require a state, participating in the Medicaid program under Title XIX, to pay for nontherapeutic abortions if it paid for childbirth. Finally, in its third opinion,[8] the Court holds that the Constitution did not require public hospitals to perform elective abortions, even though these hospitals provided care for childbirth. Because it deals directly with constitutional issues, the most important decision is the second case, *Maher v Roe.*

Justice Lewis Powell, the author of the majority opinion in the first two cases, was able to convince six of the nine Justices to accept a novel approach to the Equal Protection Clause of the Constitution, an approach which had only two weeks previously been relegated to a minority position.

At issue in *Maher* was a Connecticut Welfare Department regulation requiring the patient to submit a certificate from a physician stating that the abortion was "medically necessary" before payment would be authorized for a first trimester abortion. The plaintiffs argued that while the Constitution did not require the states to pay for abortions, it was a violation of equal protection for the state to pay for childbirth and refuse to pay for elective abortions.

To analyze this question, Justice Powell applied the "Rodriguez criteria":

1. Does the legislation operate "to the disadvantage of some suspect class or impinge upon a fundamental right explicitly or implicitly protected by the Constitution?; and if not,

2. Does the statute rationally further some legitimate, articulated state purpose?[9]

Justice Powell determined that there was no discrimination against a "suspect class" (such as race or religion) since poverty is not a suspect class, so that "[a]n indigent woman desiring an abortion does not come within the limited category of disadvantaged classes so recognized by our cases." Though the minority agreed with this conclusion, Justice Thurgood Marshall noted that nonwhite women obtain abortions at a rate nearly twice that of whites, and that almost 40 percent of minority women (five times the proportion for whites) are dependent on Medicaid for their health care. In his words, "Even if this strongly disparate racial impact does not alone violate the Equal Protection Clause...at some point a showing that state action has a devastating impact on the lives of minority racial groups must be relevant."

The Issue of Impingement

The central issue in the case then became, does the Connecticut regulation "impinge upon a fundamental right"? If it does, the state must demonstrate a "compelling interest" to uphold the requirement; if it does not, only a finding of "rationality" is necessary. Since in *Roe v Wade*,[10] the Court previously held that a state could not demonstrate any "compelling interest" to interfere with the abortion decision when abortion was safer than childbirth, the Court either had to expressly overrule *Roe* or find that the requirement did not "impinge" upon a fundamental right. Thus Justice Powell's reasoning on this point is critical.

Conceding that a woman's "right to privacy" includes her constitutionally protected right to make a decision to terminate her pregnancy, Justice Powell argues that the Connecticut regulation does not interfere directly with this right, and need therefore only meet the "rationality test." The previous abortion cases, he argues, dealt with criminal prohibitions on abortion, which represented "a complete abridgement of constitutional freedom" and an "absolute obstacle" for a woman desiring an abortion. On the other hand, the refusal to fund abortions for poor women:

> places no obstacles—absolute or otherwise—in the pregnant women's path to an abortion...she continues as before [Medicaid] to be dependent on private sources for the services she desires...[the state has thus] imposed no restriction on access to abortions that was not already there. The indigency that may make it difficult—and in some cases, perhaps, impossible—for some women to have abortions is neither created nor in any way affected by the Connecticut regulation.

Interference vs Encouragement

Justice Powell restates his reasoning, after declaring that his decision "signals no retreat from *Roe* or the cases applying it":

> There is a basic difference between direct state interference with a protected activity [abortion], and state encouragement of an alternative activity consonant with legislative policy [childbirth].

To bolster this conclusion he uses the education analogy, concluding that the state may favor public schools by funding them and simultaneously refuse to fund private education. This analogy is, of course, misleading. A pregnant woman desiring an abortion does not have the choice between a "free" one at a public hospital and a "paid" one at a private hospital. Her choice is between trying to raise the money for an abortion, legal or illegal, or having her child. Moreover, the Court's previous abortion decisions provide no support for this view.

In *Doe v Bolton*,[11] the court struck down as unconstitutional a requirement that two physicians concur before an abortion could lawfully be performed—even though this requirement "interfered" with a woman's abortion decision far less than denial of payment is likely to. Justice Brennan is more honest in noting, "This disparity in funding by the State clearly operates to coerce indigent pregnant women to bear children they would not otherwise choose to have, and just as clearly, this coercion can only operate on the poor, who are uniquely the victims of this form of financial pressure."

However, having found that the woman's right to decide to terminate her pregnancy is not infringed, all that is left for the court is to find that the provision rationally furthers a legitimate purpose. The regulation accomplishes this, the Court concludes, by furthering the state's "strong interest in protecting the potential life of the fetus," and its "strong and legitimate interest in encouraging normal childbirth." This, however, treats pregnant woman who want abortion differently from those who want children, and a more precise equal protection analysis might have found the regulation defective on this basis. One does not have to be an advocate of abortions to see that what the Court did was to base its decision on its own view of the regulation's "wisdom and social desirability" (although Justice Powell denies this in the majority opinion). The Court said it was refusing to play "super legislature," but it did just that in *Roe v Wade* and, by partially reversing itself, did it again. Since the opinion affects only the rights of the poor and minorities, the case can be read as saying that the majority of the United States Supreme Court believes that the rights of this group are less worthy of protection than the rights of the economically advantaged.

Likely Impact

What will be the likely result of these decisions? First, a serious blow has been dealt to the protection afforded minorities by the Equal Protection Clause, and its future usefulness is open to question. Second, the poor and minorities seeking redress of grievances in the courts will attempt to keep important cases away from the Supreme Court. Third, the federal government and most state legislatures are likely to restrict the abortions paid for by Medicaid to those that are "medically necessary." Some states may even decide not to pay for any abortions, although nothing in these opinions supports favoring the fetus over the life and health of the mother. Fourth, even though only 18 percent of all public hospitals now provide elective abortion services, this number is likely to decrease.

Finally, the physicians, hospitals, and clinics offering abortion services will be forced to return to a type of pre-*Roe v Wade* procedure, under which most states permitted abortions only upon the certification of two physicians that the abortion was medically necessary. Under these cases, and the decision in *Doe v Bolton*, it seems clear that the "two doctor" rule is unconstitutional. However, the Connecticut requirement for *one* physician to certify "medical necessity" has been upheld. Thus the likely ultimate result is that abortion clinic physicians and others who perform abortions will simply be forced to make such a certification before being compensated under their state's Medicaid program. The factors they should be able to consider in making such a determination should be at least as broad as those enunciated by the court in *Roe v Wade*, which included the psychological and economic impact of the pregnancy and potential child on both the woman and her family.

But this possibility is a more optimistic view than is currently warranted. Justice Marshall's prediction is probably more apt, and his description of the Court's proper role more precise: "The effect will be to relegate millions of people to lives of poverty and despair. When elected officials cower before public pressure, this Court, more than ever, must not shirk its duty to enforce the Constitution for the benefit of the poor and powerless."*

---

*The precise result has not been to deny abortions to the poor, but to delay them an average of 2 to 3 weeks because of financial problems. See Henshaw, SK & Wallisch, LS, "The Medicaid Cutoff and Abortion Services for the Poor," *Family Planning Perspectives* **16**: 170–180; 1984.

Parents, Children, and the Supreme Court

Throughout most of history, parents have had absolute authority over their children. The idea that children have rights against their parents is very recent. Not until the 1960s did states pass child abuse and neglect statutes designed to protect children from their parents. Courts also have routinely permitted physicians to provide life-saving blood transfusions to children, even over the religious objections of their parents, and have proscribed the use of nonapproved drugs, such as Laetrile, on children.

Nonetheless, the law regarding parental authority to consent to or to withhold medical treatment from children remains murky. It is replete with tension between protecting the "best interests" of the child and preserving parental desires. A choice must often be made between them.

Two 1979 decisions of the United States Supreme Court reveal the difficulty in making this choice. Though ostensibly based upon interpretations of the Constitution, both opinions are primarily attempts to determine whether the challenged statute is likely to strengthen or weaken legitimate parental authority. The cases are extreme: a daughter's decision to have an abortion against the wishes of her parents, and the parents' decision to commit their child to a mental institution.

A Minor's Right to an Abortion

In a previous decision, the Court decided that a state could not give parents an absolute "veto power" over their daughter's decision to have an abortion.[12] In the most recent decision, the Court struck down a Massachusetts statute that required the daughter to ask her parents' permission before an abortion. Only after the parents refused did the statute give her the right to petition a court to decide either that she was "mature" enough to make the decision herself or that an abortion was in her "best interests."[13]

Though eight of the nine Justices agreed that the statute is unconstitutional, four used the occasion to discuss the limits of children's rights. Conceding that children are "not beyond the protection of the Constitution" and that "neither the Fourteenth Amendment nor the Bill of Rights is for adults alone," the Court, nonetheless, emphasized parental authority:

> The unique role in our society of the family...requires that constitutional principles be applied with sensitivity and flexibility to the special needs of parents...must be read to include the inculcation of moral standards, religious beliefs, and elements of good citizenship...This affirmative process of teaching, guiding, and inspiring by precept and example is essential to the growth of young people into mature, socially responsible citizens.

The Court underlined the point with a line from a previous decision: "It is cardinal with us that the custody, care and nurture of the child reside first with the parents, whose primary function and freedom include *"preparation for obligations the state can neither supply nor hinder."* (emphasis added by the Court) This statement is followed by a discussion of the necessity of legal support for parental authority to aid parents in discharging their obligations. The Court suggested ways to correct defects in the statute (e.g., court access before parental notice) apparently believing that by so doing it was supporting the family.

The law remains, therefore, that the state cannot give parents the absolute authority to prevent their pregnant daughters from having abortions—since their daughters have a constitutional right to make this decision in consultation with a physician. The daughter's right of self-determination is placed above parental authority.

Commitment of Children

Given this precedent, is a law that permits parents to commit their children to state mental hospitals, without any requirement of notice or a right to a hearing by the child, constitutional?[14] The answer is even more completely based on the Court's view of the family rather than its view of the Constitution. Writing the opinion of a five-Justice majority, Chief Justice Burger says:

> Our jurisprudence historically has reflected Western Civilization concepts of the family as a unit with broad parental authority over minor children...parents generally have the right, coupled with the high duty to recognize and prepare [their children] for additional obligations...Surely this includes a *high duty* to recognize symptoms of illness and *to seek and follow medical advice.* The law's concept of the family rests on the presumption that parents possess what a child lacks in maturity, experience, and capacity for judgment required for making life's difficult decisions. (emphasis added)

The Chief Justice then noted limits to parental authority, citing cases where parents put the child's physical or mental well-being in jeopardy. But, he was unable to distinguish operations that are "not agreeable" to the child and "involve risks," like a tonsillectomy or appendectomy, from involuntary commitment to a mental institution for an average term of 100 to 350 days. In his sweeping words, "Most children, *even in adolescence,* simply are not able to make sound judgments concerning many decisions, including their need for medical care or treatment. Parents can and must make those judgments." (emphasis added)

Reconciling the Decisions

Why can't parents decide about abortions for their minor daughters, but can decide that their daughters (and sons) should be committed to mental institutions? Implicitly or explicitly, the Court offers three rationales, none of which is convincing. The first is that the decisions are different in nature; the second is that parental authority to commit extends only to "immature" minors; and third is that physicians can safeguard the rights of minors in cases where parents are not acting in the best interests of their children.

Justice Burger suggests that the decisions are different because parents have no "absolute right" to order commitment; the statute requires that the superintendent of the mental institution make an independent judgment that the child is mentally ill and can be helped by treatment. However, substantially the same situation exists in the abortion decision. The Court previously made it clear that there is no "right" to have an abortion—the decision must be made in consultation with a physician, and only a physician may perform the abortion. The decisions are not persuasively distinguishable on this ground.

The plaintiffs in the commitment case, JL and JR, are both very young children. Thus it could be argued that the decision to commit can only be constitutionally made without a hearing if the child is too "immature" to make any intelligent choice, since minor women can only exercise their own choice to have an abortion if they are "mature and well-informed enough" to make an intelligent decision on their own. JL was admitted to a mental institution at the age of six, shortly after his natural parents were divorced and his mother remarried. He was diagnosed as having a "hyperkinetic reaction to childhood." Two years later he was returned to his home for two months, but his parents were unable to control him and eventually relinquished their parental rights to the state. JR was removed from his family at the age of three months and declared a neglected child. He was placed in seven foster homes and finally determined to be "borderline retarded," suffering from an "unsocialized, aggressive reaction to childhood."

These two children are obviously neither mature enough to make their own decisions, nor from the type of family the Court seems to admire. Nevertheless, while the facts of this case could confine it to immature minors, the statute in question does not so limit parental authority, nor does the sweeping language of the Chief Justice. At one point, as already noted, he talks about "most children, even in adolescence..." not being able to make sound judgments; at another, of the availability of habeas corpus as a remedy, to be pursued by, among others, "the person detained in a facility..." Only mature minors could be expected to pursue this remedy on their own.

Physicians and Courts

Finally, it could be argued that the Court has entrusted physicians with protecting the constitutional rights of children in need of medical care. The language of the commitment case certainly bears this reading. Justice Burger not only uses this as his basis for distinguishing the cases, but also adds considerable language about the respective roles of courts and administrative agencies, as opposed to the role of physicians. His conclusion is that only physicians can decide if a child needs institutionalized care, and therefore that courts have no business being routinely involved in this decision:

> [M]edical diagnostic procedure is not the business of judges. *What is best for a child is an individual medical decision that must be left to the judgment of physicians in each case*...The questions are essentially medical in character...We do not accept the notion that the shortcomings of specialists can always be avoided by shifting the decision from a trained specialist using the traditional tools of medical science to an untrained judge or administrative hearing officer after a judicial type hearing. Even after a hearing, the nonspecialist decision-maker must make a medical-psychiatric decision. (emphasis added).

The Chief Justice also argues that independent physicians are the child's best defense against getting "dumped" into an institution by parents more concerned with their own best interests than those of the child. Two points about this remarkable denigration of due process deserve mention. First, the hearing would, if it were more than a charade, involve the child's representative getting another expert to examine the child and make a recommendation to the court. If the recommendation was different than that of the committing physician, either it must be concluded that the case for commitment is not as strong as the Chief Justice seems to believe it generally will be, or that "experts" either have no "specific" basis for their opinions in this field, or will testify for whatever side pays them. Either way, there seems no reason to prefer an untested expert to one required to face an impartial judge and have his expertise and opinions questioned.

Second, though Justice Burger's language reads like a carte blanche to physicians to act on what they think is in the best interests of their minor patients (referred to them, of course, by their parents), the law can only be interpreted this way by ignoring the abortion case. In that case, the four Justices writing the opinion of the Court quote Justice Stewart's previous opinion with approval in which he discusses the role of the abortion clinic physician in looking out for the best interests of the pregnant minor: "It seems unlikely that she will obtain counsel and support from the attending physician at an abortion clinic, where abortions for pregnant minors frequently take place." Indeed it does. It also seems unlikely that institutional psychiatrists will not be biased in favor of institutional care, just as surgeons are biased in favor of surgery.

An Alternative: The Court's View of the Family

None of these rationales is entirely persuasive. An alternative suggestion is that the decisions are based on the Court's view of the family, and how it can be strengthened. There is a problem with this rationale as well: it is one thing to say that parents should have, at least initially, the right to have their child committed (with the concurrence of a physician), it is quite different to say that the state, standing in loco parentis, should have the same rights. Nevertheless, this is precisely Justice Burger's argument. He concludes that wards of the state should receive the same constitutional treatment that nonwards receive, and since nonwards do not get a hearing prior to commitment, wards of the state should not get one either. In his words, "We cannot assume that when the State of Georgia has custody of a child it acts so differently from a natural parent in seeking medical assistance for the child." This argument puts theory above fact, and also undercuts almost everything the Chief Justice had to say about the importance of the family initially.

Justice Brennan, speaking for three dissenters, correctly notes that saying that a child does not have a right to a hearing prior to commitment because social workers are obligated by law to act in the best interests of children is "particularly unpersuasive....With equal logic it could be argued that criminal trials are unnecessary since prosecutors are not supposed to prosecute innocent persons." And, he might have added, that commitment hearings are unnecessary because psychiatrists are not supposed to commit children who are not suffering from mental illness that is treatable in an institutional setting.

Conclusion

The decisions are irreconcilable. But the explanation for this need not be complex. The Court is not attempting primarily to interpret the Constitution, but to decide what is "best" for the country by attempting to decide what is best for the family. The result is unsatisfactory and contradictory because the Court does not know—as perhaps none of us know—what, if anything, is wrong with the modern family, and what, if anything, can be done about it. Disputes over teenage pregnancy and the confinement of children to mental institutions are both often the product of failed families. It is a mistake to attempt to construct a model of an "ideal" family from these two tragic circumstances. The result was destined to be flawed and inconsistent.

The Irrelevance of
Medical Judgment

Bob Woodward and Scott Armstrong persuasively argue in *The Brethren* that Justice Harry Blackmun saw his task in writing *Roe v Wade* as eliminating "needless infringement of the discretion of the medical profession" and ratifying "the best possible medical opinion."[15] Blackmun calls his ten years as counsel for the Mayo Clinic "the best...of his life," where he developed a tremendous respect for "what dedicated physicians could accomplish."

Most would agree that Blackmun's opinion accomplished his goal. It permitted a woman and her physician to decide about an abortion without any state interference while abortion was safer for her than childbirth, and even later in pregnancy it forbade the state to place the interests of the fetus higher than the "life and health" of the pregnant woman. The decision not only upheld the individual rights of women, but also evidenced strong judicial support of sound medical judgment.

It thus came as somewhat of a surprise to Justice Blackmun (and three other members of the Court) that when the Court reversed Judge John Dooling's decision in the US District Court for the Eastern District of New York and decided in 1980 that denial of Medicaid funds for medically necessary abortions was constitutional, the majority did not even discuss the role of either the doctor–patient relationship or medical judgment.[16] Like all other Supreme Court decisions concerning abortion, this one is highly controversial. Most commentators are choosing sides based not on the logic of the opinion, but on their agreement or disagreement with the outcome. This may be perfectly appropriate, since this 5 to 4 decision is not one necessarily mandated by past precedent or constitutional law, but one made primarily on political grounds. Nevertheless, it is worthwhile to summarize the major arguments presented by the majority and minority, primarily because two new voices in the abortion debate carry the brunt of the argument: Justices Potter Stewart and John Paul Stevens.

164

The Majority

Justice Stewart, writing for the majority, sees the case as a straight-forward and logical extension of the Court's previous decision that the US Constitution did not require Medicaid to fund "elective abortions," even if it funded childbirth because the state could legitimately favor childbirth over abortion.*

At issue in this case is the constitutionality of the Hyde Amendment. The applicable 1980 version provides:

[N]one of the funds provided by this joint resolution shall be used to perform abortions *except where the life of the mother would be endangered* if the fetus were carried to term; or except for such medical procedures necessary for the *victims of rape or incest* when such rape or incest has been reported promptly to a law enforcement agency or public health service. (emphasis added)

The pivotal constitutional issue is: does the Hyde Amendment, by denying public funds for some medically necessary abortions, "impinge upon a fundamental right explicitly or implicitly secured by the Constitution?" All the Justices agree that the right to decide whether or not to terminate a pregnancy is a "fundamental right." Therefore, the only real issue in the case is that of "impingement." The majority simply, and without much discussion, adopts the *Maher*[17] analysis as directly applicable. There the Court had held that "encouraging an alternative" (that is, childbirth) by "unequal subsidization" did not impinge upon a woman's right because it was not "direct state interference." Stated another way, the government's refusal to fund abortions "places no obstacle" before the woman that did not already exist, and imposes no independent "penalty" on a woman's exercise of her constitutional right. Using this analysis, Justice Stewart concludes that the difference between medically necessary and elective abortions is constitutionally irrelevant, the government does not have to fund either. But he does suggest some limits.

The government cannot penalize a woman who has an abortion by withholding all Medicaid benefits from her because this would amount to a "penalty" for exercising a constitutional right. Since the majority finds no "impingement," the Hyde Amendment is only required to be "rationally related to a legitimate governmental objective" to be constitutional. This test as traditionally applied has led to

*Maher v. Roe, discussed at pp. 153–157.

the almost universal approval of statutes, since determination of such a relationship is generally considered the province of the legislature, not the courts. The result is therefore predictable. Even though Justice Stewart announces that a majority of the Court believe that the Hyde Amendment is not "wise social policy," they nonetheless uphold it on the basis that such a decision is properly one for the legislature: "By encouraging childbirth except in the most urgent circumstances, [the Hyde Amendment] is rationally related to the legitimate governmental objective of protecting fetal life."

The Dissenters

The dissents by Justices William Brennan, Thurgood Marshall, and Harry Blackmun parallel their dissents in *Maher*. But a new member has been added to the Court since *Maher*, and Justice Stevens' dissent merits careful review. In his view the case is substantially different from *Maher*. In *Maher* the right in question was Medicaid payment for an "elective" or unnecessary abortion, one that by definition did not raise the issue of maternal health. But here, as he sees it, the woman's choice is "between two serious harms: serious health damage to [herself] on the one hand and abortion on the other. "Since under *Roe v Wade* the state may not protect fetal life when its protection conflicts with the health of the pregnant woman, Stevens argues that the state may "not exclude a woman from medical benefits to which she should otherwise be entitled [that is, coverage for necessary medical care as determined by her physician] solely to further an interest in potential life when a physician, 'in appropriate medical judgment,' certifies that an abortion is necessary 'for the preservation of the life or health of the mother.'"

Such a policy is by definition not "neutral," but violates the equal protection clause by penalizing those pregnant indigents who prefer abortion to childbirth. In the words of Justice Stevens, the government may not

> ...deny benefits to a financially and medically needy person simply because he is a Republican, a Catholic, or an Oriental—or because he has spoken against a program the government has a legitimate interest in furthering...it may not create exceptions for the sole purpose of furthering a governmental interest that is constitutionally subordinate to the individual interest that the entire program was designed to protect.

Justice Marshall argues that, even if one does not see jeopardy of a woman's health as a penalty, the Hyde Amendment bears no rational relationship to a legitimate state interest. He observes that the Hyde Amendment refused funding even in cases where "normal childbirth" will not result—where the child will die shortly after birth or where the mother's life will be shortened or her health greatly impaired by the birth. Since this is the result, Marshall argues that the only rational basis for the Hyde Amendment is that it was "designed to deprive poor and minority women of the constitutional right to choose abortion." And this is, of course, not a constitutionally permissible reason for legislation.

Inadequacies of the Decision

A number of observations are in order. First, the term "impingement" and its specific application are subject to wide interpretation. The majority of the Justices did not view refusal to fund a medically necessary abortion as an impingement on the right of a poor woman to have one, or as a penalty to discourage exercising a constitutional right. The minority differed. I personally find the majority's view insensitive to reality. Justice Blackmun correctly notes, "There is another world out there, the existence of which the Court, I suspect, either chooses to ignore or fears to recognize." By the late spring of 1980, federal and state courts in fourteen jurisdictions (affirmed by four US Courts of Appeals) had ruled that states participating in the Medicaid program were constitutionally required to fund all medically necessary abortions as determined by a physician. Two federal courts had found the Hyde Amendment unconstitutional, and no court had ruled otherwise. All these judges must have been surprised at this decision.

Second, by permitting the state to place the welfare of the fetus ahead of the health and welfare of the pregnant woman, this case effectively overrules *Roe v Wade* in terms of what interests the state may properly favor over others. And even if one believes that *Roe* should be so overruled on this issue, the Hyde Amendment does not rationally protect fetal life by forbidding funding for even those cases in which the child will die during childbirth or shortly thereafter.

Third, the primary motivation behind the majority's decision seems to have been their reluctance to tell Congress and the states how to spend the taxpayers' money. But Justice Stevens responds that this scheme will cost the government considerably more money since abortion costs less than childbirth, and the harm to women whose health is injured by childbirth must also be paid for by government funds. He concludes that "there are some especially costly forms of treatment that may reasonably be excluded from the program in order to preserve the assets in the pool and extend its benefits to the maximum number of needy persons." But the Hyde Amendment is not rationally related to such a legitimate governmental policy. Refusing to fund heart transplants or artificial hearts, of course, could be.

Fourth, the lack of discussion of the role of the physician by the majority is remarkable. It seems to assume either that the physician's judgment of the medical necessity of the abortion is irrelevant or that the physician will (should?) perform the abortion without pay if he or she really believes that it is medically necessary. Neither assumption is tenable.

Finally, the decision again affects only those least able to protect themselves—the poor. Though poverty is not a "suspect classification" that requires "strict scrutiny" of the statute in question, Justice Marshall is surely correct in arguing that when a governmental enactment directly affects the very lives and health of poor women who seek to exercise a constitutional right, it is unfair and unjust to judge that statute by the same mechanistic test designed primarily to give judicial deference to legislation about competing business interests.

The funding battle for abortions for poor women has been lost in the courts. It is now up to Congress and the individual states to pass reasonable legislation on this issue, an unlikely outcome given the present political climate. After the Court's previous decision in *Maher*, one supporter wrote, "It's like bringing some of the troops home from Vietnam: we didn't save all fetuses, but we saved some." The analogy, of course, is not quite accurate. It is like bringing some troops home—those who can pay for their own plane tickets.

Roe v Wade Reaffirmed

The Supreme Court's message is clear: if states continue to pass statutes that restrict a woman's access to abortion in ways not permitted by its 1973 *Roe v Wade* decision, the Court will strike them down. Efforts to persuade the Court to overrule or curtail *Roe* have failed, and the Court's commitment to a pregnant woman's liberty interest in her decision regarding childbirth or abortion can no longer be questioned.

In *Roe v Wade*[18] the Court held that the constitution protected the woman's right to decide, with her physician, to have an abortion. As long as abortion was safer than childbirth (until about the end of the first trimester), the state could not demonstrate any interest in maternal health compelling enough to interfere with the decision. Until the fetus was considered viable, that is, had "the capability of meaningful life outside the mother's womb" (usually about the beginning of the third trimester), the interests of the mother continued to outweigh those of the fetus. At viability, however, the state's interest in protecting fetal life became compelling enough to permit the state to prohibit abortions, "except when necessary to preserve the life or health of the mother." Sometimes denoted the "trimester system," the actual dividing lines regarding permissible state regulation were two: the time when abortion became more dangerous to the pregnant woman than childbirth, and the time when the fetus became viable.

The Akron Opinion

Until the Court's 1983 decision in *Akron*,[19] decided by a vote of six to three, it was possible to argue that the Supreme Court was slowly backing away from its most controversial decision of the past 25 years. In 1977, for example, the Court ruled that the Constitution did not require a state's Medicaid program to pay for nontherapeutic abortions if it paid for childbirth*; and in 1980 that the Constitution did not require Medicaid funding of any abortions, even medically necessary ones, because failure to fund them did not place any state-created obstacle in the path of a poor woman who elected to terminate her pregnancy by abortion.**

But these cases were limited to the issue of government funding. Those commentators who, like I did, interpreted them as signaling a retreat from *Roe v Wade*, have now been proven wrong. Indeed, it would be difficult to write a stronger reaffirmation of the principles of *Roe* than Justice Louis Powell's majority opinion in *Akron*.

At issue was a detailed ordinance of the City of Akron, Ohio, that attempted to place numerous obstacles in the way of a woman who wished to have an abortion. The ordinance's provisions relating to in-hospital abortions and informed consent probably best illustrate the Court's overall position.

The Hospitalization Requirement

The ordinance required that all abortions after the first trimester be performed in a hospital. This would have been acceptable in 1973, when such abortions were considered more dangerous to a woman's health than childbirth. However, in 1983 medical evidence indicated that when the method of dilation and evacuation (D&E) is used, abortions may be safer than childbirth up to a gestational age of sixteen weeks. While continuing to put the line for the state's interest in protecting maternal health "at approximately the end of the first trimester," the Court notes that this interest no longer justifies a blanket hospital rule because: (1) the safety of second-trimester abortions has "increased dramatically" since *Roe*; (2) the American Public Health Association and the American College of Obstetricians and Gynecologists both agree that second-trimester abortions using dilation and evacuation can be safely performed in nonhospital facilities; and (3) the additional cost of an in-hospital abortion, $850–$900 versus

---

**Maher v. Roe*, discussed in "Let Them Eat Cake" at p. 153.

***Harris v. McRae*, discussed in "The Irrelevance of Medical Judgment" at p. 164.

$350–$400 for a clinic abortion, places "a significant obstacle in the path of women seeking abortion."*

The Court concludes:

> By preventing the performance of D&E abortions in an appropriate nonhospital setting, Akron has imposed a heavy, and unnecessary, burden on women's access to a relatively inexpensive, otherwise accessible, and safe abortion procedure...and therefore unreasonably infringes upon a woman's constitutional right to obtain an abortion.

The Consent Requirement

Another portion of the ordinance, mislabeled "informed consent," required the physician to personally inform the pregnant woman of a number of specific items, including the statement that "the unborn child is a human life from the moment of conception"; a description detailing the "anatomical and physiological characteristics of the particular unborn child...including appearance, mobility, tactile sensitivity, including pain, perception, or response..."; the warning that "abortion is a major surgical procedure, which can result in serious complications...and can result in severe emotional problems"; and the suggestion that "numerous public and private agencies are available to assist her during pregnancy and after the birth of her child if she chooses not to have an abortion..." Although agreeing that informed consent is an essential part of the abortion decision, the Court concludes that the relevant information can vary based on the patient's "particular circumstances" and that it should remain "primarily the responsibility of the physician to ensure that appropriate information is conveyed to the patient." The information required by the Akron ordinance, the Court concludes, is "designed not to inform the woman's consent, but rather to persuade her to withhold it altogether."

Equally decisive, the Court says, is that it intrudes "upon the discretion of the pregnant woman's physician." The state "may require that

---

*In a companion case the Court approved a state requirement that all fetal remains be examined by a pathologist because of the health rationale and the "comparatively small added cost," put at $19.40. On the other hand, an Akron requirement involving a 24-hour waiting period between consent and the abortion procedure was struck down because of the cost involved in making two separate trips to the clinic. Small costs with health rationales are permitted; burdensome and unnecessary added costs are not.

the physician make certain that the pregnant woman understands the physical and emotional implications of having an abortion," but it places unconstitutional "obstacles in the path of the doctor upon whom the woman is entitled to rely for advice in connection with her decision" when it insists "upon the recitation of a lengthy and inflexible list of information."

The Court also rejects the requirement that the physician always obtain informed consent personally, because it does not believe the state can demonstrate that such a requirement is reasonably designed to further the state's interest in the woman's health: to provide her with "the critical information and counseling from a qualified person, not [to mandate] the identity of the person from whom she obtains it."

The Dissent

Roe v Wade was a seven-to-two decision. Justice Sandra O'Connor wrote the dissent in *Akron*, joined by the original two dissenters in *Roe*, Justices White and Rehnquist. Justice O'Connor's primary arguments center on the "completely unworkable" trimester system that requires courts and legislatures "to continuously and conscientiously study contemporary medical and scientific literature" to determine the acceptability of current practice, the safety of abortion at a given period in pregnancy, and the point of viability. She argues that the *Roe* approach turns the court into a "science review board" and violates a primary principle of judicial decisionmaking: the application of neutral principles "sufficiently absolute to give them roots throughout the community and continuity over significant periods of time..."

Although this is the framework she outlines, her dissent makes clear that she is not upset with the changing nature of *Roe* (which is tied to the state of medical technology) but with its outcome. She would be no happier with a decision that supported the woman's liberty interest in abortion on some more permanent basis; her goal was a decision that permitted the states to heavily regulate or outlaw abortion altogether. Her central argument is that *Roe's* trimester framework "is clearly on a collision course with itself" and therefore should be abandoned. She correctly notes that abortion will probably keep becoming safer later in pregnancy, and may even become safer than childbirth up to and after fetal viability. As a result, states will be prohibited from enacting unreasonable and burdensome restrictions on abortion up to fetal viability. But the state will still have ability to proscribe abortion altogether after viability, even if pregnancy termination is (at some future date) always safer or as safe as continuing the pregnancy to term. At this point the *Roe* trimester test will collapse into a two-period test: pre- and post-viability.

The real issue, therefore, is fetal viability and the effect of improved technology, which may permit younger and younger fetuses to survive to "meaningful lives." Under *Roe* such technological improvements will enhance the state's ability to regulate abortion, and will require women seeking abortions to have them prior to fetal viability (as they must now) and thus earlier and earlier in pregnancy. I have difficulty seeing, however, why such a result is either discontinuous with *Roe* or socially undesirable. Perhaps Justice O'Connor is worried about the day—surely in the distant future—when it may be possible not only to conceive a child in the laboratory, but also to grow the fetus to viability in an "artificial placenta." Then, since the fetus is "immediately viable" the *Roe* framework will provide no basis for distinguishing between any of the phases of pregnancy, and a new analytical framework will be necessary. In an area in which scientific advances are central concerns, it is perfectly appropriate for both courts and legislatures to reconsider their rules in light of these advances.

Reaffirming *Roe v Wade*, the Supreme Court takes a strong stand on the side of our's being a government of law. As Justice Powell put it for the Court: "The doctrine of stare decisis [following precedents] ...is a doctrine that demands respect in a society governed by the rule of law." *Roe v Wade* was the law in 1973, has been the law since, and remains the law. Debate will continue as to its wisdom, its moral basis, and its effect. The debate as to its legal status in the Supreme Court, however, seems over.*

*The confirmation hearings on Justice Robert Bork in the fall of 1987 make this prediction appear as naive as my prior one that the court was retreating from *Roe*.

Reaffirming *Roe v Wade*, Again

Roe v Wade has two distinct aspects, and each has often been separately attacked. The first is its cornerstone and most critical holding: the "right to privacy" is a fundamental human right, broad enough to encompass the right of a pregnant woman to choose an abortion. This matter of constitutional construction continues to fuel debate. The second is that the state has a "compelling interest" in protecting the life of the fetus after viability. The capability of the fetus to survive independent of the woman will be a function of science, technology, and medical practice, and thus the contours of *Roe* can be directly affected by medical advances.

Given these two independent avenues of attack that this historic Supreme Court opinion is open to, and given that the Court has heard no fewer than a dozen major abortion cases in the 13 years since *Roe*, it is remarkable how consistent the majority opinions have been in determining what legitimate restrictions state legislatures can place on abortion decisions by women and their physicians.

At issue in the 1986 case *Thornburgh v ACOG*[20] were six specific provisions of the Pennsylvania Abortion Control Act of 1982, a title that aptly describes what the statute is all about. Because of the procedural manner in which the case got to the Court, the Court could easily have simply remanded it to the trial court for further evidentiary proceedings. The fact that it chose not to do so, and the fact that the author of *Roe v Wade*, Justice Harry Blackmun, was chosen to write the opinion, means that the majority of the Court went out of its way to once again reaffirm the principles enunciated in *Roe*.

The Court seems to have done this as a way to restate and emphasize three primary points: (1) the centrality of the "right to privacy" that protects certain private spheres of decisionmaking from governmental intrusion; (2) the chilling and brutal effect unclear criminal laws have on medical practice, and thus on the exercise of constitutional rights in the context of the doctor-patient relationship; and (3) the importance of the doctrine of stare decisis in a government based on law.

The Majority Opinion

The Court invalidated all six of the provisions in dispute before it (informed consent; printed information; reporting; determination of viability; postviability care for the child; and the requirement of a second physician at postviability abortions). The last two were struck down because they required physicians to trade off the woman's health against fetal survival without explicitly acknowledging that maternal health could always be the physician's "paramount consideration." The other four provisions were dealt with in pairs.

The "informed consent" and "printed information" provisions required that certain information be given to the woman by the referring physician, or the physician who would perform the abortion, at least 24 hours prior to the abortion, and certain other information be provided by the physician or his agent at least 24 hours prior to the abortion. The physician information included the name of the physician who would perform the abortion; the fact that there "may be detrimental physical and psychological effects which are not accurately foreseeable"; the medical risks associated with the particular method to be used; the probable gestational age of the "unborn child"; and the medical risks associated with carrying a child to term. The information that could be conveyed by the doctor's agent included: the fact that medical assistance benefits may be available for prenatal care, childbirth and neonatal care; the fact that the father is liable to assist in the support of the child; and that she has a right to review printed materials which she shall be orally informed "describe the unborn child and list agencies which offer alternatives to abortion." All of this must then be certified in writing by the woman prior to the abortion.

Though not the "parade of horribles" struck down in *Akron,** the Court nonetheless struck down these requirements using the *Akron*[21] rationale that the information is designed not to get the woman's informed consent, "but rather to persuade her to withhold it altogether." In addition, the Court objected to the "rigidity" of the requirement that imposed an "undesired and uncomfortable straitjacket" on the physi-

*Discussed in *"Roe v. Wade* Reaffirmed" at p. 169.

cian. In an obvious overstatement, the Court concluded, "All this is, or comes close to being, state medicine imposed upon the woman, not the professional medical guidance she seeks, and it officially structures—as it was obviously intended to do—the dialogue between the woman and her physician." Giving some examples, the Court described requiring this information as "cruel" in a case of a rape victim having to be told about the theoretical financial responsibility of the rapist.

The "reporting" and "determination of viability" provisions required that the physician determine whether or not the fetus was viable (if viable, the physician is required to report the basis for his determination that the abortion is necessary to preserve maternal life or health; if not viable, the physician must report the basis for this determination). In addition, the physician must report to the state department of health the following information: identification of physician and facility involved in the abortion, and referring physician; political subdivision and state in which the woman resides; woman's age, race and marital status; number of prior pregnancies; date of last menstrual period; type of procedure used; complications; length and weight of "aborted unborn child"; basis for medical emergency; date of consultation examination; whether the abortion was paid for by the patient, medical assistance, or medical insurance. These reports "shall be made available for public inspection and copying within 15 days of receipt in a form that will not lead to the disclosure of the identity of any person filing a report."

In *Danforth*[22]* the Court had approved record keeping requirements designed to "preserve maternal health" so long as patient confidentiality was maintained. The Court found these rules, however, went "well beyond" maternal health interests, especially information related to payment method, personal history, and basis for medical judgment. The Court concluded that maintenance of privacy seemed problematic, and the possibility of a breach will "necessarily" make a woman and her physician "more reluctant to choose an abortion," and thus potentially "chill the exercise" of her constitutional right:

> Pennsylvania's reporting requirements raise the spectre of public exposure and harassment of women who choose to exercise their personal, intensely private, right, with their physician, to end a pregnancy. Thus, they pose an unacceptable danger of deterring the exercise of that right, and must be invalidated.

*Discussed in "Round II" at p. 147.

Justice Blackmun's majority opinion makes it clear as he can that neither he nor the majority of the Court has backed down one inch from *Roe v Wade*, and that the states should recognize it as the law of the land and respect it:

> Again today, we affirm the general principles laid down in *Roe* and in *Akron*. In the years since this Court's decision in *Roe* states and municipalities have adopted a number of measures seemingly designed to prevent a woman from exercising freedom of choice. *Akron* is but one example. But the constitutional principles that led this Court to its decisions in 1973 still provide the compelling reason for recognizing the constitutional dimensions of a woman's right to decide whether to end her pregnancy. "It should go without saying that the vitality of these constitutional principles cannot be allowed to yield simply because of disagreement with them."... *The States are not free, under the guise of protecting maternal health or fetal life, to intimidate women into continuing pregnancies.* (emphasis added)

And, in concluding the opinion, Justice Blackmun argues that the promise "that a certain sphere of individual liberty will be kept largely beyond the reach of government" is one that "extends to women as well as to men." Regarding the abortion decison, he says, "few decisions are more personal and intimate, more properly private, or more basic to individual dignity and autonomy, than a woman's decision— with the guidance of her physician and within the limits specified in *Roe*—whether to end her pregnancy."

The Dissenters

This 5-4 decision brought three written dissents. Justice O'Connor mainly restated her *Akron* dissent.* Chief Justice Burger based his dissent (a new position for the Chief Justice) on an apparent misreading of *Roe v Wade*.

Justice White wrote a detailed dissent, which reads like a warm up exercise for what he hopes someday will be a majority opinion overruling *Roe v Wade*. Justice White dissented in all but the funding cases involving abortion, and is not about to change his mind even after 13 years of unbroken majority opinions upholding *Roe*. He explains that following precedents, the doctrine of stare decisis, is a critical "prin-

*Summarized in "*Roe v. Wade* Reaffirmed" at p. 169.

ciple of law" that prevents the exercise of "judicial will" from becoming "arbitrary and unpredictable." But, he thinks, cases that determine statutes unconstitutional "call other considerations into play." Specifically, he objects to "decisions that find in the Constitution principles or values that cannot fairly be read into the document" because these decisions "usurp the people's authority, for such decisions represent choices that the people have never made and that they cannot disavow through corrective legislation." Accordingly, he argues that the Court should correct constitutional decisions that "on reconsideration, are found to be mistaken."

He thinks *Roe v Wade* is such an opinion, and argues that it is time to overrule it. In this regard Justice White restates his argument that a woman's right to abortion is not a "fundamental" constitutional right, and therefore the state need show only a "rational basis" for legislation restricting its exercise, rather than a "compelling state interest." In deciding otherwise, Justice White accuses the majority of "engaging not in constitutional interpretation, but in the unrestrained imposition of its own, extraconstitutional value preferences."

Justice White also argues that the "viability" rule is "entirely arbitrary." In his view, the governmental interest at stake is "in protecting those who will be citizens if their lives are not ended in the womb." This interest, he argues, is not contingent upon medical science or other factors, but "is in the fetus as an entity itself, and the character of this entity does not change at the point of viability under conventional medical wisdom." If these views are accepted, Justice White correctly notes, then states may adopt "a broad range of limitations on abortion (including outright prohibition) that are not now available."

Justice Stevens spends his concurring opinion directly addressing Justice White's dissent, and his commentary is devastating to White's argument. On the issue of the fetus having the same interests throughout pregnancy, for example, Justice Stevens notes:

> I should think it obvious that the state's interest in the protection of an embryo...increases progressively and dramatically as the organism's capacity to feel pain, to experience pleasure, to survive, and to react to its surroundings increases day by day. The development of a fetus—and pregnancy itself—are not static conditions, and the assertion that the government's interest is static ignores this reality...there is a fundamental and well-recognized difference between a fetus and a human being; indeed, if there is not such a difference, the permissibility of terminating the life of a fetus could scarcely be left to the will of the state legislatures.

Conclusion

Justice Stevens, of course, underlines (again) the real conflict at issue between supporters and detractors of *Roe v Wade*. It is not a contest between those "in favor" of abortions and those opposed to them. An individual can quite easily view abortion as a horror, and still support the constitutional analysis of *Roe*, just as one can support a woman's right to an abortion, and disagree with *Roe's* analysis. The real issue is between majoritarian rule by criminal statutes, and private decisionmaking by individual pregnant women. For Justice Stevens is right; if one really believes that embyros are persons from the moment of conception, their lives can hardly be left in the hands of a majority in the state legislatures.

Justice Blackmun thus seems correct when he concludes the majority opinion by saying:

> Our cases long have recognized that the Constitution embodies a promise that a certain private sphere of individual liberty will be kept largely beyond the reach of government. That promise extends to women as well as to men. Few decisions are more personal and intimate, more properly private, or more basic to individual dignity and autonomy, than a woman's decision—with the guidance of her physician and within the limits specified in *Roe*—whether to end her pregnancy. A woman's right to make that choice freely is fundamental. Any other result, in our view, would protect inadequately a central part of the sphere of liberty that our law guarantees equally to all.

Medical
Practice

Introduction

As the title suggests, these essays focus on a variety of topics that do not fit easily into any of the other more specifically defined sections. They are, nonetheless, important enough to the overall view of the role of the law in medicine during the past decade to be included in this book.

The first three essays deal with medical malpractice and the controversies that litigation by injured patients against physicians have flamed. The last ten years opened with a medical malpractice insurance "crisis" and closed with the second such crisis still underway. The fourth essay deals with the related but seemingly intractable problem of the impaired physician. How can impaired physicians be identified so that they can be helped, and patients can be protected? States continue to license physicians, but no state requires that they be reexamined for competence any time during their careers. The current system simply does not work, and we will need to dramatically improve it during the coming decade.

The section concludes with two essays on cardiopulmonary resuscitation (CPR), a dramatic medical technique, but one that has so taken on a life of its own that physicians and others have a very difficult time not using it on every patient before they die. Indeed, the

New York Commission on Life and the Law has recommended, and New York Legislature has enacted, a law dealing only with orders not to use CPR.* This is remarkable and could only have been brought about by a medical technique that has gotten so far out of hand that physicians either no longer know when its use is appropriate, or if they know, cannot be trusted to act according to their own best judgment in deciding whether or not to perform CPR on a patient. Because it is thought by most that the reasons for this lack of common sense in performing CPR are primarily legal, a discussion of when CPR should and should not be employed can be used to judge both medicine and the courts.

The first essay, "Medical Malpractice," is partially adapted from an article for the *Encyclopedia Britannica's* 1987 *Medical and Health Annual*. All of the other essays appeared originally in the *Hastings Center Report*: "Doctors Sue Lawyers," in October, 1977; "Confidentiality," in December, 1976; "Who to Call," in December, 1978; "CPR: The Beat Goes On," in August, 1982; and "CPR: When the Beat Should Stop," in October, 1982.

---

*1987 NY Laws 818; NY Public Health Law article 29-B.

Medical Malpractice

"Medical malpractice" denotes the basis for a civil action brought by a patient against a health care provider for injuries suffered as a result of negligence. The current method for compensating victims of these occurrences is primarily a fault-and-liability system. The first principle of tort liability is that the party at fault pays for the damage inflicted upon an innocent victim. Whether the health care provider is at fault is determined in an adversary proceeding, in which both the provider and the patient are represented by legal counsel. The trier of fact, usually a jury, must decide if the defendant must pay the plaintiff for the injury. The formula for determining whether or not a physician is responsible to compensate the patient is commonly phrased in terms of whether the physician acted in the way a reasonably prudent physician would have acted in the the same or similar circumstances.

The adversary system, the jury, and the rules of evidence have all generated controversies between the medical and legal professions, and these debates often degenerate into unflattering and unconstructive caricatures of the two professions. Although these debates have entertainment value, they are pointless from a public policy perspective. For example, some physicians like to quote the famous line from *Henry VI, Part II*, "The first thing we do, let's kill all the lawyers." (IV. ii) But it is seldom appreciated that the character who spoke these words was an unschooled murderer and anarchist named Dick the Butcher who wanted to overthrow the legal system so that there would be no laws but those that he and his fellows from time to time orally decreed. Of course, physicians are not actually calling for the mass execution of lawyers, any more than they really want to overthrow the US Constitution and strip citizens of their rights of access to the courts. But strong rhetoric often obscures real problems, and the medical malpractice insurance crisis, which the American Medical Association ranks as the "number one" issue confronting its members, is a case in point.

Primarily because they are usually decided by lay juries in public, physicians have deplored lawsuits against them for almost a century and a half. In 1845, for example, physicians indicated alarm at the increase in malpractice lawsuits, and suggested alternatives to jury trials, such as committees made up of physicians, to judge such claims. In 1872, the American Medical Association recommended that physicians be appointed independent arbiters by the court to judge their peers. Physicians have also historically hated the term "malpractice" itself, a term that denotes "evil" or "bad" practice. In fact, it refers simply to a physician who has not lived up to the customary professional standard set by the actions of the "average competent physician" in the same or similar circumstance. When a patient is injured by the actions of a physician, the patient can successfully sue the physician for the harm done only if the patient is able to prove that the physician was "negligent" (i.e., that the physician breached a duty owed to the patient) and that this negligence caused the patient harm. For example, if a physician leaves a sponge or clamp in the patient's abdomen after surgery, and this necessitates an additional operation to remove it, the patient can sue for the cost of the additional operation, lost earnings, and "pain and suffering." Our society permits such suits for three basic reasons: (1) to control quality by holding doctors responsible for their actions, (2) to compensate patients for injury, and (3) to give patients an opportunity to express dissatisfaction with the care they have received.[1]

An Insurance Crisis

In the mid-1970s, there was a "crisis" in the availability of malpractice insurance that physicians and hospitals could purchase. While the cause of this crisis seems to have stemmed primarily from management and investment decisions on the part of insurance companies, the resulting solution adopted by 49 of the 50 states was to change the laws to make it harder for injured patients to sue physicians. Some also formed quasi-public agencies, called joint underwriting associations, to write insurance policies for physicians who could not obtain it from private companies. In the mid-1980s, the insurance crisis returned. In this decade the problem is not availability of insurance, but affordability. Underwriters have dramatically increased premiums, especially in high-risk specialties like obstetrics and neurosurgery, and many physicians are concerned about the amount of their income needed to pay the premiums.

Insurance companies argue that the precipitous rate increases are required by increases in the frequency and severity of malpractice litigation. They cite figures indicating sharp increases in the absolute numbers of malpractice suits filed, and in the number of million dollar plus verdicts returned by juries. The companies argue that America has become an increasingly lawsuit-prone society, and that premium increases are needed to keep pace with increases in litigation expenses and unrealistic jury verdicts. As they did a decade ago, the insurance companies seek changes in the legal system to make it harder for injured patients to seek compensation from physicians and hospitals. And even though there is no more substance to their argument than there was in the 1970's, more than 30 states enacted new medical malpractice and tort reform acts in 1986 alone. A few states, however, like Florida and New York, also increased the authority of the insurance commissioner to freeze or regulate malpractice insurance rates.

The AMA's Position

The American Medical Association (AMA) has sided with the insurance companies, and also focused its attention on "reforming the tort system." The AMA's four major recommendations for change are: (1) limitation on jury awards of non-economic damages ("pain and suffering") to $100,000; (2) requiring deductions in awards for money victims receive from any other source (eliminating the collateral source rule); (3) using mandatory periodic payments for all awards for future damages that exceed $100,000; and (4) limiting attorneys fees by modifying the contingency fee (under which lawyers are generally paid a flat percentage of the award, usually 30–50%) to a sliding scale, in which the lawyer's percentage would diminish as the total award grew.

The ABA's Position

Lawyers argue that such changes are misplaced. The American Bar Association (ABA) for example, rejected the AMA's recommendations, arguing that the real problem with medical malpractice *is* medical malpractice. Accordingly, the ABA recommends that physicians should set up much tougher methods of policing themselves to eliminate incompetent and impaired physicians, and thereby protect the public. The ABA also notes that changes in the legal system were tried in the 1970s, and failed as a method to deal with the problem, and that this demonstrates that the real problem is not with the legal system. Alternatively, the ABA argues that if the problem is with the way we handle personal injury suits in this country, then we should change the entire tort system across the board, and not just make adjustments for physicians. There will undoubtedly be proposals for tort changes that will affect *all* personal injury lawsuits. The chemical industry and manufacturing companies, for example, have a huge financial stake in limiting compensation for injury caused by their products.

As to the AMA's specific proposals, trial lawyers generally argue that limiting "pain and suffering" awards to $100,000 is unfair to severely injured victims who this amount cannot begin to compensate; requiring reduction of awards for outside insurance benefits penalizes those who carry such insurance and is a windfall to the wrongdoer; and radically changing the contingency fee will close the courtroom to the poor and middle class who cannot afford to pay a lawyer out of their own pocket to bring a complex and costly lawsuit.

Consumer Groups

Consumer groups, like Ralph Nader's Public Citizen, view the professional debate between physicians and lawyers as a sideshow. They note that it is the insurance companies which have dramatically raised the rates for insurance coverage, and argue that these companies are simply exploiting the fear of physicians in order to reap profits. Stock market values of the insurance companies, they note, more than doubled in 1985, the same year in which these companies drastically increased their premiums.

St. Paul Company, the largest insurance company involved in medical malpractice (with about 20% of the entire malpractice insurance market) is a good example. In 1984 the company suffered its first loss since the San Francisco earthquake of 1906. By drastically increasing its premiums (60% in two years) and cutting its coverage, it made an incredible comeback. Its stock went from 52 to 92 in just 12 months, and almost all of this increase was directly attributable to its malpractice insurance business. Its Chairman, Robert Haugh, explains how the system worked. The company actually pays out more in malpractice claims than it takes in, but it retains the premium money for a long time and makes money by investing these premiums, usually more than enough to both offset losses and make a profit. The company now has $1.1 billion in its medical malpractice reserves earning investment income at 16% annually. In Mr. Haugh's words, "We know what this business is all about, and we feel comfortable with it."[2] In fact, while property/casualty companies suffered a $46 billion dollar underwriting loss from 1975 to 1984, they also had about $121 billion in investment gains during this period, for a net gain of about $75 billion.

This is probably why many think the insurers are too comfortable, and need to be much more closely monitored and regulated. The president of the National Insurance Consumer Organization, Robert Hunter, for example, argues that insurance premiums are much more closely linked to the industry's profit cycle (which is determined primarily by their return on investment) than to legal liability. He notes that in 1975, the date of the beginning of the last "insurance crisis," the industry earned only 4% on its reserves. Premiums skyrocketed in 1976 and 1977 to make up for this loss. The premium increase brought sharp profit increases (similar to those now being experienced), which attracted a flood of new capital and led to premium price cutting to accumulate more cash to invest at high interest rates. When these rates fell, the industry earned only 2% on equity in 1984. As a consequence, and a repeat of 1976–1977, premium rates shot up for 1985–1986. The only difference was that other insurance lines were affected as well, including liability insurance for day care centers, cities, and many products. In his view, it is completely implausible to argue that the legal system, as represented by judges and juries, worked perfectly well from 1977 to 1983, and then all of a sudden went out of control with unanticipated high verdicts in 1984 and 1985.[3]

These consumer groups seem correct. Although the number of million dollar plus verdicts has increased to over 400 in *all* personal injury cases in 1985, these are verdicts, not payments or settlements, which are usually significantly lower. Other studies have concluded that the total number of civil lawsuits is *not* increasing per capital in the US, and that the median (middle value) jury award has actually remained constant for the past 25 years, at about $20,000.

The Economist's View

Economists have found the situation mixed. There is general agreement that the system is inefficient at compensating victims of mal-

practice, but is cost-effective as a method of quality control. Its ineffi-
ciency at compensation stems from two facts: very few people who
are actually injured by malpractice ever get into the system; and there
are very high transaction costs that go to the lawyers and insurance
companies who get more of the total premiums than injured victims
do. Studies suggest that approximately 1 of every 100 patients admitted to a hospital suffers an injury due to medical negligence. Nonetheless, only between 1 in 25 and 1 in 50 of these injured patients is
ever compensated through the current system. But even though the
compensation is very limited, as long as the potential for a suit deters
any meaningful percentage of negligent injuries, the system will be
cost-effective from the patient's viewpoint.[4]

From a systems-wide perspective, the system is actually inexpensive. Malpractice premiums for physicians *and* hospitals combined
amount to less than 1% of total expenditures on health care. The
alleged added expense due to "defensive medicine" (extra tests doctors do to protect themselves instead of their patients), cannot be documented. And although some specialists pay higher rates, the average
physician, even in 1986, still spent less than 5% of gross income on
malpractice insurance, a percentage that has not changed significantly
during the past decade. Of course, there is plenty of room to make the
system more cost-effective and more efficient.

What to Do

Some look abroad with envy. Sweden and New Zealand are seen
as lands of no lawsuits, and some have urged that we adopt their "no-
fault" systems, which simply compensates everyone for injury, regardless of cause, by paying their medical bills. What this solution
misses, of course, is the fact these countries, like every industrialized
country in the world except the US and South Africa, have a system
of national health insurance in which *everyone's* medical bills are paid
for, regardless of source of injury or illness. Likewise, Sweden has a
vast network of social support systems that are government-financed
and available to all. We *could* adopt the Swedish system for malpractice lawsuits, but we would have to adopt some form of national health
insurance and a government-financed social support system first.

In early 1986, the General Accounting Office issued a report to
Congress on "Medical Malpractice." Its subtitle, "No Agreement on
the Problems or Solutions" aptly describes the current situation. In

April, 1986, the *New York Times*, in a lead editorial, depicted the current crisis in insurance availability (all insurance, not just medical malpractice) as one with three possible solutions:

(A) crack down on the lawyers who manipulate juries to win outlandish settlements—and fat contingency fees. (B) crack down on the insurance industry, which seems unable to manage its cash flows responsibly. (C) try to teach the public that enormous liability judgments are not cost-free. (D) all of the above.

The *Times* thinks the correct answer is "D," and chided the Reagan Administration for concentrating only on (A) in its proposals to reform products liability tort law. In medical malpractice, the answer is "all of the above" as well. Specifically, the legal profession must discipline its members who file frivolous lawsuits and pursue untenable claims, and should do so vigorously. States and the federal government should regulate insurance companies so that accurate data on premiums and claims paid can be compiled, and reasonable premium raises can be announced well in advance, limited on a yearly basis, and cancellations permitted only with adequate notice.

There are also at least two additional items that should be added to the list: more effective methods to prevent incompetent and impaired physicians from hurting patients; and more effective communication between physicians and their patients. The former will demand a commitment on the part of the medical profession to police its own, and on the part of state legislatures to change the composition of medical licensing boards to include larger proportions of nonphysician members. More effective communication between physicians and patients is something we can all work on now.

Informed Consent and Malpractice Suits

Although physician concern for lawsuits is not new, the public's perception of what is possible in medicine, and indeed medicine's power over disease, has changed dramatically and in some cases naively. This perception is shared by both patients and their physicians, and has been spawned by the arrival of the "new technological age" of medicine. Technology has given us, as novelist Don Delillo puts it, "an appetite for immortality." We believe that diseases *can* be controlled, and that physicians *should* be able to do something for us when we fall ill. We want to believe that we can have it all; live our lives without regard to physical or mental dangers, and then go to the "repair shop" (the hospital) when we suffer a physical or mental "breakdown" and have it fixed. We have adopted the image of ourselves that commentators on industrialization have feared for decades: we see ourselves as machines, and physicians as mechanics. If physicians can't repair us, it must be because they lack the skill, don't know the latest techniques, or make a mistake.

The major problem with medical malpractice insurance is medical malpractice insurance. But unrealistic expectations on the part of patients, and ritualistic silence and demands for blind faith on the part of physicians, exacerbate the broader medical malpractice situation greatly. Enhancing the doctor–patient relationship by taking informed consent seriously could go a long way toward "solving" the medical malpractice problem. Other solutions deserve fair hearings, but should be judged against the three primary goals of the current tort system: compensation for injury, quality control, and responsiveness to consumers. Only changes that enhance one or more of these goals, without offsetting losses in others, deserve serious consideration. Unfortunately, although Dick the Butcher may not have realized it, there is no simple solution to this problem that plagues both doctor–lawyer and doctor–patient relationships.

Doctors Sue Lawyers
Malpractice Inside Out

Suits by physicians against lawyers who bring unsuccessful malpractice charges against them, though rare, became increasingly popular in the mid-1970s. They were hailed by the American Medical Association in 1976 as a new weapon that could "discourage the filing of frivolous or nonmeritorious cases..."[5] Such lawsuits usually allege defamation of character, malicious prosecution, abuse of process, or professional negligence.

Defamation suits are generally unsuccessful because lawyers are shielded by an absolute privilege regarding their statements so long as these statements are part of the judicial proceeding. Malicious prosecution requires, as the name implies, a showing of actual malice toward the defendant in bringing the suit. It also requires that the original suit be terminated favorably to the defendant and a showing that it was initiated without probable cause. Abuse of process does not require either a conclusion of the litigation in favor of the defendant or lack of probable cause, but does require a finding that the lawyer "abused" the judicial system by bringing a lawsuit for some improper ulterior purpose or motive, such as attempting to coerce a physician into waiving fees.

The most interesting charge is that of professional negligence. Specifically, the physician alleges that the lawyer owes both a legal and ethical duty to the physician not to bring a malpractice suit when the lawyer should know that it is "frivolous." One physician involved in this type of litigation characterized such malpractice suits as "unnecessary," making an analogy to unnecessary surgery. Leonard Berlin put his rationale less politely: "The school bully who harasses innocent children in the playground isn't stopped by reporting to his parents [the bar association or licensing authority]–he's only stopped by a punch in the face. Lawyers who abuse the court system similarly can only be stopped through the same system."*

A review of the appellate decisions suggests that while this line of attack is extremely appealing to physicians, it is not likely to reduce the incidence of "frivolous" suits.

Three Countersuits

In a Texas case a physician was granted summary judgment in his favor in a suit for failure to properly treat and diagnose injuries sustained by the plaintiff in an accident. The physician then sued both the plaintiff and his lawyer saying that they should have known that the suit was groundless and unfounded, and that the allegations made were untrue. He sought $750,000, alleging mental trauma requiring psychiatric treatment, increases in his malpractice premiums, public ridicule, and loss of earnings. The lower court granted summary judgment in favor of the attorney and his client, and the appeals court affirmed. In a very brief opinion the court made it clear that under Texas law attorneys simply could not be sued on the basis of the facts alleged by the physician, noting that he failed to properly allege a case for abuse of process, malicious prosecution, contempt of court, or invasion of privacy—the only actions the court seemed to consider as plausible under the circumstances.[6]

A New York case involved a physician who was sued for negligent treatment of a patient even though he did not see the patient at all during the course of the illness in question. The physician sought $200,000 from the lawyer for, among other things, failing to properly investigate the case, filing a frivolous lawsuit, and practicing law in "a malicious and unethical fashion with a reckless disregard for the truth or falsity of the allegations." The court quickly disposed of the notion that the attorney had any duty toward the defendant-physician in a malpractice suit:

> The courts of this State have consistently held that an attorney is not liable to third parties for the negligent performance of his obligations,

*Leonard Berlin won an $18,000 injury verdict in a countersuit that he filed against a lawyer for suing him "without probable cause." An Illinois Appeals Court later reversed, holding that he could only sue successfully if the suit was brought maliciously. *Berlin v. Nathan*, 381 NE 2d 1378 (Ill. App. 1978).

even where such negligence results in damage to third parties... for to hold an attorney personally responsible for instituting a frivolous action on behalf of a client would operate to discourage free resort to the courts for the resolution of controversies, contrary to public policy.[7]

The final case involved a surgeon who had performed an emergency hernia repair operation on a three-month-old boy. The parents alleged that she had damaged one of the infant's testicles in the operation. At trial all expert witnesses, including the two called by the plaintiff, testified on behalf of the surgeon and said that there was no evidence of damage. The plaintiff's lawyer had not spoken to his witnesses prior to trial about their testimony. The surgeon accordingly sued the lawyer for $6000 alleging he "owed an affirmative duty to her and to the public at large to refrain from filing groundless litigation," and that, with a minimum of investigation, he would have known that the suit was groundless. The court found that the lawyer's client could certainly sue the lawyer for "negligence or ineptitude," causing the loss of the case. In examining the Canons of Legal Ethics, however, the court was unable to find a clear duty on the part of lawyers to the defendants in the suits they bring. The court noted that the Canons *permitted* a lawyer "who discovers his client has no case" to withdraw from the case, but did not require it, if the client was determined to continue. The court concluded its discussion by noting (similar to the New York court) that the courts must be open to all persons for "redress of wrongs," and that permitting such a lawsuit against an attorney would have a "chilling effect ... on the basic right of a citizen to seek redress in court for what he considers to be a wrong."[8]

First Counter-Countersuit

As if these cases were not discouraging enough, the attorney in the last case filed the first "counter-counter suit" in this country. Harry Burglass is seeking $902,500 from Rowena Spencer and her lawyers for malicious prosecution, defamation of character, and "experimenting" with a new point of law at the expense of his reputation.[9]

While this suit is unlikely to be any more successful than Spencer's (or than the initial suit filed against her), it could breed yet another countersuit by the doctor. The treatment which was the basis for the initial suit against Spencer was provided in 1967, and this litigation could drag on for a second decade before it is finally settled. This case thus demonstrates that countersuits have all the problems of malpractice suits in general. They are difficult to prove, expensive to prosecute, take physicians away from their practice, and make them spend much of their time in judicial proceedings, and offer only a slim chance of ultimate success.

Limitations of Countersuits

These cases illustrate that courts are likely to permit countersuits only in cases where the motivations involved are questionable enough to support an action for either malicious prosecution or abuse of process. They are unlikely to expand the lawyer's duty to include care for the individuals against whom suits are brought because of their view that permitting such suits based on an allegation of negligence alone could significantly weaken the public's access to the courts.

Though this view is understandable, it is not unassailable. The California Supreme Court fashioned an affirmative duty on the part of a psychologist to warn a third party of his patient's violent intentions toward her.* The relationship between a plaintiff's attorney and defendant is certainly closer than that between the psychologist and the intended victim, and the policy arguments against the imposition of such a duty are no more compelling.

These cases raise again the question of *professional* loyalty. How far can lawyers go in representing their clients before they are obligated to consider the effects of their actions on both society in general and specific third parties? The courts are likely to continue to give plaintiff's attorneys the benefit of the doubt. Bar associations, licensing authorities, and the public, however, need not be so lenient.

---

*This case is dealt with in the following essay, "Confidentiality and the Duty to Warn."

Confidentiality
and the Duty to Warn

In 1967, Prosenjit Poddar, a graduate student from Bengal, India, came to the University of California at Berkeley. In the fall of 1968 he met Tatiana (Tanya) Tarasoff at folk dance lessons in International House where he resided. They saw each other about once a week until New Year's Eve, when Tanya kissed him. Poddar apparently interpreted this kiss as a symbol of the seriousness of their relationship. When he explained this to Tanya, she replied that he was wrong, and that she was more interested in others than in him. This rejection caused Poddar to undergo a severe emotional crisis during which he became withdrawn, ignored his studies, and wept frequently. His condition continued to deteriorate until he began to visit a campus psychologist. Sometime late in the summer, while Tanya was in Brazil, he confided to the psychologist that when she returned he was going to kill her.

The psychologist believed Poddar and notified the campus police, requesting that they have him committed. They briefly detained him, but he was released because he appeared rational. At the order of the psychologist's superior, a psychiatrist, no further steps were taken either to commit Poddar or to warn Tanya. In late October, Poddar went to Tanya's home to talk with her. She was not at home, and her mother told him to leave. He did, but returned later that night with a pellet gun and a butcher knife. This time he found Tanya alone in the kitchen. He said he wanted to speak with her, and she began to scream. He then shot her with the pellet gun. She ran from the house. Poddar followed, and catching her, stabbed her repeatedly and fatally with the butcher knife.[10]

Ms. Tarasoff's parents brought suit against the psychologist, his superior, the campus police, and their employer, the University of California, for failure to warn them, Tanya, or anyone who could have reasonably been expected to notify Tanya of her danger, and for negligently failing to confine Poddar. After having their suit dismissed for failing to state a cause of action, they brought an appeal to the California Supreme Court. The court concluded that the defendants, including the campus police, were immune from suit for failure to act properly to confine Poddar on the basis of a California statute. However, it did rule in this December 1974 decision that the psychologist had a "duty to warn" a potential victim. The decision was withdrawn after a tremendous adverse reaction to it by the psychiatric community.[11] It took the court an additional eighteen months to re-issue the decision in different language. The 1976 decision is similar though not identical to the one issued in 1974.[12]

The Physician as Public Protector

The court's final decision dramatically illustrates the erosion of the traditional model of the physician as "healer," and the creation of a framework that demands the physician also be a "protector of the public." In a hard-fought, 5–2 decision the court held that when a therapist predicts that his patient is a danger to another person, he has a duty to warn that person of that danger. Four of the seven justices went even further, holding that, in addition, if a therapist *should have* made such a prediction on the basis of professional standards, "he bears a duty to exercise reasonable care to protect the foreseeable victim of that danger."

The difficulty of resolving the issues involved is illustrated not only by the divided court, but also by more than a dozen law review case notes that have been written on the 1974 version of the case and which are almost evenly divided on the duty-to-warn issue.

The court's most difficult task was to define a legal *duty* to warn. There were no precedents, either in California or any other jurisdiction, that had found a physician or a therapist legally responsible for damages caused by his patient on the sole basis that he had failed to warn the injured person of his patient's dangerousness. All the cases cited by the court involved some additional negligent act on the part of the physician—for example, negligent *failure to diagnose* tuberculosis, thereby placing members of the immediate family at risk; negligent failure to disclose a diagnosis of tuberculosis *to the patient*, thereby putting his wife at risk; negligence or fraud in *informing* a patient's neighbor that smallpox is not contagious; and negligence in *informing* a family that typhoid fever or scarlet fever of a sibling would not infect them or other family members. Similarly, the only mental health case cited by the court involved a Veteran's Hospital that had arranged for a dangerous mental patient to work on a farm without informing the farmer of the man's background. The farmer consequently permitted the mental patient to come and go freely during working hours. The patient borrowed the farmer's car, and drove to his wife's residence, and killed her. Under these circumstances, the court permitted the estate of the patient's wife to sue the Veteran's Administration for wrongful death; but liability was predicated on the hospital's failure to provide adequate supervision of a patient still under its control.[13]

Confidentiality or Preserving Life?

With no precedent to rely on, the court had to determine which policy was the "right" one. As with similar issues that many would classify as questions of medical ethics, the court's ultimate decision is not likely to please everyone, and will add fuel to the arguments of some that issues like this should not be solved by the courts at all.

The principal argument in favor of imposition of the duty to warn in this case is that it might have saved a human life. This interest was described by the majority as "the public interest in safety from violent assault." The court explained:

> Our current crowded and computerized society compels the interdependence of its members. In this risk- infested society we can hardly tolerate the further exposure to danger that would result from a concealed knowledge of the therapist that his patient was lethal.

The principal argument against imposition of such a duty is that this mandated "breach of confidentiality" would destroy the integrity of the doctor–patient relationship with the result that individuals would be deterred from seeking treatment, those who were treated would not be candid with their therapist making the treatment less effective, and this would decrease treatment and increase violent crime.

A related argument was also advanced: the art of predic ng violence is very inexact, and therefore to protect themselves against liability imposed by a duty to disclose, therapists are likely to make many warnings. These warnings are likely to cause their patients to terminate treatment and possibly act out their aggressive impulses. Faced with this possibility it is likely that therapists will attempt to commit all their violent patients. This will lead to a vast increase in unnecessary civil commitments and deprivations of liberty.

Such arguments are not trivial, and the majority felt compelled to respond directly to them in their opinion. The argument that a potential "breach of confidentiality" would destroy the therapeutic relationship is refuted by the longstanding legal ability of therapists to warn potential victims of a patient's violent acts. The only difference in this case was that the warning is now being required as a *duty* instead of being available as an *optional* right. Since this "right" to warn currently exists for both medical and mental health practitioners, and has apparently not destroyed patient trust, the majority are content that this expansion will not be harmful either. Other commentators have gone further, noting that not only is psychotherapy flourishing in an age when there are many exceptions to confidentiality, but many mental patients seek therapy with the hope that something will be done to curtail their violent actions. The issue, according to these commentators, is therefore not lack of confidentiality, but lack of trust. They make an analogy to group therapy, where it is known at the outset that no absolute guarantee of confidentiality exists, and suggest that to maintain trust the therapist inform the patient of this duty to disclose at the outset of the relationship.[14]

Predicting Violence

The issue of prediction is more troublesome. The majority argues that absolute accuracy is not called for. All that is mandated by the opinion is that the therapist "exercise that reasonable degree of skill, knowledge, and care ordinarily possessed and exercised by members of [their particular profession] under similar circumstances....Within the broad range of reasonable practice and treatment in which professional opinion and judgment may differ, the therapist is free to exercise his or her own best judgment without liability; proof aided by hindsight, that he or she judged wrongly is insufficient to establish negligence." In a footnote, the majority notes that in some cases it will be difficult for the therapist to actually identify the intended victim, but that each case should be judged on its own circumstances in this regard.

While this is a narrow ruling—and has no direct application to the facts of this case since the therapist *did* in fact predict that Poddar was dangerous—it is by far the most controversial point in the decision. The basis of the dispute is that while the standard enunciated by the court is the common one used for professional negligence in almost all fields (e.g., law, medicine, architecture), its application to psychiatry and psychotherapy presents the problem that there is no "standard" mode of diagnosis or treatment, and thus no firm principles on which either a practitioner or jury can judge a therapist's performance. Moreover, the burden on the therapist to maintain his patient's confidence in general continues to exist, and improper disclosure will subject him to suits for breach of confidentiality and invasion of privacy. This has caused some to argue that the court was "insensitive when it imposed a burden of warning upon a psychotherapist who must decide which patient poses a threat to some third person, while keeping an eye on his not otherwise extinguished duty of nondisclosure where the patient is not, in fact, dangerous."[15]

Who Should Bear the Risk?

The court's ultimate decision on this issue is based on three beliefs. The first, not altogether persuasive, is that increased warnings may in fact decrease the number of involuntary commitments since a warning is a less drastic step than a commitment and may have the same protective effect. The second is that "a risk that unnecessary warnings may be given is a reasonable price to pay for the lives of possible victims that may be saved." The third, unstated but implicit in the opinion, is that while such decisions involving the weighing of therapy, privacy, and dangerousness in deciding to give warnings are extremely difficult, making these decisions is what psychiatry is all about. Professional psychiatrists and therapists therefore properly bear the burden of making such decisions. In the words of Flemming and Maximov:

> Hospitals and the medical sciences, like other public institutions and professions, are charged with the public interest. Their image of responsibility in our society makes them prime candidates for converting their moral duties into legal ones. *Noblesse Oblige*[16]

A corollary to this is that therapists are in the best position to purchase insurance against injuries suffered by innocent third parties at the hands of their patients. In less kind language, a spokesman for the American Trial Lawyers Association argued that the psychiatric profession's reaction to the initial *Tarasoff* decision should not be taken seriously. In his words, "The specter of a crippled psychiatric profession is merely the routine excess of hegemonic professionals who yearn for the immunity of legal irresponsibility."[17]

Although one can agree or disagree with the outcome of this case, it illustrates many of the difficulties in resolving controversial issues of medical ethics and public policy in the courts, and suggests that the legislative forum may be more appropriate in cases involving complex issues of fact and public policy. The shift toward making the "physician–healer" the "physician–public protector" has been proceeding piecemeal. Reporting of contagious diseases, child abuse, and gunshot wounds has been generally uncontroversial. How far we should go in requiring reporting of other conditions (such as fitness to drive), however, and what effect such requirements will have on the ability of physicians to properly serve patients, is an issue that merits serious discussion and analysis.*

---

*This case was eventually settled out of court for an undisclosed amount, estimated at $100,000. Poddar has completed serving his "life" sentence, and has left the United States. AIDS has renewed the debate about a "duty to warn" those who might be at risk from someone infected with HIV.

Who to Call
When the Doctor Is Sick

What should be done to the sick or impaired airline pilot? Physicians at the American Medical Association's Third National Conference on the Impaired Physician (1978) agreed that he or she should be immediately grounded and required to submit to monitored treatment that showed beneficial results before being permitted to fly again. Pilots who do not accept treatment should be permanently grounded. The logic seems unassailable, and, in fact, is closely in line with current Federal Aviation Administration policy. Some pilots, however, think that since a qualified copilot is always on board the plane, such "coercive" treatment for conditions like alcoholism or drug dependency is unwarranted. Not surprisingly, pilots are reported to believe that any physician who is impaired should immediately cease practice and do so permanently unless he or she is successfully treated and rehabilitated in a closely supervised environment.

Three Views of the Problem

Although almost all physicians acknowledge the problem of the "impaired physician," that is, one who is alcoholic, drug dependent, or too ill, mentally or physically, to function competently, there is little agreement on what to do about a problem that affects an estimated five to ten percent of the profession. Though there are many positions, three general views seem to predominate: (1) impaired physicians are a danger to the public, and the public has a right to see that they are disciplined to stop them and discourage others; (2) impaired physicians are sick, and the medical profession should see that they are treated in a nonpunitive and noncoercive setting; and (3) impaired physicians are both sick and a danger to patients; their

201

practices should be suspended, and they should be treated and permitted to return to practice only in a carefully controlled setting until recovery is complete.

Discipline the Deviant

The first position is probably the one held by most nonphysicians. It is based on the premise that the impaired physician can do serious and fatal damage to patients and that the public has a right to be protected against such harm. Dramatic stories—John Nork's string of unnecessary and disabling back operations in California, performed while he was under the influence of drugs, and the drug-addicted Marcus brothers of New York (fictionalized in the novel *Twins*)— have dramatized to the public how little the medical profession has done to police itself and to protect patients from the impaired physician.

Such instances have led to a number of legislative initiatives. In California, for example, by law every malpractice insurance settlement or jury verdict of more than $3000 must be reported to the state's Medical Quality Assurance Board. In addition, hospitals are required to report all denials, removals, or restrictions of hospital privileges. In New York and other states, legislation (termed the "tattletale act" by some physicians) requires physicians to report incompetent physicians to the licensing agency. Such a requirement makes the long-standing ethical obligation of the AMA's Principles of Medical Ethics a legal one as well: physicians "should expose, without hesitation, illegal or unethical conduct of fellow members of the profession."

The first position is well articulated by three physicians with the State Medical Board of Virginia:

> It is vital that the deviant, incompetent, or 'sick' doctor be reported and disciplined. If such a physician is allowed to continue to practice, he himself, his patients, and the entire medical profession are all losers. The doctor's torment is prolonged, his patients' lives are endangered, and the profession's standards are lowered. The need for medical discipline is great. The public demands it, and the medical profession *must* demand it.[18]

In the past it was almost a tautology to say that licensing boards operated solely to "protect the public" and not to punish the physician;

such a statement is no longer completely accurate. A number of states have given their licensing authorities the power to levy fines against deviant physicians. Other states, like Massachusetts and Maryland, have added the word "Discipline" to the title of the licensing board itself, and, while almost every state made the medical malpractice laws more favorable to physicians in the wake of the insurance crisis, many as a *quid pro quo*, increased the authority of the state licensing board to discipline physicians. The purpose is prevention: it is better to stop the incompetent or sick physician before he injures patients, rather than sue him after the fact.

Treat the Sick

It is probably fair to say that the second position, that the impaired physician is sick and should be treated nonpunitively, is one held by the majority of physicians. Indeed, at the Conference on the Impaired Physician, many doctors voiced tremendous hostility toward licensing boards (even though they are run almost exclusively by physicians) as clumsy, legalistic, coercive, and punitive. Their argument is that "peer review" mechanisms in hospitals are the most appropriate ways to identify such physicians. Impaired physicians should be "taken aside" by their fellow physicians and convinced in a friendly, nonthreatening manner to accept treatment. Most agreed that, though physicians hardly ever do this, they could do it if they understood the problems of alcoholism and drug abuse better and that most physicians so confronted would be grateful. All agreed that few family practitioners or internists know anything about alcoholism. Requiring yearly physicals for staff privileges would be unlikely to turn up problems of alcoholism or drug abuse because the examining physician was either a friend or was unqualified to make such a finding.

Ironically, states that require hospitals to report suspension or restriction of staff privileges to the licensing authority have given hospital staffs a new weapon to use in keeping such problems away from the licensing board. For example, a number of participants at the conference noted that in their hospital the impaired physician is taken aside and told that privileges will be maintained only if he or she immediately undergoes treatment, and that if he does not, he will be restricted *and* a full report will be filed with the state board. If a hospital reporting statute does nothing else, it enables hospital staffs to more effectively police their members.

LeClair Bissell of New York, one of the leading proponents of keeping control in the hands of physicians, outlined the characteristics of any effective mechanism: it must be rapid, fair, get the physician back to work as soon as possible, and provide monitoring of practice upon return. In describing the fourth point she underlined a problem that no one has yet dealt with effectively: the dual role of the therapist and monitor. Many state medical societies and boards place an impaired physician under a psychiatrist's care and ask him to report to them periodically about the condition of the monitored physician. In her view, the role conflict of therapist and monitor makes this impossible, and psychiatrists should be assigned one role or the other, but never both. This view was seconded by George Carroll of the Virginia State Board of Medicine, who noted that in his ten years of experience, private psychiatrists rarely had anything negative to say about physicians they were asked to both treat and monitor.

Discipline and Treat

The third position is that ways should be developed to use discipline to encourage treatment and rehabilitation. The California statute, for example, requires the board to "take such actions as possible to aid in the rehabilitation of physicians." And the position of the AMA in encouraging both state medical society programs and reporting to the licensing authority, falls into this category. Their "Model Bill on Disabled Physicians," often termed the "Sick Doctor" bill, has been enacted in one form or another in more than thirty states.

This bill provides that the licensing board can require a complete physical and mental examination of any physician believed unable to "practice medicine with reasonable skill and safety to patients due to physical or mental illness, including deterioration through the aging process or loss of motor skill, or abuse of drugs, including alcohol." After the examination, usually conducted by a special committee of the state medical society, a report is sent to the board. The physician may voluntarily request a restriction on his license, the board may take no action, or the board may begin formal proceedings against the physicians' license. The purpose of the law is to make it somewhat easier and swifter to take action against a "disabled" physician than it would be to take action for other reasons. The thrust is to encourage physicians to seek help voluntarily and to have restrictions placed on their licenses consistent with the degree of impairment.

Although the boards may want to both discipline and rehabilitate, they have neither resources nor expertise to treat alcoholism and drug dependency. Three state licensing boards that were represented on a number of panels at the conference provide an illustration. Maryland's annual budget is $17,000, Virginia's $325,000, and California's $7,000,000. In very rough terms, this gives Maryland's board about $3 a resident physician, Virginia's about $25, and California's about $150. Of course, California's board can be far more independent of the state medical society than can Maryland's. Add to this the great variety of disciplinary laws in the fifty states, and one quickly concludes the ways in which a disabled physician is treated will depend in large measure on the state in which he works.

Some Possible Steps

Many physicians had come to the conference wanting to find out what steps their hospital executive boards should take to protect both the public and the medical profession from the disabled, impaired, or incompetent physician. Though there were no answers, many approaches emerged. As a public member of a state licensing board,* I am certainly not unbiased, but it seems to me that the following steps make sense:

1. State and county medical societies should develop programs to deal with impaired physicians with the aim of identifying them and getting them into treatment or behavior modification programs *before* they become a danger to their patients. These programs can be entirely confidential and run independent of the licensing authority.
2. Hospitals should conduct similar programs with the same degree of confidentiality.
3. Once the physician is actually a danger to his patients, however, his or her conduct legitimately becomes a public matter. Though rehabilitation and treatment should still be encouraged, it is now secondary to patient protection. Therefore, all physicians and hospitals should be required by law, under penalty of losing their own licenses, to report such physicians to the licensing authority. As a

*I served as a Public Member of the Massachusetts Board of Registration in Medicine from 1976 through 1982, the first five years as Vice-Chairman of the Board and Chairman of its Complaint Committee.

corollary, hospitals should be required to report all denials, removals, or resignations of privileges, and all restrictions on privileges to the state licensing authority; and malpractice insurance companies should be required to report all malpractice settlements or payments, together with the facts, in cases where the final payment is more than a certain amount, perhaps $5000.

4. Licensing boards that do not already have it should be given the legislative authority to order physical and mental examinations on short notice.

5. Serious consideration should be given to a national licensure and disciplinary scheme that could not only provide adequate financing and uniform standards for the country, but could also get away from the local politics and friendships that prevent reporting of the vast majority of impaired physicians.

6. Medical school education and continuing education must deal with the identification, prevention, and treatment of alcoholism, drug dependency, and mental impairment.

The sick physician should be treated; the physician who continues to practice while sick and thus endangers patients should be disciplined. There is little disagreement with these conclusions. There is also little agreement as to how these objectives can be accomplished.

CPR

The Beat Goes On

Cardiopulmonary resuscitation (CPR) is the best-known emergency medical treatment in the United States. Television commercials sponsored by the American Heart Association (AHA) feature such stars as Reggie Jackson encouraging us to "help save a loved one's life: learn the life-saving skills of CPR." And in Steven Spielberg's film, *E.T.*, when the extraterrestrial's heart stops beating, we are not at all surprised that the medical team attempts CPR. By 1977, 12 million Americans were trained in CPR and 60 to 80 million more were planning to take such training.

The national enthusiasm for CPR training is based on statistics: approximately one million persons in the U.S. annually experience an acute myocardial infarction. More than 650,000 die of ischemic heart disease and 350,000 of these deaths occur outside the hospital, usually within two hours after the onset of symptoms. A significant percentage of these victims can be saved by prompt initiation of CPR.

Altogether, 60 to 70 percent of sudden deaths caused by cardiac arrest occur before hospitalization (including drowning, electrocution, suffocation, drug intoxication, and automobile accidents) and this has led to the characterization of the community as "the ultimate coronary care unit."[19]

Emergency Cardiac Care can be divided into: Basic Life Support (BLS) and Advanced Cardiac Life Support (ACLS). The latter requires the use of sophisticated medical technology (generally limited to the hospital environment); the former can be performed by almost anyone with some basic training in technique. The training in BLS generally involves an introductory lecture or film on heart disease and cardiac arrest, demonstration of the techniques for recognizing and treating cardiac arrest, and practicing resuscitation techniques on a plastic mannequin. Additional education on prevention as well as techniques for helping choking victims may also be included. Upon completion of the three-to-six-hour course, the student is "certified."

Out-of-hospital CPR raises a number of issues with legal and ethical dimensions: Should the public be encouraged to take CPR training? Should physicians be required to take CPR training? Is there any legal liability on the part of a certified CPR layperson? And in an emergency, who should be in charge of the CPR effort?

Training the Public

CPR is part of a much broader movement to get the public more involved with its own medical care. The effort is aimed at eradicating the notion that medical care takes place only in the hospital, and that individuals therefore have no responsibility other than to see to it that their loved ones are hospitalized when they need (and want) medical care. This strongly held notion obviously breaks down in emergency situations where immediate action is required to save a life: a choking victim in a restaurant; a victim of a cardiac arrest; a drowning victim.

Since it is unlikely that a trained medical professional will be available in such situations, most of these individuals will die unless a significant percentage of the public has some basic lifesaving training. Individuals who engage in certain sporting activities, including camping, canoeing, scuba diving, and mountain climbing, have long sought training in basic first aid and lifesaving for their own protection. It is time to extend our notion of "medical wilderness" to include our urban areas, and to encourage basic CPR training for the inhabitants of cities and towns for our own protection.

The primary arguments against such broad-based training seem unpersuasive: people forget, and a partially or poorly trained person can do more harm than good. Both true, but the answers to these problems are to encourage at least annual recertification, and to develop quality control mechanisms that ensure that the basics are properly taught and adequately learned. Also, since in many cases the outcome without intervention will surely be death, an attempt to do what one can is better than no attempt at all.

Physicians and CPR

There is no excuse for physicians not to know CPR. Introduced only 20 years ago, it was obviously not part of the training of many physicians practicing medicine today. Yet as CPR is the most basic lifesaving technique for the most common form of sudden death, I believe physicians are ethically required, and should be legally required, to learn this technique. It is remarkable that this proposition raises any dissent.

To most nonmedical observers, physician arguments against mandatory CPR training seem bizarre. In California, the only state that currently requires its licensees to be certified in CPR, physicians voiced three basic arguments against the regulation: (1) doctors are never called upon to do CPR; (2) the training required would be a waste of time; and (3) it could give the AHA a veto power over medical licensure. Surely these physicians can't be serious. The response of William G. Gerver seems correct from a public perspective. "Most people are amazed when they learn that physicians are not required to be certified as competent to resuscitate a person...organized medicine is turning a deaf ear to the public's expectations of us as physicians."

Similarly, when the issue came up before the Massachusetts Board of Registration in Medicine, the major argument against a rule requiring CPR training of physicians was that it would require licensed physicians to take training from nonphysicians!

After an incident in Brooklyn in which a man suffered a cardiac arrest and was pronounced dead by a physician, only to be resuscitated a short time later by paramedics, the *New York Times* editorialized that the lack of physician training in CPR: "leaves the public at the mercy of random encounters with unprepared physicians...If physicians don't all volunteer for training, state legislatures and licensing bodies ought to consider requiring it."[20]

The public reasonably expects its licensed physicians to know CPR. Many don't and it is time to make this a requirement of medical licensure. It also makes sense to require it for *all* health care professionals who are licensed by a state agency.

Legal Liability

If there is one universal rule in health law it is: in an emergency, treat first and ask legal questions later. This rule of course applies primarily to medical professionals, because the public (including judges and juries) expect and want professionals to help them and others in emergency situations. But it also applies to anyone with special training that might be of assistance to a victim who requires such assistance. Nevertheless, professionals and nonprofessionals alike are concerned that they might be successfully sued for performing emergency medical care. The short answer is that it is extremely unlikely that any suit will be filed, and almost inconceivable that a suit would be successful, so long as the emergency care was rendered in "good faith."

I know of no case where any layperson who was performing CPR in a reasonable manner has been sued successfully, and it is unlikely that any such suits will be successful in the future. It makes much more sense from both an individual and societal point of view to concentrate on properly administering lifesaving treatment rather than on the potential for a lawsuit. With the former there is a realistic chance to save a life; with the latter there is no realistic chance of a successful suit.

Some have proposed passing statutes to grant CPR-certified individuals immunity from suit when acting in good faith, and some states have actually passed such legislation. These suggestions tend to encourage a sort of legal paranoia, based on a single episode of the old Dr. Kildare television series (in which he was unsuccessfully sued for helping a victim of an auto accident at the roadside), and cocktail party anecdotes. The proper response to such misinformation is accurate information, not unnecessary legislation. People do forget, after all, and some certified individuals perhaps should not be attempting CPR under certain circumstances (for example, when a more qualified person is available). By providing individuals skilled in CPR with legal protection they do not need, we may be encouraging marginal practitioners to inflict needless harm on helpless victims who will be left without legal recourse. Individuals skilled in CPR do not need such legal protection; unskilled or inadequately trained individuals do not deserve it. We should distinguish between encouraging training and encouraging irresponsible actions.

Conflicts on the Scene

Emergency medical technicians (EMTs) commonly complain that they have begun CPR only to be interrupted by a physician demanding that he or she be permitted to take over. This is, of course, perfectly acceptable if the physician is as skilled or more skilled at CPR than the EMT, but deplorable if he or she is less skilled. Equally often, physicians complain that they were administering CPR to a victim and a policeman has demanded that he be permitted to take over the case.

This bald assertion of authority is, of course, equally deplorable, unless the policeman is more skilled than the physician.

There is no simple answer to this territorial dispute. Though I think cooperation is the most common practice, when it is not, the only prudent thing to do is retreat in the face of an armed policeman; just as the only prudent thing for the EMT to do may be to retreat or risk the wrath of the physician. Whatever one's stance, the victim will probably be better off even with a less well-trained rescuer than with two potential rescuers fighting over him or her. Rescuers should defer to those more qualified in BLS and not interfere with ongoing efforts until more qualified assistance or adequate transportation to such assistance is available.

The purpose of CPR is to prevent sudden, unexpected death, a goal that has almost universal acceptance. The technique is simple enough that most people can learn it and use it effectively, and it works. Nevertheless, it raises questions that require resolution. If they are to be properly resolved, we must concentrate on actions most beneficial to victims and potential victims. Life in our medical wilderness requires that we all lend a cooperative hand.

CPR

When the Beat Should Stop

Deciding to perform cardiopulmonary resuscitation (CPR) outside a hospital is easy because death is the only alternative, it is a noninvasive procedure, and most individuals want to avoid sudden, unexpected death. Only the first of these reasons exists in the hospital setting, where CPR becomes a much more complicated and invasive procedure, and death may be both expected and welcomed. In-hospital CPR involves the possibility of placing tubes within the trachea for artificial respiration; electrocardial shock; introducing one or more intravenous lines; administering intravenous medication to correct abnormalities in blood pressure and cardiac rhythms; and, in rare circumstances, emergency placement of a temporary pacemaker. Thus it is not surprising that decisions are made *not* to perform CPR on specific patients in every hospital in the country.

Nonetheless, recent judicial decisions and medical articles indicate that significant disagreements persist regarding the proper indications for a "do not resuscitate" (DNR) order and who should initiate such an order. Much of this confusion is caused by the assumption either that every situation is unique, so that no rules can be adopted; or that every situation is fundamentally the same, and so should be governed by the same set of rules. In fact, DNR orders are generally written for two related, but fundamentally different reasons: poor prognosis and poor quality of life.

Indications for DNR

The National Conference on CPR and ECC (Electro-Cardiac Conversion), cosponsored by the American Heart Association and the National Academy of Sciences, noted: "The purpose of CPR is the prevention of sudden, unexpected death. CPR is not indicated in certain situations such as in cases of terminal irreversible illness where death is not unexpected."[21] This is a prognosis criterion. Miles, Cranford, and Schultz, who developed DNR guidelines adopted by the Minnesota Medical Society, have suggested:

> A decision not to resuscitate is considered for a variety of reasons: a request by a patient or family; advanced age of patient; poor prognosis; severe brain damage; extreme suffering or disability in a chronically or terminally ill patient; and in some instances, the enormous cost and personnel commitment as opposed to the low probability of patient recovery.[22]

Though these reasons fall into both categories, most deal with quality of life. The three courts that have looked at DNR orders to date have not always differentiated between these two reasons for DNR orders.

Poor Prognosis

CPR may be contraindicated because it will do no good; that is, the patient will die soon anyway, and nothing can be done to stop the course of the disease. When the patient's condition is hopeless, no medical intervention is required. Although there is room for debate about the normative aspects of such a prognosis, it is reasonable to conclude that there are cases where such a decision is basically medical and can and should be made by the attending physician.

The Massachusetts Court of Appeals seems to have believed it had just such a case when it was asked to review the appropriateness of a DNR order on Shirley Dinnerstein. She was 67 years old and had been suffering from Alzheimer's disease for six years. She was getting progressively worse. At the time of the hearing she was confined to a hospital bed in "an essentially vegetative state, immobile, speechless, unable to swallow without choking, and barely able to cough ... her condition is hopeless...it is difficult to predict exactly when she will die [but] her life expectancy is no more than a year."[23] Her family, consisting of a son–physician and a daughter who lived with her prior to her admission to the nursing home, concurred in the DNR order.

The court viewed CPR as "highly intrusive" and "violent in nature," and Ms. Dinnerstein as a patient "in the terminal stages of an unremitting, incurable mortal illness." The court concluded CPR would do nothing for her and therefore was not a "significant treatment choice or election." Given her condition, which was exclusively diagnosed by her physicians, the court concluded the DNR order was

> ...a question peculiarly within the competence of the medical profession of what measures are appropriate to ease the imminent passing of an irreversibly, terminally ill patient in light of the patient's history and condition and the wishes of her family.

The notion that prognosis is a medical decision, and if the physician determines that the patient's case is "hopeless" then CPR is not indicated, seems perfectly acceptable. On the other hand, if her condition is really hopeless, and if CPR is intrusive and unproductive, what justification exists to inflict it on her regardless of the wishes of her family? The "hopeless" prognosis is more akin to pronouncement of death than to elective surgery, and the role of the family should probably be seen as irrelevant in this type of a DNR decision just as it is in a determination-of-death decision.

Poor Quality of Life

The second major reason for DNR orders is the belief that a patient's quality of life is so poor that preventing death by CPR is not justified. For example, Sharon Siebert was 41 years old, had been seriously brain-damaged in an operation five years previously, and according to her physicians had a life expectancy of 37 more years. She had the mental age of a 2-year-old with no prospects for improvement. She had to be fed artificially because she could not swallow, was confined to a bed or wheelchair, and could communicate only "slightly" and "simply." A DNR order issued by her physician, with the consent of her parents, was challenged by a friend, Jane Hoyt. Following a court hearing, the DNR order was revoked on the basis that it was proper only if Ms. Siebert herself would have wanted it, and there was no evidence that she would have. The guardian "must consider all of the factors which his ward would consider were she able, and, so far as possible, evaluate them as she would."[24]

This seems proper: when DNR decisions are based on quality of life, only the patient's own view should be relevant. Physicians obviously have no special expertise in this area. Since most DNR decisions are probably based on quality-of-life assessments, it is critical that their nonmedical nature be recognized so that such decisions are based on proper criteria and made by those qualified to assess the situation. Only in this way can we protect the patient's dignity and autonomy.

Indiscriminate Indications

Children and wards of the state usually make courts take special efforts to protect their interests. Nevertheless, the "Baby Billy" case is instructive on the DNR issue. Billy was born with serious cardiac problems and was abandoned by his mother at birth. A corrective operation failed, and physicians testified that no further medical treatment was available to the child. Survival beyond his first birthday would be unprecedented. In short, his condition was "hopeless." As in *Dinnerstein*, the court found CPR highly invasive and dangerous: "If heroic medical efforts were to be used to resuscitate the child, such efforts would be invasive and traumatic and would cause the child to suffer substantial pain and possible brain damage." It's almost as if the judges envisioned CPR in the horrific and destructive way it is portrayed in John Carpenter's *The Thing*. Under these circumstances, an argument could be made that this DNR decision was basically a medical decision.

However, the state division of child guardianship refused to consent to a DNR order, and the case wound up in court. The Massachusetts Supreme Judicial Court found the facts outlined above, but treated the case as a "quality-of-life" case (like *Siebert*) rather than a "hopeless prognosis" case (like *Dinnerstein*).

Accordingly, the test the court laid down for a lawful DNR order was not a medical determination of hopelessness, but a judicial determination under the substituted judgment test; that is, the court "must attempt to 'don the mental mantle' of the child and seek to act on the same motives and considerations as would have moved the child... in this way, the free choice and moral dignity of the incompetent person are recognized."[25]

Confusing matters a bit more, the court distinguished the Baby Billy case from *Dinnerstein* on the basis that there was "no loving family with whom physicians may consult regarding the entry of a 'no code' order..." Such consultation is, of course, relevant if the issue is one of substituted judgment for a person who has lived for a period of time with a family and made decisions that tend to demonstrate a value system upon which their likely decision regarding a DNR order could be guessed. But it makes no sense with regard to an infant. We can only make decisions for infants on the basis of an objective "best interests" test, since we have no basis to believe the infant would act idiosyncratically. Accordingly, guidelines can and should be developed by private and public bodies for this category of patients. Consultation with the court or family on either "best interests" or "substituted judgment" makes sense *only* in a quality-of-life type of assessment. If the case is medically hopeless, no such assessment is relevant.

Both the courts and physicians have confused the issue of DNR orders in the hospital more than necessary by indiscriminately using two types of indication that raise very different issues: medical prognosis and quality of life. Decisions based on the first can properly be seen as medical questions (like determination of death) if prognosis is viewed as a determination that the patient's condition is "hopeless" in the sense that CPR cannot prevent death from occurring very soon and would not be viewed as good and accepted medical practice. Of course, if the patient is competent, I would want his or her consent to such an order. But even if he or she did not consent, the patient can no more prevent death than the physician, and at some point even the patient's demand for CPR becomes a demand for "mistreatment" with which the physician need not comply.

Decisions based on quality of life, on the other hand, should remain the patient's exclusive province. He or she should be able to refuse CPR for any reason, including poor quality of life. Here, as in most other areas, the patient has a much stronger right to refuse treatment than to demand what may be considered mistreatment. The patient's guardian should likewise be able to refuse CPR for the patient if it can be demonstrated that this is what the patient would want. And where there is any reasonable question regarding the incompetent patient's wishes in a quality of life case (or, in the case of a never-competent person, what is in that person's best interests), the court is the proper forum for its resolution.

To derive the maximum benefit from the lifesaving technique of CPR, we should encourage its use by all trained individuals outside the hospital, and know when its use is no longer justified inside the hospital.*

*The need to differentiate indications for CPR has recently been more fully explored. *See* T. Tomlinson and H. Brody, "Ethics and Communication in DNR Orders," *New Engl J. Med* **318**: 43 (1988).

The Mentally
Retarded
and Mentally Ill
Patient

Introduction

One of the recurring themes of the past decade has been determining who can make decisions for an individual unable to make them, and on what basis such "substitute" decisions should be made. This issue, of course, was at the core of the "Baby Doe" debate, and the three essays on the Baby Doe regulations could just as easily have been made a part of this section. The Baby Doe regulations, like the cases of Phillip Becker, Joseph Saikewicz and Mary Hier, deal with the issue of medical treatment for a specific condition, and whether it should be given or withheld.

In the case of Phillip Becker, to whose legal saga the first two essays in this section are devoted, the issue is whether or not this young teenage Down syndrome child should undergo a heart operation. In the case of Joseph Saikewicz, the question is whether this elderly, severely mentally retarded gentleman, should be given chemotherapy for his cancer. And in the case of Mary Hier, a 92-year-old mentally ill nursing home resident, the issue is whether the gastrostomy tube supplying her nourishment, which she pulled out, should be replaced.

The focal point of the judicial pronouncements in these cases is on who should be empowered to make these treatment decisions. The Phillip Becker cases explore the limits of parental autonomy in making treatment decisions about their children; Joseph Saikewicz and Mary Hier deal with the application of the doctrine of "substituted judgment," which is that the substitute decisionmaker should attempt to decide the treatment question in the same manner the incompetent person would decide it if the incompetent person could decide it. As will be seen, this is a reasonable rule when dealing with previously competent individuals who have expressed preferences, but can be a dangerously arbitrary exercise when dealing with mentally handicapped individuals who have never been able to express their preferences with regard to treatment.

Two other major issues continue to be debated in the courts: sterilization and antipsychotic medication. Sterilization of the mentally handicapped raises the question of whether *anyone*, including a court, should be able to authorize this procedure, which cannot accurately be termed a "treatment." Antipsychotic medication, on the other hand, is a treatment. The legal issue is whether the patient should nonetheless be permitted to refuse it because of its potentially devastating side effects. If there can be said to be a trend in law, it is that courts can *only* authorize sterilization in cases where it can be demonstrated that sterilization would be in the mentally incompetent individual's best interests; and that mental patients who can understand the decision and its consequences, have the right to refuse antipsychotic medications. These latter decisions have led some psychiatrists to characterize the mental patients' situation as "rotting with your rights on." This conflict illustrates, as well as any, the professional biases of both physicians and lawyers: physicians often intent on treating illness to preserve or restore "health," no matter what the patient's wishes; and lawyers often intent on protecting the patient's liberty to decide for himself, no matter what the harmful health effects that decision may have on the patient. It is a dichotomy encountered again in the "Death, Dying, and Refusing Medical Treatment" essays, when dealing with refusals of medical treatment that almost always result in the patient's death.

All of the essays in this section originally appeared in the *Hastings Center Report*. The two pieces on Phillip Becker appeared in December, 1979 and February, 1982. The essay on "Sterilization" in August, 1981; "Refusing Medication" in February, 1980; Joseph Saikewicz in February, 1978; and Mary Hier in August, 1984.

Denying the Rights
of the Retarded
The Case of Phillip Becker

The Down syndrome child in need of treatment is a medical ethics paradigm. In a film distributed by the Kennedy Foundation, for example, a Down syndrome newborn is allowed to starve to death because his parents will not consent to surgery to correct a duodenal atresia. And it is to detect and abort fetuses affected with Down syndrome (trisomy 21) that amniocentesis is most often used. Down syndrome children are also better than dogs to practice surgery on, as William Nolen reminds us in his *The Making of a Surgeon.* He recounts a conversation between a pediatrician and a young surgeon about a septal defect repair—a major heart operation—the latter is going to perform the next day. The young surgeon says he's "not worried a bit," and wouldn't worry even if his patient died on the table. "Oh, now I get it," replies the pediatrician, "you're doing a mongoloid."

The Down syndrome child in court, on the other hand, has been a rarity—most legal commentators have assumed that courts would routinely order performed almost any treatment withheld solely because the child's parents believed that a Down syndrome child's quality of life is low. However, there is now legal authority in California for the proposition that a Down syndrome child need not receive heart surgery for a ventricular septal defect, even if such surgery would be routinely ordered in the case of a normal child.[1] Since this conclusion is unprecedented and potentially dangerous to the wellbeing of all mentally retarded citizens, it is important to understand both the facts and the opinion.

In July 1978, when the case first came to court, Phillip Becker was 11 years old. The petition alleged that his parents were not providing him with the "necessities of life," specifically a heart operation. He had never lived at home with his parents, but had been institutionalized since birth. At the time of the petition, he was residing at Schnuhr's Nursery (a home for nineteen multihandicapped children) and was attending the Rouleau Children's Center School in San Jose, California. He could write his name, had good motor and manual skills, could dress himself, was toilet-trained, and could converse reasonably and take part in school and Boy Scout activities. He was described as "near the top level" for a Down syndrome child, and there was testimony that his IQ was "around the 60 range." A school psychologist testified that he could be placed in the county's sheltered workshop following his schooling.

The Medical Testimony

In 1973 Gary Earl Gathman, a pediatric cardiologist, made a clinical diagnosis of ventricular septal defect (a hole between the two pumping chambers of the heart which elevates the pulmonary artery pressure) and recommended cardiac catherization to define the anatomy of the problem. The parents refused. In late 1977 Phillip was again referred to Gathman because Phillip required extensive dental surgery, and the anesthesiologist refused to administer general anesthesia until Phillip's cardiac problem was clarified. This time the parents consented, and Gathman performed the cardiac catherization. The results confirmed his clinical impressions and indicated that Phillip had already experienced changes in his pulmonary arteries. Gathman testified that an average life expectancy for someone with Phillip's condition, left uncorrected, is 30 years; that about one-quarter of such persons would die suddenly; and that most others would slowly deteriorate. Before they are totally debilitated, they suffer from fainting spells during which they turn blue-black because of the lack of oxygenated blood. This makes them a constant worry to their caretakers. These children also tend to be smaller, thinner, and slower than normal; they "cannot run and play and keep up with other children."

Gathman recommended corrective surgery to the parents, putting the risk of death at 3–5 percent, which he considered a "low risk." The risk was related to changes in the pulmonary artery, which would continue to deteriorate. He also noted a less than 1 percent chance that a heartblock necessitating a pacemaker would result from the surgery. Gathman indicated that he did not always recommend surgery for Down syndrome children: "For a child who has an IQ of less than 30 and is institutionalized on a permanent basis, there's very little that can be gained from surgery."

The only other physician to testify was James William French, also a pediatric cardiologist, who had been asked by Gathman if Phillip's condition was surgically correctable. He testified that he believed that it was; the pulmonary vascular disease had not progressed so far that surgery was not feasible. While he was not asked to give an opinion on this case, he indicated that he "probably would have" recommended surgery if Phillip was "otherwise normal." Though confirming most of Gathman's testimony, he put the risks of death from surgery at 5–10 percent, noting that in his personal opinion Down syndrome children were "more risky" to operate on than normal children because they are "less cooperative" and "seem more subject to infection." The witness admitted that he had recommended against surgery "in cases where the children were so severely damaged intellectually or the central nervous system so damaged that they virtually did not function or they were virtually incapacitated."

One of Phillip's teachers, a school psychologist, a program coordinator, and a probation officer each gave additional testimony. This established, among other things, that Phillip's parents visit him about twice a year (although they testified they saw him five or six times a year) and that he is a lovable child who "refers to lots of men as Dad." There was no further medical testimony, although the parents placed into evidence a letter from physician Harry E. Hartsel, who predicted that Phillip would not learn to read, write, or take care of himself; he leads "a life I consider devoid of those qualities which give it human dignity."

The Parents

The only other witnesses were Phillip's parents, who testified that although Phillip had never lived with them, they considered themselves responsible for him as part of their family. They monitor his care and once transferred him to another institution when they thought he wasn't getting proper care. They thought about the proposed heart surgery for a long time, and consulted with a number of physicians and a Jesuit priest (they are both Catholics). Their primary reason for refusing consent was that they did not want Phillip to outlive them. They believed that geriatric care in this country is terrible and that Phillip would not be well cared for after they died. "His quality of life would be poor in such a place," and "life in and of itself is not what it's all about." They also did not want him to be a burden on their other children. When asked who would be better off if Phillip were dead, Mr. Becker replied: "I think it would be best for everyone, including Phillip and the survivors. There's no useful point in extending his life beyond the natural, by means of this operation." Ms. Becker testified similarly, saying, "Geriatric care in this country at its best is not good. And I really don't want Phillip to be extended into geriatric care. As it is, Phillip will be with us our whole life. We are looking at him living thirty or forty years. We'll be seventy or eighty."

The Judge's Opinion

The testimony took a day-and-a-half, with Judge Eugene M. Premo of the juvenile court presiding. He ruled immediately after the closing arguments. In a statement that can be charitably characterized as rambling, the judge decided not to order the surgery. He found the proposed surgery "elective" and not "a life-saving emergency." He determined that there were inherent risks of the surgery which no one could control, "such as a slip of the knife, malpractice, negligence" all of which would "make this entire case moot." He described the parents as articulate, intelligent, caring, loving, and thoughtful. He spoke of two of his judicial colleagues who had Down syndrome children— one cares for the child and "the other one can't literally face the fact that the child exists." He said he didn't think he personally could handle it "if it happened to me."

The judge condemned the physician who testified that he would not recommend surgery if the child's IQ were 30 or less, denoting the IQ test as a "vehicle of terrible abuse" that could be wrong and is always arbitrary. Given what he regarded as a subjective medical opinion, the judge was kindly disposed to the Beckers. He concluded that their decision was "in the range of debatable actions." He did not want the government interfering with parental choice; he was worried "by the fact that 1984 is coming upon us." His legal conclusion: "There is no clear and convincing evidence to sustain this petition [to operate]."

The Appeal

At the urging of the Pro Life Council of California, which had been instrumental in bringing suit, the California Attorney General's Office agreed to appeal. The opinion of the appeals court, written by Judge P. J. Caldecott (and concurred in by the two other judges on the panel) is, if anything, less well-reasoned than Judge Premo's. The appeals court considered three questions: (1) was there any substantial evidence to support the judge's decision; (2) should the judge have required clear and convincing proof (or should the evidentiary standard have been preponderance of the evidence); and (3) should Phillip have been advised of his right to have counsel appointed for him at the proceeding? All three questions were quickly disposed of, the latter two on technical points.

On the first point, the appeals court noted that one physician testified that the operation on Phillip was more risky than average because of pulmonary vascular changes and because he was a Down syndrome child. In addition there was risk of nerve damage that could necessitate a pacemaker. This evidence alone showed that the court "had before it a child suffering not only from a ventricular septal defect but also from Down syndrome, with its higher than average morbidity, and the presence of pulmonary changes." And this was all the appeals court needed to conclude that the lower decision was based on substantial evidence.

The Precedent

The case stands for the proposition that Down syndrome children do not have to be given treatment if it can be demonstrated that the risks of the treatment are slightly higher than they would be in a normal child. How did this remarkable eugenic policy come to be articulated by an American court? Let me suggest some possible reasons.

First, nowhere in the lower court proceedings is there any articulated standard for decision. Courts have recently been choosing among three: the best interests of the child, substituted judgment (usually considered the same as the best interests standard when a child is involved), and reviewing the parental decision to see if it is "fair and reasonable." The lower court did not cite any of these; instead it decided that the parents can make the treatment decision as long as it is "in the range of debatable actions." This standard is reasonable if it means that the proposed medical care is debatable in the medical community, and that there is no medical consensus on treatment for this child.[2] However, as applied by the trial judge, the standard seems to be, as long as the parents sincerely believe in what they are doing (and are willing to debate it?), courts should not "second guess" them.

The appeals court did no better. It bowed to the best interests standard, but nowhere applied it. Of course, it could not. It makes no sense to argue that nontreatment is in Phillip's best interests, unless one is willing to argue that the quality of life of even high-level Down syndrome children is so low that they are better off dead, even if death will come slowly with pain and discomfort. Alternatively, the parents argued that institutional life is so poor that it would be cruel to subject Phillip to it in later life. This argument would let us justify not treating any senior citizen in an institution or nursing home because the care is so poor. Likewise, the argument that the parents do not want their son to outlive them cannot prevail. It is, in fact, statistically more likely that one or both parents will die before he does than that he would suffer harm because of the surgery. And even if this were not so, we do not (yet) permit parents upon their deaths to have their children executed.

Second, the decision was made in a judicial vacuum, with no consideration of similar cases from other jurisdictions. Two cases that could have been discussed were *Quinlan* and *Saikewicz*.[3]* The first decision permitted a parental decision to withhold or withdraw treatment, but only if the patient had "no reasonable possibility of returning to a cognitive, sapient state." The second decision found that Joseph Saikewicz, a 67-year-old institutionalized individual with an IQ of 10, was "cognitive and sapient" and that any decision not to give him chemotherapy for cancer (arguably an "elective" treatment for a nonemergency condition) could only be made by determining what decision he himself would make if he could make the decision. On the basis of either of those decisions, treatment for Phillip Becker would have been ordered in New Jersey or Massachusetts. Ironically, the appeals court, in citing a *Stanford Law Review* article by John Robertson (who argues in support of such a child's right to treatment), uses only his definition of Down syndrome.[4]

Third, the medical evidence favoring surgery could have been better presented. Only one physician was prepared to testify that the operation should be done on Phillip; the other was not ready to commit himself. Both were pediatric cardiologists, and the judge seemed to need a surgeon to assure him that the operation could be done and was in the range of the routine. Even though neither court ever used the adjectives "ordinary" or "extraordinary," both opinions clearly assume the recommended surgery is extraordinary; that risks of 5–10

Quinlan is discussed at p. 261 and *Saikewicz* at p.244.

percent mortality and one percent of needing a pacemaker are major risks on which one can reasonably decide against treatment. Incredibly, neither court balanced the surgical risks against the certainty of prolonged debilitation and death.

Conclusion

Parents have a right to their moral convictions, but the law places some limits on their actions. When a parental decision is challenged, courts have an obligation to transcend emotional considerations. Courts must make principled decisions based on clearly articulated standards. The Becker decision is an arbitrary decision based on a vague standard. It devalues human life in the name of family autonomy, and it particularly attacks the mentally retarded.*

Phillip Becker was denied his right to treatment he needed to live because he is a member of an oppressed minority group: the mentally retarded. We cannot take rights seriously for ourselves if we do not respect them in our weakest citizens.[5] Taking rights seriously is not a painless or cost-free measure, but it is the only way our values can survive. Given all this, it is extremely disturbing that the California Supreme Court decided not to review the case. On narrow, legalistic grounds involving standards of review, the case may appear uninteresting. However, it was not decided on these grounds and has a symbolic value far beyond its petty prose. It stands as a beacon cautioning us that courts too make mistakes. The mistakes must be acknowledged, reversed if possible, and challenged by other courts confronted with similar decisions.

---

*For a discussion of the "Baby Doe" regulations, promulgated in an effort to mandate treatment for Down syndrome infants, among others, see the three essays at pp. 126–142.

A Wonderful Case and an Irrational Tragedy

The Phillip Becker Case Continues

Why are the courts having so much trouble deciding what should be done with Phillip Becker? Two years ago the first series of cases involving Phillip came before the courts. The California courts decided that his parents had the right to refuse recommended heart surgery for this twelve-year-old Down syndrome child even though this would result in a slow and painful death for Phillip. That decision was arbitrary, unprincipled, emotional, and placed parental autonomy ahead of the right to life of a mentally retarded child.

Another court has now ruled in Phillip's favor. Unfortunately, the opinion is again unprincipled and emotional. I can applaud the outcome and still be appalled at the methodology. In the previous case the state sought an order for surgery, alleging child neglect. In this case, Patsy and Herbert Heath sought guardianship of Phillip, arguing that they had become his psychological parents and that it would be in Phillip's best interests to be placed in their care. They promised, among other things, to reconsider the surgery decision.[6]

In his decision, California Superior Court Judge William J. Fernandez described the case as "both a wonderful and an irrational tragedy"—wonderful because "so many people [came] forward to try to make a little boy's life better...tragic because God and nature may already have determined Phillip's future life course [i.e., his ventrical septal defect may now be inoperable]...irrational because the contestants are spending thousands of dollars and thousands of hours fighting over rights." The judge believed "that time could be better spent trying to make the last part of Phillip Becker's life happier than the earlier part."

Judge Fernandez confronted two related issues: parental autonomy and the child's right to life. Regarding parental autonomy, the judge should have considered the limits of what Bernard Dickens of Toronto describes as the modern American neoconservative doctrine of "radical nonintervention."[7] Specifically, the notion is that the interventionist activities of the liberal welfare state have failed to arrest family breakdown and violence, juvenile delinquency, poverty, etc., and that letting families alone may be a better way to promote their autonomy and that of their children. Limiting state intervention in family decisions to all but the most extreme life-and-death cases is also generally espoused by Joseph Goldstein, Anna Freud, and Albert Solnit in their *Before the Best Interests of the Child.*[8] They argue that before the best interests of the child should even be questioned, it must be demonstrated by due process of law that the family has fallen short of legally set and prenotified standards of child care. Although the 1979 *Becker* opinions were decided before their book was written, they are consistent with this noninterventionist notion.

Unfortunately, Judge Fernandez declined to discuss this critical issue, and instead launched into a discursive essay on parenting. In his words: "The issue [in the former proceedings] was too narrow, that is, the risk of surgery. The basic issue is and always has been one of parenting." In this regard the judge finds it critical to compare the role of the Beckers *as parents* to that of the Heaths.

Although the Beckers win biologically, the Heaths are the clear psychological winners. The court found that Phillip never received nurturing, constancy of affection and love, or the opportunity to develop a basic trust and confidence from his biological parents. They placed him in an institution at birth, viewed his potential for development as very low, and refused to consent to life-prolonging surgery.

"Their basic opinion of Phillip was that he was a permanently mentally retarded low IQ Down's child who would never have a hope of living in society," the court argues. Their view is that he should be permanently institutionalized, not have surgery, not have visits with the people who love him, and never develop "any permanent attachments to anyone or to have any home life." They even barred the Heaths from visiting him after the initial litigation.

In contrast, the Heaths have had a continuing relationship with Phillip since he was five years old. "Their love for him is unquestioned...they offer love and home care, tutoring, and all that Phillip may need in terms of educational, vocational and basic skills training." He would live in their home, have a private tutor, and surgery would be reconsidered. Their expectations for Phillip are great and "they will always treasure him as if he were their own son." On this basis, the judge concluded that "psychological parenting between the Heaths and Phillip exists."

The judge was no closer to the mark when he discussed Phillip's rights. Nowhere, for example, did he even suggest that the mentally retarded as a class have a right to life. Instead his analysis focused on the much narrower issue of a right to habilitation; suggesting that those mentally retarded persons who do not have the options the Heaths have now opened to Phillip may not have a right to live. Instead of using solid legal principles, the judge posited a hypothetical conversation in which he explained Phillip's options:

> THE COURT: Phillip, I am convinced...you have arrived at a crossroad in your life...Your first choice will lead you to a room in an institution where you will live. You will be fed, housed, and clothed, but you will not receive any life-prolonging medical care...You will not be given an opportunity to become attached to any person...Your biological parents will visit you occasionally...
>
> Your second choice will lead you to a private home where you will be bathed in the love and affection of your psychological parents... You will be given private tutoring...in order that some day you may enter into society and be a productive member of our community... You will have a chance for life-prolonging surgery... Best of all, your psychological parents will do all in their power to involve your biological parents in your habilitation and to unite both families together ...

The judge clearly believed Phillip would decide to live with the Heaths. But since he felt such an analysis was without California precedent, the judge went on to find that Phillip's biological parents had caused him to suffer severe emotional, physical, and medical harm.

Admitting that he was "stretching important legal doctrines to the utmost" to make a "little one's life better," the judge concluded that guardianship should be given to the Heaths, Phillip's psychological parents, without severing the parental bonds with his biological parents. The judge's stated purpose is to provide Phillip a "chance to secure a life worth living."

Since the search for a legal principle of decisionmaking is fruitless, the judge's emotional reaction to the case should be examined. Here the judge was remarkably candid. The twenty-page opinion has 26 additional pages of notes; in some the judge expressed his reactions. At one point he remarked, "What the Court is critical of is their [the

Beckers'] insensitivity to his [Phillip's] needs and the immutable label of retarded that they have placed upon their child." And later, "Obviously the Beckers' attitude is unrealistic. Life is an ever changing kaleidoscope." He recalled his feelings when he read Phillip's medical records:

> I weep uncontrollably at the struggle of this wee lad to survive. My soul reaches out to him and his laboring heart to try to give it ease, and in this time of grief, I think of Tiny Tim and what might have been but for old Marley's ghost.

Later he wondered out loud if he was not letting his emotions carry the day. He suggested that a cynic might say, "Judge, isn't that just a little bit subjective, if not maudlin and trite?" His response, of course, was in the negative. He went on to explain that "Judges are humans and not machines," and in his view this is the most critical characteristic of a judge. He even took the occasion to compare his decision to Solomon's, noting again his reactions to Phillip's medical records:

> Intuitively, I reason if I a stranger but a parent can be so overcome with grief as I read the symptoms pointing to the slow but inexorable approach of the child's death, what does the real mother feel? I am struck by the indifference towards Phillip's doom displayed by Mr. and Mrs. Becker. It can only come from a failure to associate with the child and see him on a daily basis. No true parent can watch a child's life slowly ebbing and not cry out, "Oh Lord, let the child live."

Though argued before different courts and couched as different legal proceedings, the basic issue at stake is the same now as it was in 1979: who should decide if Phillip Becker possesses a "life worth living," and therefore a right to life? Goldstein et al. argue that these are "highly personal terms about which there is no societal consensus" and suggest accordingly that this is precisely the type of decision "parents must remain free of coercive state intervention in deciding ..."[9]

In the first Becker case, Judge Eugene M. Premo agreed when he determined that the parental decision against surgery was "in the range of debatable actions" and therefore should be made only by the parents. But his decision cautions that such a loose rule may protect parental autonomy at too high a price: the lives of children. Judge Fernandez, for example, is especially persuasive when he compares the perceptions of Phillip's biological parents with those of his psychological parents regarding the type of life Phillip is capable of living:

BIOLOGICAL PARENTS
Phillip can't talk, communicate, write his name, draw, cook, or form loving attachments. He is a low Down's, not educable and has few basic skills.

PSYCHOLOGICAL PARENTS
Phillip can talk, communicate, write his name, draw, cook, and form loving attachments. He is a high Down's, is educable, and has many basic skills.

The question of life-prolonging surgery for the individual described by Phillip's biological parents may be "debatable" for some; but surgery for the individual described by his psychological parents is not a matter "about which there is no societal consensus."

Most of these characteristics are, of course, matters of fact, and the courts are excellent forums in which to prove such facts if they are disputed. Courts are less skilled at making predictions about the "best" placements for children. And this court provides an example of the problems courts get into when they try to engineer the future. Instead of finding Phillip's biological parents unfit and giving Phillip's exclusive custody to the Heaths, Judge Fernandez gave the Heaths guardianship without severing the Becker's parental rights. Now Phillip has not one set of parents, but two. This is not only unprecedented; it seems to make little sense if one is seeking a continuity of care and responsibility for the child.

Both of the lower court judges let their emotions as hypothetical parents to Phillip determine their decisions. Instead of dealing with the child's right to life, the judges concentrated on the quality of his life, so that the rulings seem to depend upon Phillip's chances to be self-sufficient. And no attempt was made to define the functions or limits of parental autonomy.

I applaud Judge Fernandez for giving Phillip a chance to live. But his emotional appeals and dearth of analysis make this case almost useless as a precedent. A tribute to the judge's humanity, it is otherwise as much of a travesty as the 1979 Becker opinions. Judges must use their humanity in judging others. But judges sit as judges, not surrogate parents, and they have a responsibility to articulate their opinions on the basis of principles. Unless they do, issues of parental autonomy will be decided by emotional caprice. All our children deserve more from the law.

Judge Fernandez should have stated simply that parental rights exist primarily to enable parents to prepare their children for adulthood and their own emerging autonomy.[10] When parents abdicate this goal, the state or others properly step in to act for them.*

*This opinion was affirmed by the Court of Appeals, and the California Supreme Court declined to review it. Successful heart surgery was performed on Phillip in the Fall of 1983 with the consent of the Heaths.

Sterilization of the Mentally Retarded

A Decision for the Courts

In the most notorious case involving sterilization of a mentally retarded person, the United States Supreme Court upheld a Virginia statute that was the basis for sterilizing Carrie Buck, a "feeble-minded eighteen year old" who was the "daughter of a feeble-minded mother and the mother of an illegitimate feeble-minded child." Oliver Wendell Holmes, speaking for the Court, argued:

> It is better for all the world, if instead of waiting to execute degenerate offspring for crime, or let them starve for their imbecility, society can prevent those who are manifestly unfit from continuing their kind ...Three generations of imbeciles are enough.[11]

After an orgy of eugenic sterilization, the pendulum has swung back to the point where this opinion represents neither good law nor good genetics. On the way, in 1942, the same Supreme Court concluded in a different context that the right to procreate was a basic constitutional right. Sterilization "forever deprived" an individual of this right, which is "fundamental to the very existence and survival of the race."[12]

At least three approaches are now possible. One, the majority approach, is for courts in states that have no specific statutes authorizing sterilization to simply declare that it cannot be done on individuals who cannot personally consent to it. Another is for courts to individually evaluate cases in which sterilization of a mentally retarded individual is contemplated, and to authorize it only if it is in the person's "best interests." And the third is for courts to adopt a "*Quinlan**-

---

*See discussion of the *Quinlan* case at p. 261.

type approach" by defining the test to be applied in making a decision to sterilize and permitting the family of the incompetent and their physician (with or without the help of a review committee) to make the decision without resort to the courts.

If any court could be expected to take a *Quinlan* approach it would be the New Jersey Supreme Court, the same court that wrote the *Quinlan* decision. Thus it will undoubtedly surprise some *Quinlan* enthusiasts that this court opted for the second choice, requiring a court to decide if an incompetent may be sterilized.

The Case of Lee Ann Grady[13]

The case that brought the issue to the New Jersey Supreme Court involved Lee Ann Grady, a nineteen year old with Down syndrome. The oldest of three children, she lived at home with them and her parents, and had never been institutionalized. Her IQ was in the "upper 20s to upper 30s range." She could converse, count, and recognize letters of the alphabet. She could dress and bathe herself. Her life expectancy and physical maturation were normal; however, her mental deficiency prevented the normal emotional and social development of sexuality. If she became pregnant, she would not understand her condition, and she would not be capable of caring for a baby alone.

Because of her sexual development, her parents had provided her with birth control pills for four years, although there was no evidence that she had ever engaged in sex or had any interest in it. At the age of twenty she was scheduled to leave her special class in the public school system. Her parents would have liked to place her in a sheltered work group, and eventually in a group home for retarded adults so that she could begin a more independent life and have a place to live after they died. They believed that dependable and continuous contraception was a prerequisite to this change; and they and their doctor sought to have her sterilized, using a tubal ligation, at the Morristown Memorial Hospital. The hospital refused without court approval, and this lawsuit followed.

The Appeal

The lower court granted the parents' application; and the public advocate and attorney general, both invited to appear by the trial judge, appealed. The appeals court began its analysis by recognizing that "sterilization [destroys] an important part of a person's social and biological identity," and rejected any notion that might assign fewer rights to the mentally retarded than to other citizens. The court declined to classify the proposed sterilization as voluntary (since she could not understand the problem or the proposed solution) or compulsory (since no one was actually objecting to the procedure on her behalf). Instead it defined a new category: a procedure "lacking personal consent because of a legal disability."

Following closely the logic of the *Quinlan* case, the court concluded that the right to prevent conception through sterilization is one of the privacy rights protected by the US Constitution, and declined to discard it for the mentally retarded "solely on the basis that [their] condition prevents conscious exercise of the choice." As in *Quinlan*, this court saw its task as fashioning a way to preserve the right of choice for incompetent persons. Unlike *Quinlan*, however, the court refused to permit the person's parents and a court-appointed guardian to make the decision. Instead it insisted that only a court can make this decision for an incompetent. The reasons for this departure from *Quinlan* are instructive.

The court gave two. First, it saw the alternatives in *Quinlan* as "much more clear cut": indefinite life in a coma versus natural death. It believed such a choice is "not conducive to detached evaluation and resolution by the person exercising it on behalf of the patient," but that the decision relies more "on instinct than reasoned calculation." Second, there was no history of abuse in "*Quinlan*-type" decisions. In contrast, many choices were open to Lee Ann Grady; and there was a history of horrible sterilization abuses.

The court noted that similar decisions in abortion and child custody cases are routinely made by courts, and held that, under its inherent parens patriae power, the courts have the authority to protect incompetents who cannot protect themselves because of an innate legal disability. This power has been applied in medical cases, including authorizations to treat children over the parents' objection, kidney and bone marrow transplant cases, and in the *Quinlan* case itself.

The court understood that its decision is *not* Lee Ann's, but

> ...it is a genuine choice ... designed to further the same interests she might pursue had she the ability to decide herself. We believe that having the choice made in her behalf produces a more just and compassionate result then leaving Lee Ann with no way of exercising a

constitutional right. *Our Court should accept the responsibility of providing her with a choice* to compensate for her inability to exercise personally an important constitutional right. (emphasis added)

Standards for Decision

Having decided that a court is a proper decisionmaker, the court considered the standards and procedures to be employed. The court stopped short of requiring a showing of "strict necessity," but did enunciate very strict criteria that must be met before sterilization can be authorized. The court must appoint a guardian ad litem to represent the interests of the ward, and "should receive" independent medical and psychological evaluations by qualified professionals. The trial judge must personally meet with the individual before concluding that the person lacks capacity to make a decision about sterilization and that the incapacity is unlikely to change. This incapacity must be proven by clear and convincing evidence. Finally, the court must be persuaded, also by clear and convincing evidence, that the sterilization is in the person's "best interests." In making this determination, the court must consider the following:

- Possibility of pregnancy
- Possibility of physical and mental harm from pregnancy and from sterilization
- Likelihood of sexual activity
- Inability of the person to understand contraception
- Feasibility of a less drastic means of contraception
- Possibility of postponing sterilization
- Ability of the person to care for a child, or the possibility of a marriage at a future date with ability of the couple to care for the child
- Evidence of relevant medical advances
- Demonstration that the proponents of sterilization are seeking it in good faith for the primary concern of the person and not their own or the public's convenience

This list is not meant to be inclusive, and "the ultimate criterion is the best interests of the incompetent person." Because the trial court did not apply the stringent "clear and convincing" standard of proof (it used the more traditional "preponderance of the evidence" standard) to the best interests conclusion, the case was remanded for further proceedings.

Protecting Individual Rights

The court found a new substantive right, the right to sterilization, and sought to protect incompetent individuals from its arbitrary use by demanding strict due process protections. This is far superior to a blanket prohibition of sterilization, and much more protective of individual rights than permitting families to make this decision on their own. The basis on which *Quinlan* is distinguished is instructive. It supports the notion that a terminally ill patient in a chronic (persistent) vegetative state has suffered a form of "death" and therefore has no good options. Since the only two choices for such a patient are both so bleak, the court thinks that "instinct" rather than logic must be called upon to make this choice. If the court's words are taken seriously, New Jersey courts might be inclined to require a court decision in cases like *Saikewicz*, where the choices are much more complex, since an individual presumably has a greater interest in preserving life than in preserving an ability to procreate.*

The second distinction from *Quinlan*, the history of abuse, seems to indicate that should abuses in other areas of medical care for incompetent persons come to light, courts will step in to protect this class of persons. If *Quinlan* was read by some as judicial retreat from protecting incompetent individuals, the New Jersey courts have clearly signaled that this "retreat" was made under very specific circumstances from which we cannot generalize.**

If a structural problem exists with the decision, it is in the role of the guardian ad litem. The court defines the guardian's role as "representing the interests of his ward." This seems inappropriate for a proceeding in which the judge, not the guardian, will make the judgment as to whether the sterilization is in the person's best interest. Since the petitioners will always be arguing in favor of the sterilization, it would make more sense to require the guardian ad litem to present all of the arguments against sterilization to the best of his ability. As Charles Baron of Boston College Law School has argued, "Without advocates on both sides of the issue to develop the record, the court is too likely to face mental set and path-of-least-resistance pitfalls ... [A] general pattern will seem to emerge from the evidence; an accustomed label is waiting for the case, and without awaiting further proofs, this label is promptly assigned to it."[14]

* The *Saikewicz* case is discussed at p. 244.
**See discussion of the *Claire Conroy* case at p. 309.

The New Jersey model is as good as anything that exists, but it stands almost alone in the country. Since most courts will not act without them, statutes should be enacted in each state to prohibit sterilization of the mentally retarded and other incompetent individuals except when a court finds, after an adversary hearing, that the sterilization is in the best interests of the individual. Such a procedure permits sterilization in cases where a compelling case can be made, and protects potential victims from abusive sterilization for the benefit of others.

Refusing Medication in Mental Hospitals

Denial of a person's liberty is such a drastic act that any discussion of other rights for the institutionalized mentally ill must consider the reason for confinement. There are two basic justifications: (1) to protect other members of society, the "police power" justification; and (2) to protect or care for those unable to care for themselves, the *parens patriae* justification. Mentally ill patients may or may not be competent to make any number of decisions, including the decision to accept or reject medications. A dilemma arises when patients refuse medication: under what circumstances can they be forcibly treated or involuntarily discharged?

Courts are just beginning to come to grips with these issues, and are finding that the constitutional "right to privacy" applies to mental patients. Therefore, competent mental patients have a constitutional right to refuse medication. This is a new notion in mental health law and apparently in mental health therapy as well. On the other hand, it is perfectly consistent with the treatment refusal cases in the medical care area, and there seems no *a priori* reason why the constitutional rights of medical and mental patients should differ.

As Judge Brotman of the United States District Court of New Jersey put it, "Individual autonomy demands that the person subjected to the harsh side effects of psychotropic drugs have control over their administration."[15] The judge also defined four factors that could be weighed by a court in forcing treatment in a non-emergency: (1) the patient's physical threat to patients and staff at the institutions; (2) the patient's capacity to decide on his particular treatment; (3) whether any less restrictive treatments exist; and (4) the risk of permanent side effects from the proposed treatment. In a further proceeding, Judge

Brotman ruled that for involuntary patients, the weighing of these four criteria need not be done by a judge, but could be done by an "independent psychiatrist" at an informal hearing at which the patient could be represented by counsel.

On the other hand, a later ruling by Judge Tauro of the US District Court of Massachusetts makes no distinction between voluntary and involuntary patients, defines emergency treatment very narrowly, and requires consent of the court or a court-appointed guardian before incompetent patients can be medicated against their will in nonemergencies.[16]

Some psychiatrists have reacted strongly to the decision. For example, Alan Stone, President of the American Psychiatric Association, called the opinion "the most impossible, inappropriate, ill-considered judicial decision ever made in the field of mental health law."[17]

The 162-page opinion involved a suit by seven mental patients against fifteen physicians to prohibit forced medication and seclusion for them and all other mental patients at Boston State Hospital, and for money damages. The judge granted the injunctions, but refused to award money damages. Both sides appealed.

The portions of the opinion regarding the use of seclusion were decided on the basis of a specific state statute and so will not be discussed. The medication issue was decided primarily on constitutional grounds. The court made a number of important preliminary findings. The first was that under the Massachusetts civil commitment statute (consistent with the law in many, but not all, states) commitment to a mental institution, even involuntary commitment, is not equivalent to an adjudication of incompetence. A mental patient retains the right, under Massachusetts regulations, "to manage his affairs, to contract, to hold professional...licenses, to make a will, to marry, to hold or convey property, or to vote...." Second, the court found that antipsychotic drugs, such as Thorazine, Mellaril, Prolixin, and Haldol, are powerful chemical agents with serious and severe potential side effects. One of these is tardive dyskinesia, a usually irreversible neurological side effect that includes involuntary motor movements, particularly of the face, lips, and tongue.

The defendants argued that mental patients are *de facto* incompetent to make treatment decisions and that the state must therefore act as *parens patriae* and make such decisions. The court disagreed, and found that most mental patients "are able to appreciate the benefits, risks, and discomfort that may reasonably be expected from receiving psychotropic medication." The court found this "particularly true" in patients who had experienced the medication, and noted that the "therapeutic alliance" between psychiatrist and patient demanded an understanding and acceptance of the treatment program by the patient. If a physician wanted to treat a patient against his will, a probate court had to be petitioned for the appointment of a guardian. If the probate court found the patient unable to give informed consent to treatment, it would appoint a guardian who could give consent.

Both parties agreed that psychiatrists could medicate in an emergency, but differed as to what constituted an emergency. The patients argued, using the language of the seclusion statute, that an emergency existed only when there was a "substantial likelihood of personal injury to the patient, other patients, or staff members." The defendant physicians, on the other hand, argued for a much broader definition of "psychiatric emergency," including suicidal behavior, assaultiveness, property destruction, extreme anxiety and panic, bizarre behavior, acute or chronic emotional disturbance having the potential to interfere with the patient's ability to function on a daily basis, and the necessity to decrease the chances of further suffering or rapid worsening of the patient's condition. The court found this definition "too broad, subjective, and unwieldy," and adopted the seclusion statute's definition.

The court agreed both that Boston State Hospital had "a duty to provide treatment" and that the patients had "a right to receive treatment." However, the court disagreed that the duty to provide treatment carried with it "an implicit right to impose such treatment contrary to a patient's expressed wishes," specifically their wishes not to be forcibly injected with psychotropic medication in a nonemergency situation, or where there are less drastic or less invasive alternatives available.

As in the New Jersey case, the court found that forced medication violated the patient's "constitutional right to privacy" in the sense of self-determination. It accordingly required physicians to obtain informed consent prior to administering psychotropic drugs "that may or may not make the patient better, and that may or may not cause unpleasant and unwanted side effects."

The court then added an unnecessary alternative basis for its decision: the First Amendment. This will produce both confusion and criticism. The interest at stake, as defined by the court, is "the right to produce a thought." "Realistically, the capacity to think and decide is a fundamental element of freedom." Without thought, there can be no ideas; without ideas, no communication. Since communication is a constitutionally protected right, thought and idea formation should also be protected. This novel argument stretches the First Amendment too thin. Our freedom of thought is restricted in elementary and secondary schools, yet the state can force us to attend them; it is restricted in prisons and mental institution; it is restricted in the military, in most careers in government service and by the mass media. The judge would probably argue that all these examples are either voluntary, involve minimal invasions, or are cases in which a compelling state interest exists to overcome the right to produce ideas. The response is that this "right" is at least as "broad, subjective and unwieldy" as the concept of psychiatric emergency proposed by the defendants in this case. It was not needed for the decision, and would have been better left unsaid.

As previously mentioned, the court makes no distinction between voluntary and involuntarily committed patients; the distinction is between competent and incompetent patients. Only the former have a right to refuse treatment; the latter's decisions are made by a court-appointed guardian. The defendants argue, nonetheless, that at least if voluntary patients refuse treatment, the hospital should be able to remove them from the institution. The argument is that a patient volunteering for admission implicitly agrees to the hospital's treatment program, "and may not second guess the institution staff by picking and choosing the type of medication to be used." In this view, the admission agreement amounts to a contractual waiver of any constitutional right to refuse medications. The court rejects the argument on the basis that there is no clear-cut evidence that the patients understood their constitutional rights at the time they signed their admission papers, and accordingly could not knowingly and voluntarily waive them. The implication, however, is that hospitals can condition voluntary administration on the patient's agreement to participate in certain types of treatment, and failure of a patient to cooperate under these circumstances could be grounds for involuntary discharge.

Since the physicians violated the constitutional rights of the patients by forcibly medicating and secluding them, should they not be required to pay money to the patients for harm done them? Following the strength of the first part of the opinion, the damages section is surprisingly weak. The court finds that the repeated violation of the patients' constitutional rights were made in "good faith" and that the physicians are entitled to a "good faith immunity defense." In the court's words, "The defendants desired only to help the plaintiffs."

The plaintiffs also asked for damages for assault and battery, malpractice, and treatment without informed consent. As to the former, the court distorts their claims since there is no doubt that what the defendants did could properly and traditionally be labeled battery. Instead the court says: "To impose liability on physicians for good faith, non-negligent [intentional?] touchings and restraints would impede if not immobilize the administration of a mental hospital." Malpractice claims were rejected because of the "barely adequate" resources the physicians had available to them at Boston State: "Like front-line surgeons, they were required to work with what they had." Likewise, the court found the medication practices "reasonable," saying that they were "consistent with accepted medical standards." All of this bowing to "accepted medical standards" and "good faith" seems very unconvincing when the entire thrust of the opinion is to mandate changes in "accepted medical practice" so that it will conform with the demands of the US Constitution. It would have been preferable for the judge to have simply found that the defendants had violated the rights of their patients, but awarded only nominal damages, or made the ruling apply only to future cases.

It is important to understand what this opinion mandates before predicting, as some psychiatrists have, the immediate demise of institutional treatment of the mentally ill. It means that "competent" patients who are not substantially likely to injure themselves or others "may" refuse medications. They need not. Indeed, during a two-year period when this rule was temporarily in force, the defendants could only identify twelve out of 1000 patients who refused their medications for "prolonged periods," and most of these changed their minds within a few days. If this is any indication of the extent of the problem, having a guardian appointed for those in this group who are incompetent, and making other treatment arrangements for those who are not, is not very burdensome. Some of the patients in the suit were being forcibly medicated or secluded for masturbating in the halls, or walking the wards naked. Forced medication for these activities is forbidden by the opinion. While this behavior may be disruptive to other patients, an alternative to forced medication (perhaps a separate ward for all who fall into this category) will have to be found.

When patients refuse treatment, but cannot be released to the community (e.g., because they are a danger to others outside the institution), then the mental hospital becomes the equivalent of a prison: incarceration without treatment. This seems to be the real concern of institutional psychiatrists. But we must pick one option: either institutionalize and forcibly treat or institutionalize and let the competent patient decide whether to be treated or not. The pendulum is now swinging from the first option to the second; only the development of medications with less severe side effects and more efficacy is likely to cause a reversal.*

*This case went on to be heard by five more courts, including the US Supreme Court. The right of mental patients to refuse treatment, except in clearly defined emergencies, was upheld, and this is consistent with a trend in the law. In 1987, the First Circuit Court of Appeals upheld an award of attorney's fees to the lawyers for the patients. *Rogers v. Okin*, 821 F.2d 22 (1987).

The Incompetent Person's Right to Die

The Case of Joseph Saikewicz

Death is a natural process and a uniquely personal experience. If pressed to categorize it, most would probably term the major controversies surrounding it ethical, rather than medical or legal. Nevertheless, there is an increasing trend to ask the courts whether life-sustaining treatment should or should not be withheld from patients who are unable to make this decision themselves. Judges are asked to decide this question, not because they have any special expertise, but because only they can provide the physicians with civil and criminal immunity for their actions. In seeking this immunity, legal considerations quickly transcend ethical and medical judgments.[18]

Until now, the *Quinlan* case* was the most important case on this issue. However, the case of Joseph Saikewicz[19] is legally more significant. Saikewicz was sixty-seven years old, and profoundly mentally retarded, with an IQ of ten (a mental age of approximately three years). He had been institutionalized all his life, and was unable to communicate verbally.

On April 19, 1976, he was diagnosed as suffering from acute myeloblastic monocytic leukemia, an invariably fatal disease. The only known treatment is chemotherapy, which offers a 30 to 50 percent chance of a remission lasting from two to thirteen months. The treatment itself has serious side effects, including pain, discomfort, pronounced anemia, bladder irritation, loss of hair, bone marrow depression, and, in rare cases, death. Because Saikewicz would not understand the reason for the pain he would experience as a result of these side effects, physical restraints would probably have been necessary

*Discussed at p. 261.

during the weeks in which the drugs were administered and daily blood transfusions performed. Left untreated, he would live for several weeks or perhaps months, after which he would probably die without pain.

One week after the diagnosis, the superintendent of the institution at which Saikewicz was a resident petitioned the probate court for the appointment of a guardian empowered to make the necessary decisions concerning his case. The judge appointed a guardian *ad litem* and two physicians testified against treatment; the judge ordered that no treatment be administered. The order, dated May 13, was based on the judge's findings that the factors favoring treatment (chance of increased life expectancy and the fact that most people with this disease do accept treatment) were outweighed by factors against treatment (the patient's age, inability to cooperate with treatment, side effects, low probability of remission, immediate suffering, and quality of life possible after a remission).

An immediate appeal was taken. The Supreme Judicial Court affirmed the decision on July 9, saying that a full opinion would follow. On September 4, Saikewicz died of bronchial pneumonia, apparently without pain or discomfort. In November 1977, more than fourteen months after his death, the court issued its full opinion. Saikewicz's death gave the court time to consider the issues carefully without having to rush to a decision because of the necessity of rendering emergency treatment (as is often true in such cases), or having to watch the patient and family suffer during the proceedings. It is often said that "hard cases make bad law." The fact that Saikewicz was dead made this case easier, and one would expect to find a statement of both what the law is and should be in this area. The court has not disappointed those with such high expectations.

The court's opinion deals with three issues: (1) the right of any person, competent or incompetent, to decline potentially life-prolonging treatment; (2) the legal standards that control the decision whether potentially life-prolonging, but not lifesaving, treatment should be administered to an incompetent person; and (3) the procedures that must be followed in arriving at that decision.

Patients Rights

On the first issue, the court determined that "the substantive rights of the competent and the incompetent person are the same in regard to the right to decline potentially life-prolonging treatment." Though it thinks it "advisable to consider the framework of medical ethics which influences a doctors decision as to how to deal with the terminally ill patient," the court notes that such considerations are viewed for "insights" and not as "controlling."

In reviewing these, the court stresses and approves of the distinction between "curing the ill and comforting the dying." The patient has a right to privacy "against unwanted infringements of bodily integrity in appropriate circumstances," and only in cases where the state has a strong interest can the individual's decision to refuse treatment be overridden.

State Interests in Treating

From a review of the cases, the court identifies four potential state interests: (1) "the preservation of life"; (2) "the protection of the interests of innocent third parties"; (3) "the prevention of suicide"; and (4) "maintaining the ethical integrity of the medical profession."

The court finds the first interest, the preservation of life, the most important, but not absolute, especially in a case where life will "soon, and inevitably, be extinguished," and makes its strongest statement on individual autonomy:

> The constitutional right to privacy, as we conceive it, is an expression of the sanctity of individual free choice and self-determination as fundamental constituents of life. The value of life as so perceived is lessened not by a decision to refuse treatment, but by the failure to allow a competent human being the right of choice.

The court argues that the second interest, protection of innocent third parties, was "considerable," but finds it unnecessary to discuss it, because the patient's only two living relatives did not want to be involved. The question of suicide is summarily dismissed in a footnote that argues that the act of refusal is not suicide; the state's only interest is in the prevention of irrational self-destruction, not rational decisions to refuse treatment when death is inevitable. The court further finds refusal of necessary treatment in appropriate circumstances "consistent with existing medical mores," noting that such considerations, in any event, are subordinate to the patient's right to decide his own fate.

Accordingly, the court concludes that the only state interest in this case is the preservation of life, and the court's duty is to balance this interest against the individual's interest to be free to reject unwanted bodily intrusions. The judgment of the probate court is accordingly affirmed.

Decision Standards

The next issue is the legal standards to be applied. Here the court agrees with *Quinlan* that equity requires that a mechanism exist to permit incompetent patients the same right to refuse treatment that competent patients have "because the value of human dignity extends to both." Any such decision must be based solely on the "best interests" of the incompetent patient. Significantly the court finds, contrary to *Quinlan*,* that "statistical factors indicating that a majority of competent persons similarly situated choose treatment (as all agreed was the case here) [does not] resolve the issue." In the court's words:

Individual choice is determined not by the vote of the majority but by the complexities of the singular situation viewed from the unique perspective of the person called on to make the decision. To presume that the incompetent person must always be subjected to what many rational and intelligent persons may decline is to downgrade the status of the incompetent person by placing a lesser value on his intrinsic human worth and vitality.

The court makes the following distinctions between Saikewicz's condition and the average rational person: "Unlike most people, Saikewicz had no capacity to understand his present situation or prognosis." He would never understand the reason for the pain inflicted upon him by the chemotherapy. Therefore his situation should be compared to the competent individual who is simply told that something very painful will be done to him, over a long period of time, for reasons he will not be told and for an end he could never understand.

*That court found that a decision to disconnect Karen Ann Quinlan's respirator would be proper because "this decision should be accepted by a society the overwhelming majority of whose members would, we think, in similar circumstances, exercise such a choice in the same way for themselves or for those closest to them." See discussion of *Quinlan* at p. 261.

In short, the court rejects the objective "reasonable person" standard (usually applied, for example, in informed consent cases), and opts instead for a subjective test, the goal being to "determine with as much accuracy as possible the wants and needs of the individual involved." This is the doctrine of substituted judgment, trying to determine what the incompetent person would do if he or she could make the choice. It is adopted "because of its straightforward respect for the integrity and autonomy of the individual."*

The task of the probate judge is, therefore, to ascertain the incompetent person's actual interest and preferences. While the court approved the specific competing considerations taken into account by the probate court, it rejected any diminution of a patient's rights based on "quality of life." "To the extent that this formulation equates the value of life with any measure of the quality of life, we firmly reject it....The supposed ability of Saikewicz, by virtue of his mental retardation, to appreciate or experience life had no place in the decision." The court finds, however, that the probate court did not use the phrase "quality of life" improperly, but used it only in reference to the pain and disorientation that the chemotherapy was likely to cause Mr. Saikewicz.

Determining Proper Procedures

The final issue is how the question is to be procedurally decided. The Supreme Judicial Court finds the probate court a proper forum for immunity determinations and declares that such cases should be presented to it for resolution. The probate court should further charge the guardian *ad litem* to present all the arguments in favor of administering life-prolonging treatment, so that all viewpoints are aggressively represented in an adversary proceeding. Only after such a proceeding should the judge reach a decision.

The Supreme Judicial Court invites, but does not require, the probate court to make use of the views of ethics committees, attending

---

*This doctrine can work well when the formerly competent patient has made his wishes known, but seems misplaced as applied to Mr. Saikewicz because he never had the opportunity to make *any* decisions himself. An objective "best interests" test is more appropriate. But, because of testimony that competent people always take chemotherapy, that test would have required either ordering treatment, or denying it primarily on the basis of mental retardation.

physicians, and other medical experts. It rejects outright the immunity-granting role of ethics committees envisioned by the *Quinlan* court:

> We take a dim view of any attempt to shift the ultimate decision-making responsibility away from the duly established courts of proper jurisdiction to any committee, panel, or group, *ad hoc* or permanent .... The New Jersey Supreme Court concluded that "a practice of applying to a court to confirm such a decision would generally be inappropriate, not only because that would be a gratuitous encroachment upon the medical profession's field of competence, but because it would be impossibly cumbersome."

The court's final language on the respective roles of ethics committees and courts in granting legal immunity for actions is as strong as any in the opinion:

> We do not view the judicial resolution of this most difficult and awesome question—whether potentially life-prolonging treatment should be withheld from a person incapable of making his own decision—as constituting a "gratuitous encroachment" on the domain of medical expertise. Rather, such questions of life and death seem to us to require the process of detached but passionate investigation and decision that forms the ideal on which the judicial branch of government was created. Achieving this ideal is our responsibility and that of the lower court, and is not to be entrusted to any group purporting to represent the "morality and conscience of our society," no matter how highly motivated or impressively constituted.

This decision will be the subject of heated and continued debate. It is both wide-ranging and carefully-reasoned. It is also an example of an aggressive judicial opinion in an area that some other courts have been eager to avoid. The Supreme Judicial Court has made it clear that it considers judges the only proper people to grant legal immunity in life and death cases and seems to welcome the challenge that these cases present. It argues strongly in favor of using the adversary process—a process highly favored by lawyers, but generally scorned by members of the health care professions. The decision also quite properly puts ethics committees in their place, as advisors, not decisionmakers, and is likely to be followed by other courts on this issue.

Autonomy Affirmed

Primarily, however, this is a decision about patients rights to human dignity and autonomy, and it is here that the decision is likely to have its widest impact. The court finds all patients have the absolute right to refuse life-sustaining treatment that will not cure or preserve life. It further argues that such decisions must be based on what is important to the individual patient, and not on what a majority of "reasonable persons" might do.

The court rejects the rationale of many previous courts in the informed consent area, which permits juries to find in favor of plaintiffs only if the information not provided by the physician would have caused a "reasonable person" to refuse to undergo the proposed treatment (for example, would a disclosure of a 5 percent complication rate persuade the average reasonable person, *not* the individual plaintiff, to refuse an ulcer operation?). This court finds that self-determination is paramount and cannot be delegated. Therefore its logic compels a conclusion that courts and juries should be asked not what the "reasonable person" would do in such circumstances, but what "this particular patient" would have done. It is only such a subjective test, like the substituted judgment test proposed by the court, which protects the patient's right to make important treatment decisions.

This aspect of the decision will likely be the most controversial. How can one ever know what another person would or would not do? How can we ever know what Joseph Saikewicz would have decided? Such questions may be unanswerable. But they are the proper questions. A correct resolution of them may be more likely to come from a judicial decision after an adversary proceeding, in which all interested parties have fully participated, bringing in all their own perceptions, beliefs, and biases, than from the individual decisions of the patient's family, the attending physician, an ethics committee, or all these combined. By using the admittedly difficult machinery of a legal proceeding, we can promote the incompetent person's "right of privacy and self-determination" and ensure that it is not used as a license to kill unwanted patients.

The Case of Mary Hier
When Substituted Judgment Becomes Sleight of Hand

If a 92 year old, mentally ill woman, who has been a resident of a psychiatric hospital for the past fifty-seven years, pulls out her gastric feeding tube, and resists reinsertion, should it be surgically replaced? To answer this question, we need more facts. But determining what facts are relevant, and their influence on the ultimate decision, depends on the standard applied to make the decision.

The subjective, or substituted judgment standard calls for the decisionmaker to make the decision that would be made by the incapacitated person if that person could competently express his or her wishes. It's a legal fiction, but it serves the highly desirable goal of promoting self-determination. The objective, best interests standard, on the other hand, promotes well-being by attempting to determine what is "best" for the person.

Both standards can be indeterminate in hard cases. Nonetheless, uncritical use of the substituted judgment standard can at times lead to a decision that most of us would consider not only inconsistent in terms of patient interests, but also a dangerous precedent for others similarly situated. When the evidence of what the previously competent patient would want is persuasive, for example, where the patient has signed a specific "living will" or designated a proxy to make decisions, we should follow these wishes to honor our commitment to self-determination. But in cases where the patient has never been competent, or has never expressed his or her desires while competent, application of the substituted judgment standard often serves simply as a pretext to permit otherwise unjustifiable actions. The case of Mary Hier is a vivid illustration.

The Case of Mary Hier[20]

At the time her case was heard in court, Ms. Hier had recently been transferred from a New York psychiatric hospital to a Massachusetts nursing home. New York State public assistance continued to pay for her care. She had no relatives who could be contacted. In addition to believing that she was the Queen of England, she suffered from senile dementia. Because of a hiatal hernia and a cervical diverticulum (a pouch in her esophagus that prevents placement of a nasogastic tube), she was unable to take food orally.

The nursing home had applied to the court to have a guardian appointed with authority to consent to the administration of psychotropic drugs (Thorazine) and the performance of a gastrostomy to replace her feeding tube, which she had recently pulled out and would not allow to be replaced.

The probate judge, Thaddeus Buczko, found that Hier was mentally incapable of making necessary decisions about her medical treatment. He authorized the administration of Thorazine, but not the surgery to replace the feeding tube (the nursing home had subsequently withdrawn its request to have this surgery authorized).

In order to reach this conclusion, the probate judge applied the substituted judgment doctrine, attempting to "don the mental mantle of the incompetent and substitute itself as nearly as possible for the individual in the decision making." The judge found that Hier objected to the injection of Thorazine by needle, but that its administration would decrease her agitation and increase her communicativeness. He concluded that she would accept it if she were competent. The judge also found that Hier objected to the gastrostomy, that alternative methods of feeding were not available, that her prognosis was "poor" with or without it, and that it involved "high risks." Giving "significant weight" to her stated preference, the judge ruled that she would refuse the gastrostomy if competent.

The Appeals Court

Because the Thorazine decision was not contested, the Massachusetts Appeals Court focused on a single issue: Would Ms. Hier, if competent, reject the gastrostomy procedures? It found that the probate court's conclusion that she would reject the gastrostomy, "cannot be said to have been erroneous." Its analysis, however, offers slim support for this conclusion. The appeals court concluded that the following factors were properly taken into consideration to determine Hier's desires:

- The proposed operation is intrusive and burdensome

- Hier repeatedly and clearly indicated her opposition to procedures necessary to introduce tube feeding

- The benefits from performing the gastrotomy are diminished by her repeated history of dislodgments, and such dislodgments probably cannot be prevented except by using physical restraints

- Physicians who evaluated her condition made "thoughtful recommendations that surgery was inappropriate in her case."

These conclusions are insecure both factually and analytically. Two physicians who evaluated her did recommend against surgery. But Mayo Johnson did so only because "the patient has made it clear to me that she does not want tube feedings...." The other physician, Randolph D. Maloney, recommended against surgery primarily because he thought enough resources had already been expended on Heir:

> [We] are also dealing with a patient that is requiring *an enormous amount of professional time both medically and legally*....I now feel that enough time has been given to this patient that these efforts could be better directed towards other endeavors....There comes a time when *it is economically untenable to proceed* with this type of treatment and it has become *inordinately expensive* in this situation.... (emphasis added)   (affidavit)

Thus one physician treated her as a competent adult, the other as a resource allocation problem. Neither opinion adds much to a substituted judgment determination. Moreover, the court's truncated burdens-benefits analysis requires much more substantiation. We need to weigh not only the potentially formidable problems of performing surgery on this patient (which are graphically portrayed) but also the burdens of *not* performing surgery. These are never discussed, and this omission alone makes the conclusion suspect. Ms. Hier will die of starvation if she is not fed. This may be the proper outcome, but we need to know a lot more about how it will occur and the pain and suffering it will cause before we can reasonably reach this conclusion.

Past Precedents

The court summarily rejected the conclusion of the *Storar* case in which New York's highest court refused to use substituted judgment to let a 52 year old mentally retarded man suffering from terminal bladder cancer refuse blood transfusions he found disagreeable and distressful. *Storar* is overly simplistic, but without sufficient attention to the case itself, the Massachusetts Court makes almost exactly the same fatal error as the New York court.* In *Storar,* the court considered bladder cancer and blood loss as two completely separate medical conditions, one of which was treatable and the other not. This simple-minded view led the court to order treatment for the blood loss without regard for the incurable cancer that was its cause. By treating these two inextricably intertwined conditions separately, the court arrived at an artificial and unsatisfactory conclusion.

Both *Hier* courts made the same mistake, although less dramatically. They treated Ms. Hier as suffering from two conditions: a mental illness and an aversion to feeding tubes. The courts ordered Thorazine for her mental illness, but rejected surgical intervention to feed her because she indicated a preference not to have it. In fact, her mental illness and her removal of the feeding tube are probably causally related; and this relationship deserves exploration. The appeals court at least acknowledges the relationship in its concluding paragraph by noting that the court-ordered Thorazine treatment may "cause her to become more acquiescent towards surgery." Indeed.

But more important for a burdens/benefits analysis is an exploration of the finding of Kenneth Slater, the psychiatrist who examined her, that the administration of Thorazine via a feeding tube "would be safer and less painful than administration by injection." Since it was the needle injections, we are told, that Hier objected to, and not the Thorazine itself, the trial judge should at least have considered whether a single surgical procedure, which would permit her to be both fed and medicated, would be less invasive and objectionable than twice-daily injections.

* The *Storar* case is discussed at p. 280.

Can Substituted Judgment Serve?

The fatal flaw in the opinions is a reliance on the substituted judgment standard in a case in which substituted judgment should have been considered identical to the best interests tests (since we have no competent expression by Hier that her personal beliefs or desires are idiosyncratic). In an almost self-congratulatory tone focusing on autonomy, the court actually uses the right to refuse treatment to *deny* treatment to a patient whom some physicians find troublesome. This sleight of hand is performed by declaring that Ms. Hier is incompetent to make medical decisions for herself; nonetheless, her ambiguous, incompetent statements and actions are used as a pretext for ruling against treatment. This approach, unfortunately, accomplishes the very thing that Massachusetts hoped to avoid by adopting the substituted judgment doctrine: it devalues the life of the mentally ill. This is how it works.

In the case of Joseph Saikewicz, the Massachusetts Supreme Judicial Court first applied the substituted judgment doctrine to the treatment setting to conclude that a 67-year-old severely mentally retarded man could refuse cancer chemotherapy because the treatment would produce adverse side effects that would make him "experience fear without the understanding from which other patients [the majority of whom opt for chemotherapy] draw strength."* Of course, the only reason Saikewicz would not be able to understand the treatment is *because* he is mentally retarded. Thus using his potential fear of treatment as the primary rationale not to offer it to him can be seen as a disguise for the real reason: his mental retardation.

The other case relied upon by the *Hier* court, that of Earle Spring, is similar. Spring was a 78-year-old senile nursing home patient on kidney dialysis. He occasionally resisted transportation to the kidney dialysis center, and this was used as the primary evidence that he did not want to continue dialysis. His actions, however, were inherently ambiguous, and similar to those of many profoundly incompetent, senile patients in a nursing home. It is unfair to label him on the one hand profoundly incompetent, and on the other to attribute deeply held motives to his seemingly random disruptive behavior.**

But no case illustrates better than this one the use of substituted judgment as a pretext by construing the actions of incompetent pa-

*The *Saikewicz* case is discussed at p. 244.
**The case of *Earle Spring* is discussed at p. 285.

tients as indicative of competently held beliefs. The key fact is that Hier pulled out her gastrostomy tube. What do we make of it? Since she is severely mentally ill, has been institutionalized for fifty-seven years and has never indicated any desire not to eat (indeed, she steals food from other patients), we could easily conclude that we can learn *nothing* about her personal preferences from this act. On the other hand, if we do not want to replace the gastrostomy tube surgically (and surgery is required only because the dislodged tube was not noticed soon enough to be replaced manually), we can follow the court's lead and construct a whole scenario about Hier's preference. This is fiction, but it reads well.

The sole *fact* found by the probate judge is that Hier pulled her feeding tube out *once*. The judge also found that her three gastrostomy operations in New York "suggest" that she "repeatedly pulled these tubes out," but there was no direct evidence of this. Even if she did pull the tube out twice in ten years in New York, and once in six months in Massachusetts in ways that required surgery to replace it, what do these actions signify? She had, for example, intravenous lines at the time of the hearing. She also resisted their insertion, but there was no evidence that she ever pulled them out. The appeals court, however, pictures not an incompetent mentally ill patient, but a competent, dying patient in agony:

> Ms. Hier's repeated dislodgments of gastric tubes, her resistance to attempts to insert a nasogastric tube, and her opposition to surgery all may be seen as a *plea for privacy and personal dignity* by a ninety-two-year-old person who is seriously ill and *for whom life has little left to offer*. (emphasis added)

On the surface, the court is promoting the rights of the mentally ill. Beneath the mask, it is making a profound statement about the resources that should be expended on the aged, mentally ill patients in nursing homes. Because it reads *Saikewicz* as saying we don't have to give older, mentally retarded patients chemotherapy because they can't understand it, and *Spring* as saying that we don't have to give disruptive, senile nursing home patients dialysis because they don't like it, it concludes that it is simply following past precedents by ruling that we don't have to feed senile, mentally ill nursing home patients who pull out their feeding tubes because they don't want to be fed that way.

Back to Best Interests

The good intentions of the *Saikewicz* court have been perverted: what was originally constructed by the Supreme Judicial Court as a shield to prevent discrimination against the never-competent patient has been transformed into a sword to justify gross discrimination against them. The *Saikewicz* court correctly noted: "The best interests of an incompetent person are not necessarily served by imposing on such persons results not mandated as to competent patients similarly situated."[21] The court was wrong, however, in thinking that the substituted judgment standard could do a better job than the traditional best interests standard to protect incompetent patients who had never been able to express their preferences competently. As the Mary Hier case illustrates, what is required is not a new standard for never-competent patients, but a broader interpretation of the relevant factors in determining the patient's best interests.*

*On July 3, 1984, the guardian ad litem, Robert Ledeux, took the case back to the original probate judge and got him to reverse his position after he heard from seven additional medical witnesses. Successful surgery was performed at St. Elizabeth's Hospital in Boston to reinsert the gastrostomy tube on July 4. The lower court's reversal does not, of course, affect the judicial precedent created by the appeals court decision. Ms. Hier is still alive in 1988, and still believes she is Queen of England.

Death, Dying, and Refusing Medical Treatment

Introduction

In one of his *Letters from the Earth*, Mark Twain has Satan observe that, before hell was invented, "death was sweet, death was gentle, death was kind...when man could endure life no longer, death came and set him free." In a country where more than 80% of us will die in hospitals, it is increasingly unlikely that (whether we believe in heaven or hell), death will be "sweet," "gentle" or "kind." Death will almost always be accompanied by technologically driven attempts to stave it off, attempts that ultimately must be futile. When should medicine let death come, and when oppose it with all the instruments at its disposal? The essays in this section chronicle the law's struggle with this question over the past decade—from the case of Karen Ann Quinlan to the case of Paul Brophy. It is a troubled chronicle; one that sees physicians and lower court judges alike frightened to take any responsibility for terminating medical treatment; frightened at times even to follow the wishes of competent patients. In many of these cases, "judging medicine" is of only secondary interest: more central is judging judges who attempt to judge medicine. The judges often do not fare well. Indeed, some of the most unprincipled and callous judicial decisions of the past decade appear in this section, most notably the decision by Judge Lawrence Waddington in the case of William Bartling; Judge Warren Deering's decision in the case of Elizabeth Bouvia; and Judge David Kopelman's in the case of Paul Brophy. The only kind thing that can be said of these trial court rulings is that all of them were overturned in strongly-worded appeals court decisions. Of course, these and many other cases in this section should probably not have been in court in the first place. It is often only the fear and incompetence of hospital counsel that drives hospital administrators and physicians to seek court relief that serves only to make hospital counsel feel more comfortable.

This section also explores cases in which I have had more personal involvement than most of the others discussed in this book. With my friend and colleague Leonard Glantz, for example, I wrote *amicus* briefs for the appeals courts in the cases of Earle Spring (for the American Society of Law & Medicine); Brother Fox (for Concern for Dying); Claire Conroy (for Concern for Dying); and Paul Brophy (for Concern for Dying). In addition, I assisted Attorney Richard Scott with the hearing in the William Bartling case, and wrote the appeal brief with him and Leonard Glantz, again for Concern for Dying. And while I was not formally involved in any of the Elizabeth Bouvia cases, I met with her on two occasions following my first essay about her Riverside Hospital lawsuit, and discussed preparation of the appeal of her High Desert Hospital case with her attorney, Richard Scott.

My viewpoint in this section should be clear: I believe competent individuals should have the right to refuse *any* medical treatment, regardless of the consequences to themselves. Distinctions made on the basis of prognosis, type of treatment, nature of the disease, etc. are all excuses to permit physicians to force treatment on unwilling individuals. Only a focus that is exclusively on the patient's wishes is worthy of the law. And only a law that upholds the right of the individual patient in this regard is worthy of support.

All of the essays in this long section were originally published in the *Hastings Center Report*. The *Quinlan* essay in June, 1976; "After Saikewicz" in June, 1978; "Brother Fox" in June, 1980; "Help from the Dead" in June, 1981; *Spring* in August, 1980; "Elizabeth Bouvia I" in April, 1984 and "Elizabeth Bouvia II" in April, 1986; "Feeding Tubes" in February, 1986; *Conroy* in April, 1985; and *Bartling* in December, 1984.

The Case of
Karen Ann Quinlan
Legal Comfort for Doctors

The case of Karen Quinlan captured the interest of both the general public and the legal and medical professions. There was widespread dissatisfaction with the New Jersey Superior Court's decision in November 1975.[1]

In March, 1976 the New Jersey Supreme Court, in a unanimous decision written by Chief Justice Richard J. Hughes, overturned that ruling.[2] And while many will agree with the court's apparent view that some way had to be found to permit Karen's parents to have their daughter removed from the respirator, there will be equally widespread dissatisfaction with this court's approach. Rather than search for narrow grounds (applicable to this particular case) on which to base its conclusion, the court chose instead to use the malpractice "crisis" as an excuse to employ expansive and ill-defined language to deliver one of the most wide-ranging, and potentially dangerous, opinions in medical jurisprudence.

The New Jersey Supreme Court, in one decision, finds a new right of privacy, permits it to be exercised by proxy, develops new criteria for terminating life-sustaining treatment, develops a new procedure for applying the criteria (including an ethics committee), recognizes the "ordinary–extraordinary" distinction as legally useful, applies the principle of the double-effect, *sub silencio*, to homicide, and absolves all those participating in terminating the care and treatment of Karen Quinlan of any civil or criminal liability.

The law of the case is this: Joseph Quinlan's request to be named his daughter's guardian was granted. The court further ruled that should physicians of his choosing conclude "that there is no reasonable possibility" of Karen's ever emerging from her present comatose condition to a "cognitive, sapient state," and that the life-support apparatus now being administered to Karen should be discontinued," and should he and Karen's "family" concur, *and* should an "ethics committee or like body" of the hospital in which Karen is then a patient agree "that there is no reasonable possibility of Karen's ever emerging from her present comatose condition to a cognitive, sapient, state," then the respirator may be withdrawn, "and said action shall be without any civil or criminal liability therefore on the part of any participant, whether guardian, physician, hospital or others."

The court arrives at this conclusion after reviewing the testimony presented before the lower court, the Harvard criteria for brain death,* and the 1957 statement of Pope Pius XII on extraordinary care. Following this factual recital, which accounts for more than half of the entire 59 page opinion, the court proceeds to dwell with extensive case citation on issues that are clear, and with an embarrassing lack of citation or logic on issues that are central and extremely difficult. For example, in denying the Quinlans' contention that any state interference with their decision is a violation of the free exercise of religion clause of the First Amendment, the court finds it necessary to cite fourteen cases. On the much more central and difficult question of whether or not the decision to decline medical treatment, under certain circumstances, is protected by one's constitutional "right to privacy," the court provides almost no analysis, but instead simply states its conclusion: "Presumably this right is broad enough to encompass a patient's decision to decline medical treatment under certain circumstances, in much the same way as it is broad enough to encompass a woman's decision to terminate pregnancy under certain conditions."

The court then tries at length to explain that no matter how ill-defined the "right to privacy" is, the state could demonstrate no "compelling interest" to justify interfering with it. The court notes that in previous cases where treatment has been ordered (such as the blood transfusion cases) the patient's chances for a normal life were usually excellent and the bodily invasion involved in the treatment was "minimal." In the court's words:

> We think that the State's interest *contra* weakens and the individual's right of privacy grows as the degree of bodily invasion increases and the prognosis dims. Ultimately there comes a point at which the individual's rights overcome the state interest. It is for that reason that we believe Karen's choice, if she were competent to make it, would be

*See "Defining Death: There Ought to be a Law," at p. 365.

vindicated by the law. Her prognosis is extremely poor...and the bodily invasion very great...

This distinction may, however, be only semantic as almost all the blood transfusion cases involved invasive surgery as well. In a 1971 decision in *John F. Kennedy Hospital v. Heston*[3], this same court had written, "It seems correct to say there is no constitutional right to choose to die." The court does not even acknowledge this language, but says only of that case, which also involved a 22-year-old woman, that it can be distinguished because she was "apparently salvable to long life and vibrant health." No mention is made of the state's interest in protecting life, regardless of the person's present medical condition or prognosis.

Having determined in this case that Karen herself would have the right to refuse further treatment if she had a lucid moment and could completely express this wish, the court next had to decide that someone else could refuse treatment for her. The court concludes that Karen's rights should not be denied her simply because she is no longer in a position to exert them, and therefore that her "guardian and family" should be permitted to "render their best judgment...as to whether she would exercise it in these circumstances." Again there is no analysis to support the conclusion that the parent and guardian must act under this rule, termed the doctrine of "substituted judgment," which has been used in transplant cases. The other tests courts have adopted in kidney and bone marrow transplant cases involving minor donors are: what is in the "best interests" of the child; and, is the parental decision "fair and reasonable" under the circumstances? Perhaps the substituted judgment test is the appropriate one here, but the court neither intimates that other possible tests exist and have been found more appropriate in similar circumstances by other courts, nor gives any guidance to the parents as to how they might determine what Karen would have wanted.

Karen could decide this question herself. The court, however, is not completely comfortable with permitting her guardian and family to decide it for her alone. Therefore they place two additional qualifications on its exercise: a prognosis by the attending physicians, and a concurrence in the prognosis by an "ethics committee."

The court first argues, with little analysis, that refusal of "artificial life support or radical surgery" is distinguishable from "self-infliction of deadly harm" and should neither be considered suicide by the patient nor dereliction of the attending physician's duty. If the view of the medical profession concerning its duty to treat is different (as Judge Robert Muir of the Superior Court had concluded when he deferred to "standard medical practice"), the court finds that it could re-examine this view in light of "the common moral judgment of the community at large. The law, equity and justice must not quail and be helpless in the face of modern technological marvels presenting questions hitherto unthought of." The court accordingly decides to "re-evaluate the applicability of the medical standards projected in the court below" to determine if they were so well reasoned and consistent as to constitute "an ineluctable bar to the effectuation of substantive relief for the plaintiff in the hands of the court."

The court's conclusion, of course, is that the medical standards are not such a barrier. The main reason seems to be the court's belief that the physicians' motivations for refusing to take Karen Quinlan off the respirator were not based on medical judgment, but self-interest: their fear of potential malpractice suits and criminal liability under the homicide statutes. The court goes out of its way to say it believes the testimony of the physicians in this case that potential liability did not affect their decision not to follow the parents' wishes. Nonetheless, the court's opinion indicates that it believes almost all, if not all, physicians are influenced by the "brooding presence of such possible liability," and that therefore standard medical practice is sometimes dictated by the fear of legal repercussions.

In this context the court proceeds to do what it had previously said it was not going to do, apply Pope Pius XII's "ordinary–extraordinary" distinction. The problem the court defines is how to get physicians to make the "ordinary–extraordinary" distinction on the basis of medical standards uncontaminated "by self-interest or self-protection concerns which would inhibit their independent medical judgments for the well-being of their dying patients." The solution is as extraordinary as it is vague and, I believe, unwise and unjustified.

The court quotes at length and with great approval a suggestion by pediatrician Karen Teel, that physicians be legally permitted to share responsibility for such decisions and advises that "ethics committees composed of physicians, social workers, attorneys, and theologians ... review the individual circumstances of ethical dilemmas..." with the attending physician.[4] The reason is to help relieve the physician's concern with legal liability.

In one of the most remarkable statements in American jurisprudence the court finds that "the most appealing factor in the technique suggested by Dr. Teel seems to us to be the diffusion of professional responsibility for decision, comparable in a way to the value of multi-judged courts in finally resolving on appeal difficult questions of law." The court further notes that such a system should help screen out cases "contaminated by less than worthy motivations of family or physician," and that "in the real world...the value of additional views and diverse knowledge is apparent."

The court gives almost no guidance as to the composition of the ethics committee, and no guidance at all on the procedures it must follow. The only guidance the court does give on criteria is inherently contradictory. In the final order, the court specifically says that the role of the ethics committee will be to determine if "there is no reasonable possibility of Karen's ever emerging from her present comatose condition to a cognitive, sapient state." This issue is, of course, one of the few relevant ones in this case on which an ethics committee will be of absolutely no help; it is purely a medical prognosis question, and calls for expert medical consultants.

Read another way, the statement is equally disturbing. Specifically, one could infer from this sentence that the ethics committee should be consulted on the definitions of "reasonable possibility" and "cognitive, sapient" life. Using a "reasonable possibility" test, for example, the court has greatly expanded the traditional "hopeless" standard for such decisions. As to "cognitive and sapient," one could read this as being the first time any United States court has sanctioned the use of a "quality of life" criterion for making such decisions. Read together with the rest of the opinion, this could be taken as a signal to let ethics committees re-evaluate the care and treatment of defective newborns, the retarded, and the mentally ill. Moreover, since such a re-evaluation would be made by a group in which no one was personally liable, either criminally or civilly, decisions to terminate the treatment of these individuals would presumably be much easier to make than they are now. The idea seems to be that all involved (including the court) will feel more (ethically?) "comfortable" with decisions thus arrived at for which no one individual is seen as responsible and for which no individual can be held legally accountable.

The "diffusion" of responsibility argument is also disturbing in this context. When everyone is "responsible" for such a decision no one is, and the decision becomes much easier to make. Experiments with ordinary people who were persuaded to apply what they believed to be deadly force to an innocent party by a researcher who assured them that it was "necessary" for the experiment,[5] should be a sobering reminder that diffusion of responsibility can lead to arbitrary and lethal decisionmaking even without granting individuals immunity from civil and criminal penalties. How much more dangerous and inappropriate is it to diffuse responsibility and simultaneously grant complete immunity from the consequences of any decision so reached? The value of an "ethics committee" is in a teaching or advisory role. As a decisionmaker it is likely to prove ineffective, acting either as a rubber stamp or a debating forum.

The opinion should stand as a monument illustrating how far a court will go to encourage physicians to do what it wants them to do in a particular case. Its loose language and potential for abuse demand exposition and discussion. Above all, it should be limited to the facts of this case and not imposed on other patients whose rights it may destroy rather than protect.*

---

*The case has, in fact, been limited in New Jersey, where "ethics committees" were soon renamed "prognosis committees" and composed exclusively of physicians to apply the *Quinlan* criteria. In 1986, I had the privilege of commenting on the *Quinlan* case together with its author, Governor and former Chief Justice Richard Hughes. Governor Hughes remembers the opinion well, and told the audience at the conference held to commemorate the "10th Anniversary" of the opinion, that the Court felt it had to act very quickly, and that his wife insisted he cancel or postpone a planned trip to Japan to write it to help relieve the burden from the Quinlan family. He also made it clear that the procedural aspects of the opinion were not taken terribly seriously by the court, but were primarily seen as a way to resolve this case. Karen Quinlan died in June, 1985. *See: New York Times,* March 30, 1986 at 8E.

No Fault Death

After Saikewicz

The social mandate of the physician is distinct from that of the judge. Yet in the wake of the *Saikewicz** opinion, physicians have reacted both by renouncing what they view as unjustified judicial incursion into their domain, and by demanding that their own social mandate be extended to permit them to make life and death decisions with legal immunity. Never has the argument for professional irresponsibility been made in such an inappropriate context.

Hospital lawyers in Massachusetts have generally taken an extremely conservative and arguably unreasonable view of the court's decision. This view has resulted in cases going to court that do not belong there, in inaccurate and harmful instructions to hospital personnel regarding orders not to resuscitate, in a high degree of hostility to the opinion on the part of the medical profession, and in a call for "remedial immunity legislation" that would grant physicians immunity for terminating treatment under certain circumstances. Each merits discussion.

Cases Following *Saikewicz*

The Massachusetts Supreme Judicial Court (SJC) concluded in *Saikewicz* that only a court could grant legal immunity to physicians who wished to terminate treatment to an incompetent patient. The following six cases were heard in Massachusetts courts during the year after it issued the opinion.

*See discussion of the *Saikewicz* opinion at p. 244.

Case 1: R. Pietrowicz was a 56-year-old woman, dying of Huntington's Disease, who had been hospitalized for five years. She had most recently been placed on a respirator, after which her condition deteriorated rapidly; she suffered cardiac arrest, extensive brain damage, and went into a deep coma. Two of her brothers had previously died of Huntington's Disease, and she had previously made known her desire not to be kept alive on a respirator. The probate court agreed that the respirator could be removed. She died 35 hours later.

Case 2: Baby Huber was a one-year-old child who had been severely burned, and was brain-dead under the Harvard criteria. The fire had allegedly been set by the putative father whose lawyers did not want the respirator disconnected because of the potential homicide charge against their client. The court dismissed the case for lack of jurisdiction.

Case 3: W. J. Norton was twenty years old, and was dead under the Harvard brain death criteria, following an automobile accident. Parents and physicians agreed on this. The court accordingly dismissed the case, saying it had no jurisdiction over the subject matter.

Case 4: Kerri Ann McNulty was a newborn whose mother had had German measles during pregnancy and who had multiple defects including cataracts, probable deafness, probable mental retardation, and a congenital heart defect that had to be surgically corrected if she were to survive. The parents would not consent to the surgery, which also carried a high mortality rate. After a hearing the judge authorized treatment, concluding, "I am persuaded that the proposed cardiac surgery is not merely a life-prolonging measure, but indeed is for the purpose of saving the life of this child, regardless of the quality of that life."

Case 5: Katherine Jennifer Demiano was a newborn who was admitted to the hospital at about six weeks of age with seizures and thereafter lapsed into a coma. She met all the Harvard criteria for brain death except for a spinal reflex. After a hearing, the judge authorized removal of the respirator.

Case 6: Chad Green was a two year old with leukemia. His parents had begun the child on a chemotherapy regime that promised a 50 to 80 percent chance of a cure after the three-year treatment. The parents

discontinued treatment, and the hospital brought a "care and protection" suit to compel continued treatment, the absence of which would surely lead to the child's death. The SJC ordered continued treatment.*

It should be observed that, of these six so-called "*Saikewicz* cases," only one (Case 4) is the type of case the court was talking about in *Saikewicz*. Three of them deal with brain death, and the SJC had previously accepted this definition of death; nothing in the *Saikewicz* case implied that physicians were under an obligation to treat a corpse as if it is a living human being. The Chad Green case (Case 6) is a traditional "child neglect" case, and the relevant legal standards for such a case were in no way changed by *Saikewicz*. Finally, *Pietrowitz* (Case 1) illustrates an inappropriate shifting of medical decisionmaking to the courts, since the only issue was one of medical prognosis and treatment possibilities in the case of a person who was terminally ill of an incurable disease who had made her wishes known.

The scope of the *Saikewicz* case is properly limited to an incompetent patient (adult or child) for whom a life-prolonging treatment is available, is used on some patients as standard medical procedure, and is opposed by the family or guardian because they do not believe the treatment is in the best interests of the patient. Therefore, while hospital attorneys and some physicians have argued that the courts would be "flooded" with cases as a result of *Saikewicz*, in fact only one such case has actually gone to court.

Legal Advice to Hospitals

Legal advice to hospitals is difficult to document, but from some published statements of physicians and speeches given by hospital attorneys, it appears that many physicians have been led to believe that they can no longer write "do not resuscitate" (DNR) orders on patients without going to court. Physicians should be aware that what their lawyers are telling them is that, if they want to discontinue treatment of a patient "and want to have a court-certified statement that they will be civilly and criminally immune for such action," then and only then must they go to court.

*After an additional case was brought, the SJC ordered laetrile discontinued as well. This caused the parents to flee the jurisdiction, and Chad died in Mexico shortly thereafter.

Since there has *never* been a criminal indictment for discontinuing treatment of a terminally ill adult, and since, under these circumstances, the only important question is prognosis and feasible treatment, the risk of prosecution is almost nonexistent and the principle questions are medical in nature. Accordingly, I would argue that it is the physician's duty as a professional to make such determinations (but only with consent, if the patient is competent) and act on them without seeking immunity. The proper procedure in these cases is well set out in the medical literature in an article cited by the court.[6]

On the other hand, discontinuing treatment of a handicapped newborn or a mentally retarded adult, not because further treatment is futile and the case hopeless, but because the physician and family do not believe that the patient's "quality of life" justifies further treatment, is *not* a medical decision at all. Such a case raises the type of questions that should be resolved in a court of law, and not by private and unreviewable decisionmaking.

The tragedy is not that such cases will go to court, but that physicians and hospital administrators are responding to families of terminally ill patients for whom there is *no hope* by saying that they cannot be put on "no code" status without a court order. Families are thus forced either to hire an attorney and go to court to ask that the doctors be granted immunity for withholding useless care, or to watch their relative continually and fruitlessly resuscitated. The family's (and patient's) suffering in this instance is tragic and needless.

Hostility from the Medical Profession

Not atypical of many physicians in Massachusetts are the remarks of two as quoted in the *Boston Globe*:[7]

> I was flabbergasted when I first heard about it. The court seems to be saying that a judicial body is in a better position to make life-and-death decisions than doctors are. The doctors I've talked to just think it's awful...patently ridiculous.

> *Physician One*

> It's an impossible burden on us. We're resuscitating the 85 year old
> with brain hemorrhages, the patients dying of metastatic cancer. We
> often have no machines left for young people.
>
> *Physician Two*

Nor is discontent limited to practitioners. In an editorial in the *New England Journal of Medicine*, the editor, Arnold Relman, voiced his dismay at what he views as an insult to the medical profession. Noting that the court leaves "no possible doubt of its total distrust of physicians," he continues: "This astonishing opinion can only be viewed as a resounding vote of 'no confidence' in the ability of physicians and families to act in the best interests of the incapable patient suffering from a terminal illness." For the judges, whom he assumes are simply uninformed, he suggests "a guided visit to a large acute-care hospital particularly to pediatric and adult intensive care units...."[8]

While the hostility is real, it is based on a complete misunderstanding of the case, a misunderstanding for which the legal community in Massachusetts must take the blame. Insofar as the hostility represents resentment of the court's conclusion that certain types of decisions (that is, nonmedical decisions) should not be made by physicians, little sympathy is due. One wonders what these physicians thought the law was before *Saikewicz*, and on what basis they had been making decisions to terminate treatment.

Legislative Immunity

To "correct" the decision, the Massachusetts Medical Society and the Massachusetts Hospital Association would like to see some private mechanism established that would grant physicians immunity from prosecution and civil suit without requiring them to go to court. In this respect they are following the logic of "Good Samaritan" statutes, which physicians of the 1960s insisted were necessary because, without a prior guarantee of "immunity," they would not render even necessary emergency treatment.

The medical community is arguing that, rather than take an almost nonexistent risk for terminating treatment on a terminally ill incompetent patient for whom there is no hope of a cure, they will actively and uselessly continue fruitless, painful, and extraordinary treatment unless they are granted legal immunity for terminating it. What an incredible self-indictment of the profession! I am aware of almost no critic of the medical profession who would make such a sweeping condemnation of the current state of medical ethics and personal responsibility in medicine as the Massachusetts medical community is making of itself. Policemen are not immune if they kill someone in the line of duty; nor are soldiers who argue they were "just obeying orders," or airline pilots who crash their planes, or any other professional whose job involves the possibility of a negligent homicide. Physicians have traditionally argued that they are special because they routinely deal with "life and death." The modern argument seems to be: "We are still special, but we won't make any medical decisions that involve life and death unless we are guaranteed immunity for those decisions." Should we, as a society, give unreviewable life-and-death decisionmaking authority to a group of individuals who are afraid to take responsibility for their decisions?

I agree with Professor Robert Burt of Yale Law School who has written regarding neonatal decisions:

> Because these decisions, dispensing life and by necessary implication dispensing death, press against our most basic communal identities, I think it proper that society impose an extraordinary burden of care-taking on these physicians. The possibility of criminal liability should force these physicians to give of themselves, to identify both with the family and with the newborn child as if the suffering of each were the physician's own. Professional norms demand that all physicians care for their patients in this emphatic way.[9]

The *Saikewicz* case has now become almost as important for the ways in which the medical and legal professions have reacted to it as for what the court said. The issues discussed here are far from resolved, and will be addressed in professional journals and meetings, as well as public forums, for some time to come. Through this open process, I hope that consensus will be reached. It is, however, unlikely that "no-fault death" will be seen by many as a proper resolution in the long run.[*]

---

[*]The immunity statute died a relatively quick death in the Massachusetts Legislature. The next case considered by the SJC, the *Earle Spring* case (discussed in "Quality of Life in the Courts" at p. 285) underlined the points made in this essay, and further limited potential criminal liability on the part of physicians to acts "grievously unreasonable by medical standards."

Brother Fox and the
Dance of Death

When a judge decides to play legislator on issues like the termination of treatment in which there are strong and competing social values and case law is embryonic, the result is likely to be very unsatisfactory. Nevertheless, judges do not seem able to resist the temptation to make final pronouncements in this intoxicating area of the law.

The Brother Fox case shows what happens when judges attempt not only to decide the specific case before them, but also to decide how all future cases like it should be dealt with. And even though the Brother Fox case hasn't provided "final answers," its seventy-three pages do provide a fruitful field on which to begin discussion. The decision of the intermediate appeals court[10] is a philosopher's dream, and should find its greatest use as a discussion-provoker, much like its predecessors: *Quinlan* and *Saikewicz*.

Brother Fox's Story

Brother Joseph Charles Fox joined the Catholic Society of Mary in 1912 when he was 16 years old. At the age of fifty-seven he met Father Philip Eichner and remained his close friend for the remainder of his life. In August 1979, at the age of eighty-three, he sustained an inguinal hernia while gardening. During otherwise uneventful corrective surgery, Brother Fox suffered a heart attack. He sustained substantial brain damage, slipped into a coma, and was placed on a respirator in an intensive care unit.

Father Eichner called in two neurosurgeons to examine Brother Fox. Both agreed that there was no reasonable possibility he would ever regain consciousness. Father Eichner accordingly asked the hospital authorities to discontinue the respirator. Had they agreed, the

273

case would have ended at this point. But they refused, saying they would comply only if a court order was obtained. Father Eichner then went to court to seek appointment as Brother Fox's committee (New York's term for a guardian) with authority to discontinue life-support systems.

At trial Father Eichner testified that Brother Fox had discussed both the *Quinlan* case and Pope Pius XII's *allocutio* with him, and had stated that he wanted no "extraordinary means" used to sustain him if he were ever in a condition similar to that of Karen Quinlan. The surgeon who operated on him testified that Brother Fox was in an "irreversible" and "permanent" coma, and "would remain a vegetable." A neurosurgeon testified similarly. The District Attorney, who was invited to intervene by the Court, called on two physicians who testified that it was possible that his condition had not "absolutely stabilized." One case in the medical literature reported a patient recovering from a similar condition.

The trial court concluded from the testimony that there was "no reasonable possibility" that Brother Fox would ever return to a "cognitive and sapient" state; and that were he himself competent, he would "order a termination of the life-supporting respirator." On December 6, 1979, it accordingly granted the petition. The District Attorney appealed. On January 24, 1980, while the panel was considering the case, Brother Fox died.

The appeals court, which issued its ruling on March 27, began by reaffirming "the undeniable right of a terminally ill but competent individual to refuse medical care, even if it will inexorably result in his death." The problem, as in *Quinlan*, was how to make decisions for an "incompetent" patient in a chronic, persistent vegetative state. The court saw three issues: medical, legal, and procedural. The medical issue is prognosis, and the court determined that the medical prognosis should be proven "in court" using the "clear and convincing" standard of proof.

The legal issue is substituted judgment, and the court insisted that life support should not be discontinued, even on patients in a chronic vegetative state, *unless* it could be demonstrated that they themselves would make such a decision if they could. On this issue the court opined that a "living will" executed at a time "when the patient contemplated the catastrophic medical possibility that actually befell him" would make the decision easy because all could be sure they were just "carrying out the stated wishes of the patient and therefore, face no moral dilemma." The court further found that Brother Fox's express statements were "highly probative of his choice" and "entitled to great weight." It accordingly affirmed the lower court's decision. Instead of concluding the opinion at this point, however, the court decided to set up a procedure for deciding all future cases involving individuals in chronic vegetative states. The court determined that before life-sustaining treatment could be removed, "both" the medical prognosis test and the legal substituted judgment test must be met, and these tests "must" be reviewed by a court "on an 'individual,' patient-to-patient basis," similar to the way that requests to transfuse blood to a child against the wishes of his parents routinely go to court.

The Brother Fox Procedure

1. The physicians attending the patient *must* first certify that he is (1) terminally ill; and (2) in an irreversible, permanent, or chronic vegetative coma; and (3) that the prospects of his regaining cognitive brain function are extremely remote.

2. Thereafter, the person to whom such a certification is made (family member, friend, or hospital official) *may* present this prognosis to an appropriate hospital "prognosis committee" of no fewer than three physicians with specialists relevant to the patient's case.

3. The committee *shall* either confirm or reject the prognosis by majority vote of the members.

4. Upon confirmation of the prognosis, the person who secured it *may* commence a court action for appointment as the committee of the incompetent, and for permission to have life-sustaining measures withdrawn.

5. The attorney general and appropriate district attorney *shall* be notified and given an opportunity to intervene and arrange examinations by their own physicians if they so desire.

6. A guardian *ad litem* shall be appointed to represent the best interests of the patient.

7. The court *shall* determine that the prognosis is accurate by "clear and convincing" evidence.

8. The court *shall* determine that the patient would decide to terminate extraordinary life-support measures if he were able to make this decision himself (apparently by a preponderance of the evidence).

9. An appropriate order for the discontinuation of extraordinary measures *shall* be entered and no participant shall be subject to

criminal or civil liability in the event death follows the termination of these measures.

The court acknowledges that such a procedure "may appear cumbersome and too time consuming to accommodate the need for speedy determinations in cases where termination of treatment is proposed." Nevertheless, the court believes such procedures are necessary to protect the terminally ill patient involved. The mixing of "musts" and "mays" will confuse many, but the court seems to be saying that the procedure is required *only if* one wants to terminate treatment, and not that the procedure itself is optional (even though there is no crime of "not going to court" and no new sanctions).

The Price of Progress

It is important to stress the positive in novel areas of the law where courts are attempting to reconcile conflicting values. Tragic choices will never be happily made. The decision can be seen as a step forward: first, hospital officials would not have permitted the removal of life-support systems without it. There is now a mechanism to implement the wishes of a patient in a condition like Brother Fox's. Second, the committee used to confirm the prognosis is now clearly termed a "prognosis committee" (and not an "ethics committee" as in *Quinlan*) and is composed entirely of physicians. Third, a clearer-than-usual distinction is drawn between medical prognosis questions and legal questions of substituted judgment. Fourth, a written "living will," and prior oral statements of preference, are given strong judicial weight. And fifth, the decision is a very narrow one, applying only to patients like Brother Fox (and Karen Quinlan) in a "permanent or chronic vegetative coma" from whom "extraordinary" means of care can be removed.

Unfortunately, these advances are purchased at a very high price. Questions of medical prognosis and substituted judgment are treated as if they are fundamentally different. While both arguably mix fact and value positions, it seems appropriate to permit physicians to make decisions based primarily on scientific fact as long as the "criteria" on which these decisions are made are socially approved. For example, we routinely permit physicians to pronounce people dead.* Simi-

*See discussion in "Defining Death: There Ought to be a Law" at p. 276.

larly, the *Quinlan* court essentially permitted families and physicians to apply a narrow prognosis criteria without court intervention. I think this is sound, and as long as society finds prognosis a relevant variable, we should permit physicians to determine it themselves.

The will of the patient is something quite different. The court acknowledges that the patient is the best source of information, and the "best solution" may be to get all competent individuals to express their preferences regarding treatment in such circumstances. Without this, we are left to the speculation of the patient's family and friends. It seems appropriate that this speculation be subjected to court review in cases where there are real choices to be made concerning the patient. On the other hand, judicial review is unnecessary when, according to medical criteria, the patient is terminally ill, and in an "irreversible, permanent or chronic vegetative coma" from which recovery is "extremely remote."

In such cases I would permit the patient's legal guardian to decide the issue without further court review (requiring the guardian to use the substituted judgment standard if possible, and the best interests standard if he were unable to determine the patient's actual preferences). It is appropriate for the court to set the ground rules, but it does not seem either desirable or necessary for the court to routinely review medical prognosis before the fact. And in the narrow medical confines of this case, it does not seem to add sufficient protection for the patient to require review of the guardian's determination either. On the other hand, where, as in cases like *Saikewicz*, there is a real choice between painful treatment with a chance of remission, and no treatment with a "peaceful" death, court review seems perfectly appropriate.

The reintroduction of the notion of "extraordinary" technologies is also unfortunate. The concept surfaces here only because Brother Fox, a devout Catholic, discussed *his own views* on treatment in the terms utilized by Pope Pius XII. But many will overlook this point, and assume that courts are willing to make a legally meaningful distinction between ordinary and extraordinary. In this case the court had to use this language in order to carry out Brother Fox's desires: but unless adopted by the patient himself, such a distinction is legally irrelevant.

Philosophers will likely enjoy an examination of some of the court's language about patients in a permanent vegetative state, for example, the possible call for a new definition of death: "As a matter of established fact, such a patient has *no* health and, in a true sense, no life, for the State to protect." The court follows this by calling up the specter of a slippery slope: "The State's interest in preservation of the life of the fetus would appear to be *greater* than any possible interest the State may have in maintaining the continued life of a terminally ill co-matose patient ... [their] claim to personhood is certainly no greater than that of the fetus." These statements are simply wrong as a matter of law: the fetus is *not* a person for the purposes of the United States Constitution; the person in a coma, permanent or not, is. The court's own opinion acknowledges this implicitly by setting up a much more elaborate (although not constitutionally mandated) scheme to termi-nate the life of a comatose person than is needed to terminate the life of a fetus.

The Brother Fox case is under appeal, and New York's highest court, the Court of Appeals,* has the opportunity to take a giant step forward by taking a step back from the intermediate appeals court decision. Judges should permit the law to develop on a case-by-case basis in this area. This case is like the *Quinlan* case, and should not be confused and conflated with *Saikewicz*. By deciding it on its facts and leaving it at that, the Court of Appeals will do us all a service.

---

*The Court of Appeals opinion is dealt with in the following essay, "Help from the Dead."

Help From the Dead

The Cases of Brother Fox and John Storar

If a "human rights prize" was awarded to those who have done the most to advance the rights of all of us to refuse life-sustaining medical treatment, that prize could reasonably go to parties in some of the most important litigation in this area: Karen Ann Quinlan, Joseph Saikewicz, Earle Spring, and now, Brother Joseph Fox and John Storar. None benefited personally from the state supreme court opinions that bear their names. Karen Quinlan continues to live long after the court approved removal of her respirator.* Both Joseph Saikewicz and Earle Spring were dead before the courts articulated their right to refuse treatment. And now the New York Court of Appeals has ruled that were Brother Fox alive today, his respirator could be removed; and if John Storar were alive today, his blood transfusions would have to be continued.[11] The fact that each was dead before the high court ruled need not terribly concern the courts, since these cases serve as an occasion to articulate legal principles that affect us all.

Brother Fox

Brother Fox, an 83-year-old member of the Society of Mary, was being maintained on a respirator in a permanent vegetative state. He had previously expressed orally a desire not to be maintained by "extraordinary means" (following the teachings of Pope Pius XII) if he were in a situation similar to that of Karen Ann Quinlan.

The New York Court of Appeals concluded that Brother Fox had a common law right to refuse treatment, and that such a refusal survives the incompetency of an individual if there is "clear and convincing evidence" of the refusal. In Brother Fox's case the court concluded

---

* She eventually died in 1985.

that his oral declarations were "solemn pronouncements" and not casual remarks. Finding that Brother Fox had personally refused treatment, the court found it unnecessary to take up either the proxy issue or the constitutional right of privacy. The court concluded simply that a competent adult could refuse treatment and that his wishes must be respected: "Even in emergencies it is held that consent will not be implied if the patient has previously stated that he would not consent."

Substantively, the court reaffirms the right of a competent person to refuse treatment, and extends this right by giving legal weight to declarations that the person would refuse treatment if he or she were to become incompetent. The only issue is the expressed desire of the formerly competent patient. Since an oral declaration on this matter is sufficient, a written declaration, such as a living will, is also sufficient. The court rejects the requirement of a complex court review mandated by the lower court before such a decision can be implemented. But it is willing to hear cases in which parties are uncertain of the desires of the patient: "Responsible parties who wish to comply with the law...need not act at their peril."

John Storar

The Brother Fox case was consolidated with a quite different case, and they were decided together. Though Fox was a "*Quinlan*-type case,"* John Storar was a "*Saikewicz*-type case,"** and therefore presented much more complex issues. John Storar was a profoundly retarded 52-year-old resident of a state facility who had a mental age of approximately eighteen months. His closest relative was his mother, a 77-year-old widow who resided near the facility and visited him almost daily. In July, 1979, he was diagnosed as having cancer of the bladder, and his mother was appointed his legal guardian to consent to radiation therapy, which produced a remission. Internal bleeding began again in March, 1980, and his bladder was cauterized in an unsuccessful attempt to stop the bleeding. At this point the cancer had metastasized to his lungs and he was considered terminal and inoperable. In May, the physicians asked his mother for permission to administer blood transfusions. She reluctantly agreed, but in June requested that the transfusions, which were given approximately

---

* *Quinlan* is discussed at p. 261.
** *Saikewicz* is discussed at p. 244.

every two weeks, be discontinued. The director of the center where Storar lived then brought a petition before the court seeking authorization to continue the blood transfusions. The mother opposed the petition. All the witnesses at the hearing agreed that Storar could not comprehend what was happening to him, that he had irreversible bladder cancer, and that even with the transfusions he could only live three to six months. Without them he would eventually bleed to death.

Storar found the transfusions disagreeable, and was distressed by the increased blood and clots in his urine after a transfusion. He had to be sedated and restrained at times before the transfusions, and received regular doses of painkillers.

Other experts testified that proper treatment would be only to administer painkillers. Mrs. Storar testified that she just wanted her son to be comfortable, and since he obviously disliked the transfusions and tried to avoid them, she believed that he himself would want them discontinued. The lower court held that under these circumstances Mr. Storar's right to refuse treatment could be exercised by his mother, because she was in the best position to determine what he would want. An intermediate appeals court affirmed.

But the New York Court of Appeals reversed the decision on the grounds that there was no realistic way to determine what John Storar himself would want done. Asking that question was, the court said, like asking, "If it snowed all summer, would it then be winter?" The court concluded that since John Storar was mentally an infant, he must be afforded the same rights as an infant. Since parents (even Jehovah's Witnesses) cannot refuse life-saving blood transfusions for their children, the court decided that Mrs. Storar could not refuse to have blood transfusions administered to her son.

Though the court did not explain its reasoning, it apparently viewed its mission as protecting the rights of the mentally retarded to receive treatment. But what explanation it did give only serves to call the conclusion into question. The court, for example, characterized Mr. Storar as suffering from *two* threats to his life: bladder cancer and loss of blood. The first was incurable, the second treatable. Making an analogy from blood to food, the court concluded that blood must be used to prevent death from the second cause: "A court should not in the circumstances of this case allow an incompetent patient to bleed to death because someone, even someone as close as a parent or sibling, feels that this is best for one with an incurable disease."

Discussion

Both parts of this decision are important; *Eichner* (Fox) because it is clear and well-reasoned; *Storar* because it is cryptic and contains no analysis worthy of the name. *Eichner* states that prior directives of formerly competent patients can be given legal effect with almost no fear of criminal or civil liability, and that prior court review of the directive is not required. Hospitals should adopt procedures that will help physicians determine whether prior written or oral directives were actually made. Some questions that should be considered are: the circumstances of the declaration, whether it was written and witnessed (although neither is required, each increases the likelihood that the declarant was serious in his intention), whether it was consistent with the patient's value system, and whether there are any conflicting family interests. If the declaration is found to be genuine, it should (and I would argue it *must*) be given effect.

On the other hand, while Brother Fox, like *Quinlan*, involved a patient in a persistent vegetative state, the case does not tell us how the New York courts should deal with the actual *Quinlan* case, since Karen Ann Quinlan never made any declaration concerning her own future desires sufficient to convince a court. Since the New York court found it unnecessary to address this issue, a real question remains. Would the New York courts treat a Karen Quinlan like a Brother Fox (by letting a proxy indicate her probable wishes) or like a John Storar (by requiring continued treatment)?

Storar is much less satisfying because it consists primarily of conclusory statements rather than legal analysis. The case seems to uphold the principle that children and mentally retarded individuals must *always* be given medical treatment that can preserve their life, even for a limited time. Though superficially very appealing, that principle immediately raises a host of difficult questions. Must Tay-Sachs infants be continuously resuscitated if their hearts can be made to beat again, even for a limited period of time? Can orders not to resuscitate ever be written on any incompetent patient in the absence of a prior directive? How severe must pain and suffering be before life is not worth living; or can they never be judged that severe for a child, a mentally retarded person, or a currently incompetent person who has not previously expressed his or her wishes?

The court's heart was surely in the right place, but its mind, unfortunately, did not follow. It very properly wants to protect children and other incompetent persons from those who would deny them life-sustaining medical treatment for reasons other than the patient's best interest. But it fails to recognize that there may be times when such treatment only prolongs suffering and is itself cruel; and it fails to suggest any test that parents, families, or lower courts can apply to decide if it is ever legally permissible to withhold life-sustaining treatment from this group of patients.

This failure is especially distressing in light of recent important decisions in two adjacent jurisdictions, New Jersey (*Quinlan*) and Massachusetts (*Saikewicz*), that dealt with this very issue. Both these courts indicated that the proper manner in which to make such a decision was to try to put oneself in the position of the incompetent person, and do what he or she would do if able to make the decision. This "substituted judgment" standard strikes me as the proper one to apply when dealing with an individual who was previously competent and who therefore had the opportunity to make many decisions and develop a value system that a close friend or family member could reasonably understand.

But the court, as Paul Ramsey has properly noted, is quite correct in rejecting this type of analysis as "unrealistic" when applied to an individual like John Storar who never made any decisions and never evidenced much more than a desire to survive and avoid pain. What the court seems to ignore is that, by refusing to approve an alternative test, it effectively deprives incompetent persons of any "right to refuse treatment" and *forces* them to be treated under all circumstances. Surely this is too blunt a ruling. Traditional legal doctrine permits parents and guardians to make decisions consistent with the "best interests" of their wards. The court would have been on much firmer ground had it enunciated this doctrine, and, if it believed that abuses needed to be prevented, then gone on to specify the parameters of best interests as applied to seriously ill incompetent persons. Some of the factors that might be considered in such a best interest decision are prognosis, pain, discomfort, the degree to which pain and discomfort can be controlled, ability to communicate and experience pleasure, and alternative treatments.

It would be tragic if the court's ruling in *Storar* has an effect opposite to the one intended. An overly cautious reading of the case by those who fear legal liability could subject children and incompetents to treatments that are ineffective and prolong agony. Such a scenario was played out in Massachusetts after *Saikewicz*.* It can be

---

*See discussion in "No Fault Death: After *Saikewicz*" at p. 267.

prevented, however, if the case is read and interpreted very narrowly, as the court indicated it should be. The court treated John Storar "exactly" like a minor Jehovah's Witness who needed blood. And on this issue the court is unambiguous: under no circumstances can children be allowed to bleed to death. The analogy is to food, not to mechanical ventilators or major surgery. Even though this analysis is defective, the court should be taken at its word when it indicates that it will not set down broad principles to deal with all cases, but will decide only the issue before it. In view of all the legal and social implications of the opinion that the court did not discuss, this seems a reasonable reading. Guardians and parents should be permitted to continue to make decisions involving more painful and drastic treatment than blood transfusions by applying the traditional best interests test. In cases of doubt or conflict, the courts should be consulted. Only by bringing a diverse group of cases involving incompetent patients to the attention of the courts will substantive rules be developed that protect the lives of incompetent patients without depriving them of the opportunity to have treatment terminated when treatment is no longer in their own best interests.

Quality of Life in the Courts

Earle Spring in Fantasyland

The term "quality of life," like "right to life," means many things to many people. It can conjure up notions of genocide or of a good death; it can be favorably compared with quantity of life, and unfavorably contrasted to protecting the equality of life. Perhaps the term is, like euthanasia, so misused and misunderstood that it should simply be banned from our lexicon. But no matter how one comes out on this question, the issues of normalcy, social worth, and resource allocation are, overtly and covertly, playing an increasing role in court decisions regarding the care and treatment of various categories of patients. The latest, in a series that includes *Saikewicz, Quinlan, Becker*, and *Eichner*, is the case of Earle Spring, a senile patient suffering from chronic kidney failure in a nursing home.[1][2]

The case provides an opportunity to explore how courts, though they heroically attempt to avoid using the term, inevitably have to deal with the issues it represents. It is suggested that their reluctance to speak to such issues directly leads them, at times, to give untenable rationales for some of their decisions. This "journey to fantasyland" may be necessary for courts because their decisions are precedents for other people similarly situated, and they therefore do not have the luxury of making a decision that applies to only one person.

At the time his wife and son sought court approval for removal of Spring from kidney dialysis, he was 78 years old. After hearing their testimony and that of the kidney specialist, the probate court agreed that treatment could be discontinued. Complicated appeals followed for almost a year, and a final opinion by the Massachusetts Supreme Judicial Court (SJC) was not issued until May 13, 1980, about a month after Spring died.

The Court's Opinion

The court's opinion came as a major disappointment to many hospital administrators and their advisors who had hoped for a blueprint describing in detail when they did and did not have to seek court immunity. Instead the court restated its position in *Saikewicz*,* although in much clearer language: "Our opinions should not be taken to establish any requirement of prior judicial approval that would not otherwise exist."

The court also added generous language on criminal liability which should reassure Massachusetts doctors:

> Little need be said about criminal liability: there is precious little precedent, and what there is suggests that *the doctor will be protected if he acts on a good faith judgment that is not grievously unreasonable by medical standards.* (emphasis added)

Since these seem to have been the two issues that most upset and concerned the medical profession, clarification of these points must be viewed as a significant, positive development.

But the most disturbing aspect of the case is the court's loose language on the enormous issue of the "quality of life" of nursing home patients. The court found the conclusion that Spring "would, if competent, choose not to receive the life-prolonging treatment" was "not clearly erroneous." By so doing, it dismissed any reference to quality of life, simply saying, "The problem of impairment of 'quality of life' associated with Saikewicz's mental retardation has no analog in the present case." But the case seems to have been decided on quality of life considerations.

Spring himself, for example, had *never* stated any preferences regarding medical treatment he might require after he became incompetent, and the relevant evidence provided by the family members is virtually nonexistent. Most dealt with the fact that, while competent, Spring had led a vigorous, active life, which he was no longer able to do. But it is almost always true that activity declines as people age, and this alone does not mean that people want to cease living. The only specific opinion rendered came from his wife, who stated that, based on their long years of marriage, she believed "he wouldn't want to live." No evidence was offered as a basis for this conclusion.

* *See* discussion of *Saikewicz* at p. 244.

Since on this reading the decision cannot properly rest on Spring's own preferences, it must rest on some view (his family's or the court's) of what people like Spring want or "deserve." In this regard, the court took the view that Spring's kidney condition was "irreversible and incurable." But this is true of all patients who suffer from chronic kidney failure and are not candidates for a transplant. However, the advent of hemodialysis has made this condition controllable. An analogy can be drawn to diabetes, another irreversible and incurable disease, which is also controllable. If Spring were senile and a diabetic, should we permit the withholding of insulin merely because of the "irreversible and incurable" nature of this disease? If the answer is no, it seems to follow that a distinction between "lifesaving" and "life-prolonging" is also a meaningless one, since both dialysis and insulin could be viewed as either.

The Court's Approach to Senility

Even more telling is the court's approach to senility. The probate court's conclusion that Spring need not be continued on dialysis seems based largely on its view that senility is an "incurable, permanent, and irreversible illness," and its conclusion that no treatment could restore Spring to a "normal, cognitive, integrated, functioning existence." Unfortunately, the SJC adopted this language in describing Spring when it stated, "The treatment did not cause a remission of the disease or restore him even temporarily to a *normal, cognitive, integrated, functioning* existence, but simply kept him alive." (emphasis added)

The same statement, rephrased, reads, "If one is not and cannot be returned to a normal, cognitive, integrated, functioning existence, but can simply be kept alive, treatment can be stopped." Such a standard, if one can call it that, is so vague and sweeping as not to have any legal pedigree at all. For example, it encompasses almost all severely mentally retarded persons, and all senile demented persons, in or out of nursing homes. It argues that only life which is "normal" (however that is defined) and involves "integrated functioning" (however that is defined) is worthy of legal protection.

Phrased another way, there are some categories of people who are so abnormal or ill-functioning that the state has no interest in seeing to it that their lives are preserved. This is probably true, but such a class must surely be carefully, rigidly, and narrowly defined.

Quinlan and Saikewicz

A similar approach, but with a much more narrow definition, was enunciated in the *Quinlan* case.* There the court said, without ever once using the phrase "quality of life," that there is a certain class of patients who can be treated differently from others. Specifically, if the guardian, family, and physician of a patient, and a hospital ethics committee all agree that there is "no reasonable possibility of returning to a cognitive, sapient state," then life-support measures may be withdrawn with legal immunity. This carefully circumscribed test can be viewed like a judicial expansion of a brain death criteria, in the sense that if the patient meets the court-defined criteria, there is no legal requirement to continue treatment. *Quinlan* can thus be viewed as the first full-fledged quality-of-life-decision (in the sense that it defined a category of patients who need not be treated).

Saikewicz specifically declined to directly define the severely mentally retarded as a class that did not have to be treated with cancer chemotherapy. On the other hand, by adopting the substituted judgment test for a group of patients (that is, the severely retarded) to whom it can never apply, to protect their autonomy (something they never had), the court actually reached the same conclusion: severely mentally retarded patients do not have to be provided with cancer chemotherapy as long as the therapy is painful or distressing.

The court forced itself to use a criterion that could not be applicable to Saikewicz himself because of its fear of explicitly adopting a quality-of-life standard, which "demeans the value of the life of one who is mentally retarded." The SJC described the term "quality of life," which had been used by the lower court judge in *Saikewicz*, as "vague and perhaps ill-chosen" and stated "to the extent that this formulation equated the value of life with any measure of the quality of life, we firmly reject it." However, the SJC *did* accept the concept as applied to an *individual*. Specifically, it could be taken into account by Saikewicz himself (read, by the judges) "as a reference to the continuing state of pain and disorientation precipitated by the chemotherapy treatment."

And, though in *Spring* the same SJC argues that it is not defining the senile patients who live in our nation's nursing homes as a class who do not deserve expensive medical treatment, this in fact seems to be what is at work behind the scenes. Though discounted by the court,

* The *Quinlan* case is discussed at p. 261.

the physician involved focused his own testimony on the quality-of-life issue. He said that he himself decides whether or not to discontinue treatment on the basis of "whether a person is a *real person, whether the person is happy to be alive, whether other people around him or her are happy to have him alive.*" (emphasis added)

The quality-of-life issue was also explicity stated by the Appeals Court: "To what extent should aggressive medical treatment be administered to preserve life after life itself, for reasons beyond anyone's control, has become irreversibly burdensome?"[13] The question, of course, is "burdensome to whom?" Certainly Spring's senile condition was troublesome, and perhaps even burdensome, to his family, physician, and the nursing home personnel. But the issue is, does senility alone make a person's life so burdensome that medical treatment can justifiably be withheld from him? The *Spring* decision can be read as an indication that the SJC's answer to this question may be affirmative.

But perhaps this is the best we can do. Perhaps openly acknowledging that we are willing to treat the retarded or the senile differently from the "normal" patient is so offensive to society's view of the equality of citizens that its explicit declaration is impossible. If this is true, and if we still want to be able to make a decision not to treat a senile or retarded person in the same way as a "normal" person, then we do need "make-believe" reasons that we can feel comfortable with. Substituted judgment seems to be one such invention from Fantasyland. Whether our fairy tale experience with it will have a happy ending, however, is far from certain. What seems more certain is that, at least on a macro level, society is likely to enter Tomorrowland with much more explicit decisionmaking regarding resource allocation. In the words of a dissenting Justice in a hypothetical case set in the year 2002:

The energies of the National Health Agency should be directed toward the young and the middle-aged and toward making life more enjoyable and richer. It should not be directed toward prolonging the agony of death and miseries of old age...we should allocate resources toward medical and health measures that make our lives worth living, rather than those that prolong lives that are not worth living.[14]

When Suicide Prevention Becomes Brutality

The Case of Elizabeth Bouvia

Elizabeth Bouvia challenges us all. This articulate, 26-year-old woman's desire to starve herself to death in a hospital enkindles strong emotions. The press speaks about "suicide appeals" and "right-to-die appeals"; the physicians and the hospital speak of their legal liability and "medical ethics"; and Ms. Bouvia speaks of her right to privacy.

Society is simply no match for this woman, and we know it. Unable to respond with either reason or compassion, we resort to slogans. Perhaps the courts cannot help her. Perhaps we cannot do better than respond with the full force of the state while claiming to be on the side of "life." The stakes are high for either publicly condemning or approving Bouvia's actions, but there is a middle road that offers both her and us a safer journey.

Elizabeth Bouvia's cerebral palsy has left her with virtually no motor function in any of her limbs or other skeletal muscles. She retains some limited control over the movement of her right hand (sufficient to operate an electrically powered wheelchair), and enough control over her facial muscles to eat when someone else feeds her and to speak. She has completed a BS degree in social work, has been married for more than a year, has attempted to have a child, and has lived independently with the help of her relatives or a personal care attendant. But in 1983 her husband left her, she dropped out of school, and she lost her state assistance for transportation. Realizing that she would never be able to find employment, she concluded that she could never live without almost total reliance on someone else.

On September 3, 1983, she asked her father to drive her from Oregon to the Riverside County General Hospital (she had previously lived in Riverside, was a California resident, and had been a Medi-Cal recipient) where she arranged for a voluntary psychiatric admission on the grounds that she was suicidal. Her plan was to get admitted to a place where she would "just be left alone and not bothered by friends or family or anyone else and to ultimately starve to death." She had allegedly attempted suicide on at least one, and possibly more, previous occasions. At Riverside she refused to eat solid food. After her attending physician threatened to have her certified as mentally ill and dangerous to herself so she could be force-fed, she phoned the local newspapers in an attempt to get legal assistance.

Shortly after, the American Civil Liberties Union entered the case, with doctor-lawyer Richard Scott in charge. He persuaded her to continue to take nourishment while he applied for a court order restraining the hospital from either discharging her or force-feeding her. A hearing on these requested orders was held in December.

At the hearing, Bouvia testified that she was no longer willing to live completely dependent upon other people: "I hate to have someone care for every personal need...it's humiliating. It's disgusting, and I choose to no longer do that, no longer to be dependent on someone to take care of me in that manner....I am choosing this course of action due to my physical limitation and disability."

The chief of psychiatry at the hospital, Donald E. Fisher, testified that he would force-feed her with a nasogastric tube even if the court ordered him not to. At closing arguments, attorney Scott argued that her decision was "exactly medically and morally analogous to the patient deciding to forego further kidney dialysis." The county argued, on the other hand, that the question was whether an individual had a right to commit suicide in a county hospital, with "the forebearance of the medical personnel."

The Decision for Force-Feeding

Judge John H. Hews announced his decision the morning after closing arguments.[15] He essentially accepted the county's presentation of the case and, as Scott has argued, "bought the Chicken Little defense" that the moral sky would fall in on Riverside Hospital if they did not force-feed Bouvia. Specifically, the judge declared that Elizabeth Bouvia was fully competent, and that her decision was "rational" and "sincere" and stemmed from her physical disability and dependence on others rather than from recent experiences.

He believed, however, that permitting her to starve to death in Riverside Hospital would "have a profound effect on the medical staff, nurses and administration" of the hospital, and "would have a devastating effect on other patients within Riverside Hospital and other physically handicapped persons who are similarly situated in this nation." Thus the judge determined that the decision should be based on weighing Bouvia's right of privacy to end her life against the feelings of other members of society who might be offended by the act.

To strike the balance in favor of unnamed others, the judge found that the decisive point was that Bouvia was "not terminal," but "had a life expectancy of 15 to 20 years." He concluded, "The established ethics of the medical profession clearly outweigh and overcome her own rights of self-determination." Therefore, "forced feeding, however invasive, would be administered for the purpose of saving the life of an otherwise non-terminal patient and should be permitted. *There is no other reasonable option.*" (emphasis added)

Reasonable Alternatives?

The judge's decision begs the question: Is there a reasonable option? In the adversary proceeding played out in California no one seemed to search for reasonable options. The county, in fact, consistently took the most extreme position. It continually threatened to eject Bouvia from the hospital by force, and leave her out on the front sidewalk, hoping someone would pick her up and take her away. Almost from the beginning the county and hospital made it clear that they did not care whether she lived or died but, because of their own fear of potential legal liability, would not let her die at Riverside General.

Attorney Scott made an early offer to try to reach an accommodation before the case became politicized and publicized, but the county was not interested. Nonetheless, his position in court was also extreme: surely this case is *not* like the kidney dialysis cases. There is no way a person in kidney failure can survive without medical intervention; Bouvia can survive simply by eating.

Moreover, refusing treatment is not the same as demanding specific services, such as the administration of pain killers, that physicians might properly consider mistreatment. The law is (and should remain) clear on the right of an individual to refuse medical intervention; the "right to die" is much more problematic and does not have the legal pedigree associated with the more narrow right to refuse treatment. Bouvia did not seek to refuse medical interventions she found personally intolerable; she sought to end her life because she found living intolerable.

Although this case is different from others, there are applicable similarities and lessons. The first is that the often-quoted "terminally ill" qualification for refusing treatment makes no logical or moral sense. Courts have not previously required individuals to be terminally ill before they were permitted to refuse treatment, and should not do so now. Karen Ann Quinlan, for example, was never "terminally ill" and could well live the fifteen to twenty years Judge Hews found determinative in this case.* Likewise, Jehovah's Witnesses who refuse blood transfusions are not terminally ill, nor are patients who refuse kidney dialysis, or those who refuse to have their gangrenous feet or legs amputated. They are likely to die of their illness only if they refuse blood, dialysis, or surgery. Thus, the patient's right to refuse medical intervention, not the terminal prognosis, is and should be the focal point.

Second, some helpful principles could be applied to confine the issue more narrowly. Two seem especially appropriate: the physician may not abandon a patient in need of care, but the physician may arrange transfer if unable, because of moral conviction or lack of skill, to care for the patient; and though the patient has a right to refuse treatment, the patient does not have a right to demand mistreatment.

Once Bouvia was properly accepted for treatment (as all admit she was) the physicians and hospital had a duty to continue to treat her with her consent until she discharged them, no longer required treatment, or was properly transferred to other appropriate caregivers. Since she did not discharge her physicians and continued to need care, the primary issue in the case is transfer. Apparently steps were taken to attempt to transfer Bouvia to another institution, but these were unsuccessful for a variety of reasons, including society's failure to provide reasonable alternative settings for patients like Bouvia.

Under these circumstances, what medical interventions can be forced on an unconsenting, competent adult, when that adult is properly in the care of a physician who believes he has a moral obligation to force treatment on the patient, and who has been unable in good faith to transfer the patient to another physician or appropriate institution? I believe the answer should be "none." Competent adults should have the right to refuse any medical interventions. Their phys-

---

*Karen died in 1985, approximately ten years after her ventilator was removed. See discussion of the *Quinlan* case at p. 261.

icians have a legal and moral duty to honor their refusals, because the physical and psychological impact of the physical invasion on the patient is greater than the psychological impact on the doctor, and because of the fiduciary qualities inherent in the doctor–patient relationship.

Five state interests have been suggested that might dictate forced treatment, but only one of them has any applicability here. The five are: value of life, protection of innocent third parties, upholding medical ethics, protecting the prison system, and the prevention of suicide. The value of life is, of course, enhanced by liberty and "lessened not by a decision to refuse treatment, but by the failure to allow a competent human being the right of choice."[16] The protection of "innocents," usually referring to minor children, is here taken to mean other patients and the handicapped in general. This seems far too broad, and other competent handicapped persons are best protected by respecting their individual autonomy.

As to upholding medical ethics, even if Fisher was correct (and there were impressive medical witnesses who testified that following Bouvia's wishes would be consistent with medical ethics), "the right to bodily integrity and control of one's fate are superior to institutional considerations."[17] The prison issue is obviously inapplicable here.* Suicide is the only state interest left.

Suicide—Prevention or Brutality?

Suicide was therefore properly the focus of the *Bouvia* case, but not in the way it was addressed. It is inadequate to dismiss Bouvia's self-destructive desire as suicide and thereby remove her right to refuse medical intervention. If we simply accept her decision, it is argued, we devalue all severely handicapped persons. But surely there is a middle ground between "simply accepting" her decision, and violently force-feeding her.

Attempted suicide is not a crime, but we do not condemn individuals who thwart suicide attempts by dragging people off ledges, preventing them from jumping off bridges, or rescuing them from drowning. There seems to be a societal presumption that individuals who attempt to kill themselves must be incompetent. Since one cannot

---

* See discussion in "Prison Hunger Strikes: Why Motive Matters" at p. 363.

properly assess competence on a ledge or on a bridge, reasonable steps to prevent a public suicide seem justified.

Similarly, it seems perfectly reasonable for health professionals at a psychiatric unit of a general hospital to try to prevent the suicide of a young woman who asks to be admitted because she is contemplating suicide. Reasonable steps can and should be taken to assess her competence and dissuade her from self-destruction. But where do reasonable and required steps end, and unreasonable and unconscionable actions begin?

Bouvia's competence has not been challenged. Donald Fisher, however, has indicated that he will not seriously consider her request to starve to death until she has at least a six-month history of virtual freedom from trouble. This is unreasonable because it sets an impossible condition. Surely no one would argue that force-feeding her for the rest of her life, or even for a period of years, is anything but horrible.

Hospitalization for one to two days had served to dissuade Bouvia from continuing previous alleged suicide attempts. The only asserted rationale for force-feeding her now is that she will "change her mind." This is, of course, possible. The question is, how long is it reasonable to force-feed a competent adult with a nasogastric tube in the hope that the person will decide to eat voluntarily?

I do not believe competent adults should ever be force-fed; but efforts at persuading the individual to change his or her mind, and offering oral nutrition should continue. If a court determines, however, that invasive force-feeding is required "in the hope that [she] will realize that there is hope in life and that now because of the action taken by her she can be a symbol of hope to others similarly situated if she changes her purpose," then to avoid permitting hospitals to become hideous torture chambers, some reasonable limit must be placed on this "treatment."

California law for the mentally ill limits involuntary hospitalization for suicide attempts to thirty days; certainly Bouvia should not be required to endure force-feeding longer than this. After that period of time (and a week seems more reasonable) what is at stake is not medical care and treatment, but a contest of wills and power. But the match is so one-sided, with the full weight of the medical establishment and state brought to bear on an almost completely paralyzed young woman, that continuing it brutalizes us all. Four or more "attendants" wrestle her from her bed in the morning and restrain her while a nasogastric tube is rudely forced through her nose and into her stomach.

Is such brutal behavior required by medical ethics? Where are nursing and medical students schooled in the martial arts of restraint, forced treatment, intimidation, and violence? If we will not refrain from force-feeding Bouvia because we do not believe she has a "right to die," we should refrain from that action because it perverts the very meaning of care and treatment. Medical care must be consensual or it loses its legitimacy.

We can all grieve for Bouvia and express sorrow at her plight without adding further violent injury to her. As her attorney put it in his summation:

> We do believe that this world will be a poorer place without that tortured but resilient and cheerful lady with so much potential, with so much to offer others. But Elizabeth Bouvia has no duty to make that offer nor does she have the obligation to endure what to her is unendurable.*

---

*Elizabeth Bouvia's story is continued in the following essay.

Elizabeth Bouvia

Whose Space Is This Anyway?

After losing both in the hospital and in the courtroom, Elizabeth Bouvia fled to Mexico on April 7, 1984, to seek her death. She was soon persuaded that Mexican physicians and nurses would be no more sympathetic to her plan than those at Riverside, and so returned to California. Because of the brutal force feeding she had endured at Riverside, she was afraid to return there. Since no other facility would admit her unless she agreed to eat, she resigned herself to eating and entered a private care location. There she remained without incident, for more than a year.

On September 22, 1985, she was transferred to LA County–USC Medical Center where a "morphine pump" was installed for pain control. She was a model patient, and her stay there was uneventful. Two days before Christmas she was transferred again. The new facility was named High Desert Hospital, but the attitudes and actions of the staff were almost identical to those she had experienced at Riverside. Both medicine and law are awkward with Bouvia, but medicine has treated her more humanely. During the past two years she has been a patient in four US health care facilities. Two let conflicts degenerate, and ended up forcing their wills on her. But the other two had the flexibility to care for her in a reasonable and compassionate manner. Both lower courts that have reviewed her case to date have abandoned her. Her Riverside trial, where she claimed a right to commit suicide by starvation in a hospital, presented a difficult judgment. But the decision involving forced treatment of Bouvia at High Desert Hospital despite her voluntarily eating is legally vacuous, and encourages the medical staff to treat Bouvia as their pet, rather than as a citizen with constitutional rights.

The "Riverside Hospital Syndrome"

Almost from the day she arrived at High Desert Hospital, Elizabeth Bouvia was labeled a problem patient whose will and spirit had to be broken. Her physicians not only adopted the "Riverside Hospital Syndrome" by insisting that she be force-fed if they thought it necessary; they also arrogantly pursued a fantasy "right to rehabilitate" Bouvia to a physical and mental state that the doctors considered "ideal" without regard to her own wishes.

The medical director, for example, determined that her "ideal weight" would be between 104–114 pounds (for her 5 foot height); since she weighed only 70 pounds, he determined she should be force fed until she reached her "ideal weight" or voluntarily consumed at least 1400 calories daily (compared with the 500–700 she had been eating). His progress notes, written on the day forced nasogastric feedings commenced (January 16, 1986), are instructive:

> *Since she is occupying our space, she must accede* to the same care which we afford every other patient admitted here, care designed to improve and not detract from chances of recovery and rehabilitation. (emphasis added)

Patients at High Desert, it seems, have no individual personalities, just fungible bodies. In addition to being force fed, Bouvia's "care" consisted of being forcibly removed from bed and placed on a gurney out in the corridor to "socialize"; being warned that if she didn't do what the hospital staff told her to do she would not be permitted to smoke in her room; and being threatened with the withdrawal of the morphine. The seemingly impossible had happened: Elizabeth eating had become news; Elizabeth eating was nonetheless being force fed; and Elizabeth eating was, with her ACLU lawyers, back in court.

The Lawsuit

A lawsuit seeking monetary damages and a restraining order to remove the nasogastric tube and discontinue its use was filed almost immediately. After three hearings, Judge Warren Deering issued an opinion denying a preliminary injunction to stop force feeding.[18] Like the doctors at High Desert, the judge was convinced that all parties were back at Riverside two years ago. The intervening time period did not matter: the judge believed Bouvia was still trying to starve herself to death. Though he never bothered to meet her, he judged that her "refusal of medical treatment and nutritional intake is motivated not by a bona fide exercise of her right to privacy but by a desire to terminate her life." On the basis of watching a six-minute videotape deposition Bouvia gave with the nasogastric tube inserted, he determined that although she "claims considerable discomfort" from the tube, "she seems to tolerate the device reasonably well."

The judge also chose to ignore any medical evidence inconsistent with his own views or those of the High Desert Hospital staff. For example, he ignored the medical director's own expert consultant on nutrition who wrote in Bouvia's chart on February 1, 1986, "a *modest weight gain* to range of from 75 to 85 lbs. *might be desirable ...*" (emphasis added) This is a much more reasonable assessment of a patient with no body movement and thus no muscle mass; a patient who is never going to reach an ideal body weight for either beauty or performance. *Outside* medical experts, called upon to make it clear that Bouvia's liquid diet *was* medically reasonable, were brushed aside. Their conclusions need not even be considered, the judge decided, because they were based primarily (he believed) on "studying the plaintiff's medical charts..."

The court's holding strips Bouvia of any remaining human dignity by denying her basic rights of citizens, and puts her at the callous whim of physicians who have consistently demonstrated their indifference to her feelings and wishes:

> The court finds and determines plaintiff at defendant facility has a right to refuse any medical treatment *except where*:
>
> *In the opinion of the medical staff at defendant hospital* plaintiff's refusal would to a reasonable medical certainty directly result in a life-threatening condition, or... (emphasis added)

In short, Bouvia can refuse anything her doctors are willing to let her refuse, and nothing they are unwilling to let her refuse so long as they are willing to say such refusal would result in a life-threatening condition. This holding is untenable as a matter of law, determining as it does that neither the US Constitution nor the laws of the State of California apply in the "space" called the High Desert Hospital. It directly contradicts previous California rulings,[19] and would, for example, permit resuscitation and kidney dialysis without the patient's consent.

Moreover, it is clear from the context that the medical staff of High Desert, unencumbered by "outside experts," can set and judge the adequacy of their own standards for treatment and "reasonable medical certainty." This invitation to arbitrariness and indeterminacy represents judicial lawlessness at its worst, and will likely also lead to medical practice at its worst. It is, for example, most improbable that the standards of medical care are higher at High Desert than they are at LA County, which cared for the identical patient with the identical condition in a far different manner.

A court has never before so encouraged physicians to make up their own standards for use in their own "space," uninhibited by the learning and usages of the medical profession at large. In short, the judge decided that neither the law nor reasonable medical practice could or should influence the medical staff's actions at High Desert Hospital.

Benefiting Elizabeth Bouvia

This opinion is on appeal and may have been reversed by the time this is published. But what are we to do about Elizabeth Bouvia? Can we still rationalize our brutality toward her on the grounds that we are protecting "community values" by "upholding the value of life"? Even if this was at stake two years ago, it is not at stake today. Bouvia says she wants to try to live independently, in an apartment, with appropriate medical and nursing care. Judge Deering doesn't believe it, and he is probably not alone. But she cannot exercise this option in any event, because society is unwilling to provide the funds for Bouvia to live independently.

We seem to prefer to force her to live in public hospitals she has not chosen, and to have her body invaded and manipulated based on her caretaker's views of rehabilitation, including "ideal" body weight and mental attitude. And, at High Desert at least, if she does not accept these nonnegotiable, arbitrarily determined, nonreviewable terms, she will be force-fed and her pain medication withdrawn. These actions pervert the very meaning of community. The community could benefit both itself and Bouvia by making funds available (either directly through the State of California, or indirectly through private fund raising) for her to live independently.

Finally, since two of the last three medical facilities that treated Bouvia succeeded in providing quality care without public fanfare or resort to the court system, the real problem seems to be High Desert Hospital, not their famous patient.*

*The case was reversed in a strongly worded opinion: 179 Cal.App. 3d 1127 (2d Dist. 1986). In retaliation it seems, the hospital cut Bouvia's morphine supply. Following suit to prevent this, she was transferred back to LA County–USC Medical Center in May, 1986. As of early 1988, she remains at LA County, although discussions aimed at providing her a place to live independently have begun.

Do Feeding Tubes Have More Rights Than Patients?

The Case of Paul Brophy

Karen Quinlan has become a symbol and a part of our language. Millions have said, "I don't want to be like Karen Quinlan," meaning that they don't want their dying artificially prolonged when they are in a permanent coma or a persistent vegetative state. Paul Brophy was one of them. In discussing the *Quinlan* case with his wife, he said, "No way do I want to live like that ..."

Through no desire of his own, Paul Brophy may become as symbolic as Karen Quinlan of the fight for the rights of patients to control the use of medical technology on themselves. A 48-year-old Easton, Massachusetts firefighter and emergency medical technician, Brophy lies in a persistent vegetative state in a hospital bed, kept alive only by a surgically implanted gastrostomy tube that permits liquid nutrients to be dripped from a plastic bag directly into his stomach. On October 21, 1985, Judge David Kopelman permanently enjoined the hospital, its physicians, and staff "from either removing or clamping" the gastrostomy tube.[20] The case is on appeal. The judge's opinion stems directly from his belief that mechanical ventilation (at issue in *Quinlan*) is categorically different from gastrostomy tube nutrition.

In March 1983, Paul Brophy suffered a subarachnoid hemorrhage as a result of a ruptured aneurysm. He regained consciousness, and after waiting two weeks for the cranial swelling to subside, he underwent a craniotomy. Brophy never regained consciousness after the operation, and was transferred to the New England Sinai Hospital in Stoughton, Massachusetts where he remains. On December 21, 1983, his wife authorized transfer to an acute care hospital for the placement of a gastrostomy tube, after which he was returned to Sinai Hospital. More than a year later, after hospital doctors, nurses, and administrators refused to remove the tube, Mrs. Patricia Brophy asked a court to authorize the withholding or discontinuation of all medical treatments from her husband, including artificial nutrition and hydration.

What Paul Brophy Wanted

At trial, evidence was presented that if Brophy were able to speak, he would want the artificial nutrition terminated. The evidence included not only the *Quinlan* statement, but also his statement to his brother, after dragging a man from a burning truck, "If I'm ever like that, just shoot me, pull the plug." The victim of that accident lived approximately three months, and Brophy received a commendation for bravery. He threw the commendation in the wastebasket and explained to his wife, "I should have been five minutes later. It would have been all over for him."

Within twelve hours after being admitted to the hospital for his aneurysm, he pulled himself up to a half-sitting position to kiss his daughter. She scolded him for not lying still, as the doctors had ordered. He responded, "If I can't sit up to kiss one of my beautiful daughters, I may as well be six feet under."

On the basis of this and other testimony, the trial judge concluded that if Brophy were "presently competent [his preference would be] to forego the provision of food and water by means of a G tube, and thereby terminate his life."

That should have been the end of the case. But instead the judge ignored substituted judgment and applied his own best interests test. The judge thought this was appropriate even though he knew what Brophy wanted, and even though Brophy's wife, to whom he has been married 27 years, wanted the treatment terminated; and his seven brothers and sisters, his 91-year-old mother, and his parish priest (the family is Catholic) concurred.

Brophy's Medical Condition

About half of the testimony related to Brophy's medical condition, and most of the remainder to "medical ethics." Though there was some quibbling among the medical experts, there was basic agreement, and the judge concluded that Brophy "has been in a chronic persistent vegetative state from the time of his brain surgery on April 6, 1983 to the present." Ronald Cranford, a neurologist, provided the most concise definition. He testified that in a coma the patient's eyes are closed, but in a persistent vegetative state the patient is "awake but unaware," has sleep–wake cycles, but is unconscious and unable to interact purposefully with his environment. There is virtually no chance that such a patient will ever regain consciousness.

Brophy and Quinlan were both in persistent vegetative states, and both courts found they would reject treatment if they could. But the *Brophy* trial court wanted to distinguish the cases, and decided to focus on the gastrostomy tube to do so. First the judge set the prognosis aside by saying, "Apart from the injury to his brain, Brophy's general state of health is relatively good."(!) Then he argued that the *"Proper focus should be on the quality of treatment furnished...to* Brophy, and not on the quality of Brophy's life." (emphasis added) The judge made two unsupportable distinctions: there is a distinction between starting and stopping, because tube feeding itself is noninvasive; and unlike other medical treatments, if a person is deprived of nutrition, the person always dies.

The Gastrostomy Tube

The judge concluded that the creation of the stoma for placement of the gastrostomy tube "is an intrusive and invasive surgical procedure." Accordingly, it can be rejected by a patient or the patient's guardian. On the other hand, once it is in place, utilizing it "requires no special knowledge," is "a nursing procedure more than a medical procedure," "requires minimal maintenance," is "neither uncomfortable nor painful"; "is not likely to produce any major complications"; and "is neither invasive nor intrusive."

The judge was so intent on concentrating on the tube that he asked Brophy's attending physician, LaJos Koncz, to "bring a tube similar both in length and width to the one utilized by Mr. Brophy" to court with him. The next day the judge showed both a fascination with and a knowledge of the tube. Koncz testified about blowing up a soft bubble to maintain the tube in position. The judge asked, "What is your estimate as to the diameter of the bubble?"

The Witness: Less than an inch. I don't know—I believe that.
The Court: I would suggest half an inch once inflated. [Taking the

tube and examining it].

The Witness: Yes, three quarters of an inch at the most. I would suggest for you to feel the consistency.

The Court: I assume it's going to be offered as an exhibit?

Previously the guardian *ad litem*, appointed by the court to represent Paul Brophy, had testified that he didn't see the gastrostomy tube as "any more invasive than a set of dentures." In contrast, the judge described a respirator as "a sophisticated mechanical device which must operate 24 hours a day blowing air into a patient's lungs, thereby mechanically moving the patient's diaphragm." Thus, though the judge concluded that it is acceptable to withdraw mechanical ventilation, he decided that it is not acceptable to withdraw gastrostomy nutrition.

The judge's key finding was that with other medical technologies such as CPR, kidney dialysis, mechanical ventilation, and chemotherapy, orders to withhold or withdraw are "not written with the specific intent of terminating the patient's life inasmuch as some patients are capable of surviving without the aforesaid treatment, i.e., Quinlan's surviving the removal of a respirator."

Unlike these types of treatment, the judge concluded, "I find that denying a patient food and water...will *inevitably in each and every instance guarantee and cause the death of the patient*." (emphasis added) To emphasize this point, the judge also found as a fact that the purpose of withholding nutrition from Brophy would be precisely "to terminate his life" rather than to spare him pain or discomfort.

Although these arguments form the core of the decision, they do not justify it. First, by concentrating on the tube, rather than on Brophy, the judge seems almost to afford medical technologies that are simple, noninvasive, and without any moving parts, rights of their own. Of course, no technology, even one as primitive as a gastrostomy tube, exists in a vacuum. It requires a human being to introduce it into another human being, and its usefulness to the patient, *as evaluated by the patient*, is the critical legal issue. Its design, length, softness, or diameter should not control the outcome.

Second, by concentrating solely on the introduction of liquid nutrients through the tube *after* it has been surgically implanted, the judge makes two mistakes. The first is that more "burdensome" and "invasive" things are happening to Brophy than the four times daily introduction of nutrients into his gastrostomy tube. As the judge himself notes, but does not follow up, Brophy requires seven and one-half hours of nursing care daily, which consists, among other things, of shaving, mouth care, grooming, turning him and positioning him in bed, and caring for his bowels and bladder. Though none of these may seem particularly burdensome in terms of pain or suffering, they are invasive and burdensome in terms of his personal privacy and *his own view* of human dignity, and they are all part of his "treatment."

Third, by concluding that Brophy could refuse the introduction of the gastrostomy tube in the first place because it is an invasive surgical procedure, the judge undercuts the logic of his argument requiring continuation. Specifically, continuation of gastrostomy feeding once started is required because if it is stopped, the patient will necessarily die. But, of course, if gastrostomy feeding is required, and the initial surgery is not performed, the patient will also necessarily die, and in the *same manner*. If this is the argument, starting and stopping must both be required, or they must both be optional with the patient.

If starting is optional and stopping is impossible, patients are also forced into paradoxical situations. If a gastrostomy tube can never be removed or clamped once it is put in place, then the only strategy that patients and their families can employ is always to refuse all gastrostomy tube placements if they believe that they will want their use discontinued when and if it becomes clear that they cannot recover consciousness. This is bad medicine and bad law. We should encourage trials of therapies, which can and should be discontinued when their purposes cannot be attained or they cease to have any usefulness to the patient.

Nor is it persuasive to argue that removing other forms of treatment does not inevitably lead to death, and is not done with the intent of terminating life. First, the judge did not find that patients must always be provided nourishment (they can refuse gastrostomies, for example). Second, the judge specifically rejected use of a nasogastric tube, intravenous feeding, and hyperalimentation, as involving more invasiveness and risk than the gastrostomy tube, seeming to hold that any of the other three methods could be terminated. Third, the judge approved the entry of a "no code" order, approved withholding intravenous antibiotics if Brophy should develop a life-threatening infection, and would have approved withdrawal of a ventilator or kidney dialysis.

An individual may not suffer cardiac arrest, but if one does, and CPR is withheld, he or she will die a "preventable" and absolutely predictable and intended death. Likewise, a patient with end-stage kidney disease who requires dialysis will certainly die of intended renal failure if it is discontinued. And, of course, the reference to Karen Quinlan is disingenuous. Both the trial court, and the New Jersey Supreme Court, rendered their opinions *because* they believed she would die. The fact that she did not die in no way changes the court's conclusion that removal of her "life-support system" would not constitute either homicide or suicide. As that same court made clear in *Conroy,*[*] this conclusion does not change when the issue is removal of artificially-introduced nutrition. Food and water may be "basic," but deprivation of air is at least as basic, and withholding it will lead to death as surely, and much more swiftly.

The Technology-Distinction Trap

The judge fell into the technology-distinction trap. Like some of the witnesses, he confused the *use of the gastrostomy tube* with the *provision of nutrition* itself. In this regard the judge seemed to adopt the flawed logic of the guardian *ad litem*:

> The removal of the G tube is not comparable to cessation of dialysis or removal of a respirator because removal of the aforesaid artificial mechanisms permit the illness or injury to run its course. Nutrition, however, is not a need required by Brophy as a result of his illness, but rather, it is a need common to all human beings.

The response is obvious. A gastrostomy tube is *not* "a need common to all human beings"; it is an intervention required by Brophy *because of his injury*; and if it is withheld or withdrawn, Brophy will die as surely from the "natural course" of his illness as Karen Ann Quinlan would have had she been removed from her ventilator (instead of being carefully weaned from it).

Medical Ethicists in the Courtroom

Medical ethics experts have been permitted to testify in previous cases, but there is probably no other case in which they have so dominated the proceedings. Four philosophers and three physicians testified as experts on medical ethics. I assume their legal role was to help the judge determine if the state's compelling interest in protecting the

*[*]The *Conroy* decision is discussed at p. 309.

"ethical integrity of the medical profession" outweighed Brophy's right to refuse treatment. However, the judge failed to control their tendency to ramble and try to "one up" each other, once on the stand. Their testimony thus was more confusing than useful. When it was over, the judge acted like a referee in an ethical debate, and declared the winners. His 143rd and final finding of "fact" was: "This Court is in substantial accord with the ethical viewpoints of Reverend John Connery, Professor Patrick Derr, and Professor Arthur Dyck."

To complete the confusion, one of the judge's crucial "conclusions of law" is: "It is *ethically inappropriate* to cause the preventable death of Brophy by deliberate denial of food and water, which can be provided to him in a noninvasive, nonintrusive manner, which causes no pain and suffering, irrespective of the substituted judgment of the patient." (emphasis added)

The judge's role should be to define the *legal* rights and duties of the parties, not to make a determination as to which "expert" view of medical ethics corresponds to his own. Judges have no more business telling medical ethicists what "medical ethics" is than medical ethicists have telling judges what the law is.

Conclusion

As this case illustrates, medical technology as simple as a gastrostomy tube can become as removed from any curing or caring purpose as an artificial heart. It is technologically possible to maintain patients in persistent vegetative states for decades on artificial feedings, mechanical ventilators, and kidney dialysis, but to what purpose when they have expressed their belief that for them this would be a fate worse than death?

Paul Brophy committed no crime. Yet for the sake of a primitive medical technology, he has been sentenced to spend perhaps decades confined to a hospital bed in a persistent vegetative state. As the Massachusetts Supreme Judicial Court put it in the *Saikewicz* case, "The value of life is lessened not by a decision to refuse treatment, but by a failure to allow a competent human being the right of choice."[21] People have rights, not technologies: the focus in this case should have been on Brophy, not his gastrostomy tube.*

*On appeal, the Massachusetts Supreme Judicial Court reversed this opinion, holding that Mr. Brophy's wishes should control. *Brophy v. New England Sinai Hospital*, 497 N.E.2d 626 (Mass. 1986). He was thereafter transferred to Emerson Hospital where, eight days after feeding was discontinued, he died.

The Case of Claire Conroy
When Procedures Limit Rights

The New Jersey Supreme Court has concluded that there is no analytical difference between terminating artificial feeding and discontinuing a mechanical ventilator, an action it permitted in the 1976 Karen Ann Quinlan case. In *Quinlan*, the court wanted to get Karen off a ventilator, and devised a method designed to make physicians feel comfortable about removing the machine. However, by focusing on the physicians and their fear of liability, the *Quinlan* court ignored Karen's right to refuse treatment and issued an ill-defined and potentially dangerous opinion that permitted Karen's guardian, family, and physician to do almost whatever they wanted to her with legal immunity.* This "Catch 22" approach to treatment refusals, in which incompetent patients are given a substantive right to refuse treatment that must be procedurally exercised in such a way that it will either never be exercised, or can be exercised in a way inconsistent with their wishes, also forms the basis of the 1985 Claire Conroy decision.[22]

The Facts and the Opinion

Claire C. Conroy was a single, 84-year-old woman who resided at Parkland Nursing Home in Bloomfield, New Jersey. She suffered from severe organic brain syndrome, chronic decubitus ulcers, urinary tract infection, heart disease, hypertension, and diabetes. She was unaware of her surroundings, had primitive brain functioning, no cognitive ability; and there was no expectation that her condition would ever improve. Medical testimony about her ability to experience pain was inconclusive. She had never married and had few friends.

*The *Quinlan* case is discussed at p. 261.

In August 1979, after she was adjudicated incompetent, her nephew and only surviving relative, Thomas C. Whittemore, was appointed her legal guardian. He placed her in the nursing home. On July 21, 1982, Conroy was admitted to Clara Maas Memorial Hospital for treatment of her gangrenous leg. Amputation was recommended to prevent death, but Whittemore refused consent, on the basis that Conroy would not have consented were she competent. The surgery was not performed, but Conroy survived. Shortly after her admission to the hospital a nasogastric tube was inserted to facilitate feeding. It was removed on October 18, but reinserted on November 3 because of Conroy's inability to take sufficient nutrition by mouth. After the reinsertion, her guardian asked that the tube be removed. Her attending physician refused.

On November 17, Conroy was transferred back to the nursing home, where the attending physician again refused to remove the nasogastric tube. The guardian accordingly filed suit to compel its removal. The trial court authorized removal, but the Appellate Division reversed the decision, stating that termination of feeding would be homicide.[23] Conroy died on February 15, 1983, with the tube still in place. The guardian *ad litem* brought the case to the New Jersey Supreme Court, which in January 1985 declared termination of *any* medical treatment, including artificial feeding, on incompetent patients lawful so long as certain procedures are followed.

The 83-page opinion, written by Justice Sidney Schreiber (who is retired, but continues to participate in cases heard before his retirement), is especially vigorous and persuasive in summarizing the law as it applies to competent adult patients. The court bases its decision on an individual's common law right to accept or reject *any* medical treatment under the doctrine of informed consent, which is designed to protect "the patient's ability to control his bodily integrity," and encompasses "the right to decline to have any medical treatment initiated or continued."

Unlike *Quinlan*, the court decided not to address the constitutional dimensions of the right to refuse treatment, because it is, "in any event, embraced within the common law right to self-determination." Given the current and likely future composition of the US Supreme Court, this may be viewed as a wise strategy. The court concludes:

> We have no doubt that Ms. Conroy, if competent to make the decision and if resolute in her determination, could have chosen to have her nasogastic tube withdrawn.

The court goes on to note that it had overstated the role of prognosis in *Quinlan* by acknowledging that "Ms. Conroy's right to self-determination would not be affected by her medical condition or prognosis ... A young, generally healthy person, if competent, has the same right to decline life-saving medical treatment as a competent elderly person who is terminally ill."

The court also examines and rejects any legal distinction "between *actively hastening death* by terminating treatment and *passively allowing a person to die...*, between *withholding* and *withdrawing* life-sustaining treatment; between *ordinary* and *extraordinary* treatment; and perhaps most importantly, any distinction *between artificial feeding and other medical treatment.*" (emphasis added) This is the most significant and well-articulated portion of the opinion: competent patients have the right to refuse *any* treatment, be it feeding, ordinary medical treatment or anything else, and have the right to demand that any ongoing treatment be discontinued, be it feeding, ordinary medical treatment, or anything else. This court takes self-determination seriously.

But since Claire Conroy was not competent, the case could not end here. Accordingly, it is necessary to set up some additional substantive rules, and a procedure to apply them, so that previously competent patients can have their right to refuse treatment exercised by proxy.

Tests to Determine Course of Action

Since what is at stake is the right to self-determination, the goal of decisionmaking for incompetent patients must be to determine what decision they would themselves make if competent. Such an expression, the court notes, could be embodied in a written document, like a "living will," but could also be an oral declaration or a durable power of attorney. A person's wishes might also be deduced from religious beliefs, or a consistent pattern of conduct: "Of course, dealing with the matter in advance in some sort of thoughtful and explicit way is best for all concerned."

The court also acknowledges that it erred in not considering Karen Quinlan's own statements to friends in assessing her wishes, since "such evidence is certainly relevant." If the individual's wishes can be determined, those wishes are controlling. If, however, such evidence is "inadequate" to determine the individual's wishes, other tests must be used. The court articulates two best interests tests.

Under the "limited objective test" life-sustaining treatment may be withdrawn if "there is some trustworthy evidence that the patient would have refused the treatment, and...it is clear that the burdens of the patient's continued life with the treatment outweigh the benefits of that life for him." Under the "pure objective test" (that is, when there is *no* evidence about what the patient might want), "the net burdens of the patient's life with the treatment should clearly outweigh the benefits that the patient derives from life...[and] the recurring, unavoidable and severe pain of the patient's life with the treatment should be such that the effect of administering life-sustaining treatment would be inhumane."

This very strict test, which centers on severe pain, is adopted because of the majority's belief that "when evidence of a person's wishes or physical or mental condition is equivocal, it is best to err, if at all, in favor of preserving life"; and not to adopt this conservative posture "would create an intolerable risk for socially isolated and defenseless people suffering from physical or mental handicaps."

But the court goes too far in insisting that "even in the context of severe pain, life-sustaining treatment should not be withdrawn from an incompetent patient who had previously expressed a wish to be kept alive in spite of any pain that he might experience." This statement is too blunt, and like most of its discussion of best interests, much too tilted toward insisting on treatment even in cases where it might not be medically indicated, might be contrary to an objective view of reasonable medical action, and might actually either amount to "mistreatment" or be inhumane under the circumstances. Even competent patients have no right to demand that their physicians deliver bad, abusive, or inhumane treatment.

Procedures in Nursing Homes

Since Claire Conroy was a resident of a nursing home when this case was brought, and died there, the court spends all its procedural energy on nursing home patients. In fact, the court describes its primary duty as protecting patients who "fit" the "Claire Conroy pattern" —incompetent, permanently impaired, and elderly—from abusive withdrawal of life-sustaining treatment. It is therefore not surprising that the procedure developed by the court to effectuate the termination of life-sustaining treatment is so rigorous that it will seldom, if ever, be employed.

The procedural portion of the opinion deals almost exclusively with New Jersey's 1983 amendments to its elder abuse statute, which charges the Office of the Ombudsman for the Institutionalized Elderly to guard against and investigate allegations of elder abuse. Using this framework, the following procedure is mandated before life-sustaining treatment can be withdrawn from an incompetent nursing home patient:

1.   A court must determine that the patient is incompetent to make the medical decision, and appoint a guardian to make it. If the patient already has a general guardian, a court must nonetheless determine if the patient is competent to make this particular decision, and if not, if the general guardian is the appropriate one to make the treatment decision.

2.   The guardian should notify the Office of the Ombudsman if the guardian believes withholding or withdrawal of treatment is appropriate under any of the three tests described above.

3.   The Ombudsman must treat every such notification as a possible case of "abuse" and conduct an immediate investigation.

4.   As part of the investigation the Ombudsman should obtain information about the patient's condition from the attending physician and nurses, and should also appoint two other physicians, unaffiliated with the nursing home, to confirm the patient's condition and prognosis.

5.   If this medical information is consistent with any of the three tests, and the guardian, attending physician, and Ombudsman agree, then, absent bad faith, all participants may withhold or withdraw treatment with legal immunity (provided that if either of the two best interests tests are utilized, "the patient's spouse, parents, *and* children, or in their absence, the patient's next-of-kin, if any, must also concur ...").

In this scheme the guardian will make the decision based on one of the three tests, and court involvement will be limited to determination of competence and appointment of the guardian.

Problems with the Approach

There are a number of problems with this procedural approach. First, it seems wholly unnecessary where the individual has written an explicit living will and designated a proxy through a durable power of attorney to carry out his or her wishes. In this case the patient's wishes are known and the person designated by the patient should be obliged to carry them out; the additional "protections" mandated by the court work to the patient's detriment by potentially thwarting, and certainly delaying, their fulfillment.

Second, as Justice Handler argues persuasively in dissent, the "pure objective test" that focuses exclusively on "recurring, unavoidable and severe pain" as the *only* basis for a withholding or withdrawing treatment is too narrow. Other considerations like prognosis and restraints, in addition to pain, may objectively make continued treatment more burdensome than beneficial to the patient.

Third, to require the Ombudsman always to treat a proposed withdrawal or withholding of treatment as potentially abusive will likely discourage families and others from using this mechanism, even if doing so would fulfill the clear wishes of their nursing-home-bound relative.

Fourth, although the court acknowledges that a major problem in nursing home care is finding a physician who will go there, it requires not one but three physicians to examine the patient and report on his or her condition. If any are needed, one outside medical consultant should suffice.

Finally, although many of the flaws in the *Quinlan* case have been clarified, the court persists in delegating its immunity-granting authority to third parties. In *Quinlan* the delegation was to an "ethics committee" (which the court has properly renamed a "prognosis committee"). In *Conroy,* the court delegates immunity-granting authority to a group that must concur, but need never meet: the guardian, attending physician, two consulting physicians, the Ombudsman; and, where either of the two best interests tests are used, the patient's family or next-of-kin.

These problems seem to flow from the court's overly narrow attempt to limit this case to nursing home residents, and its broad equation of artificial feeding and all other medical treatments (which requires its cumbersome process to be applied whenever any life-sustaining medical treatment is withheld or withdrawn). In its first footnote the court states that its "holding is restricted to nursing home residents." This is not logically possible. It neglects the fact that none of the relevant medical decisions regarding Claire Conroy was made in the nursing home at all, but during her four-month stay at the Clara Maas Hospital, where amputation was successfully refused, and where her nasogastric tube was inserted, removed, and reinserted, and where her guardian first requested that it be permanently removed. The dichotomy between nursing home and hospital is not only artificial and misleading in Ms. Conroy's case; it is artificial in the treatment course of almost every elderly nursing home resident. Almost all will be transferred to hospitals when they require invasive treatment, and a large number of them will initially enter the nursing home from a hospital. The five reasons provided by the court for differentiating between hospitals and nursing homes suggest a procedure that might be different in degree, but these differences are not so extreme as to require a procedure that is different in kind. The reasons given are the patients' average age; their lack of surviving parents, siblings, or children; the limited role of physicians in nursing homes; general understaffing and reports of inhumane treatment in nursing homes; and the less urgent types of treatment decisions that are made in nursing homes. Of these, the patient's age, family status, and needed treatment are all unaffected by the patient's physical setting.

This leaves the lack of physician contact and the general bad impression one has about nursing homes as justifications for different kinds of procedures. Neither is sufficiently persuasive. All but one of the previous nontreatment cases that have reached appeals courts have originated in hospitals, and this seems to be the setting in which patient wishes are most frequently ignored. More treatment decisions per patient per day are made in hospitals, but the court gives us no reason to assume that they are made so much better. Indeed, there are many "good" nursing homes and patients in them who have loving families and caring physicians. Nonetheless, this overly broad opinion requires these patients to be treated as if they are in a disreputable setting where a withholding or withdrawal of treatment is likely to be abusive.

By mistakenly focusing on the nursing home setting instead of the patient's wishes and interests, the court tends to ignore the plight of Claire Conroy, just as the *Quinlan* court concentrated on physician liability and ignored the wishes of Karen Ann Quinlan herself.

Possible Resolutions

The dissenting justice opines that "perhaps most important, this decision may encourage individuals to use living wills." It certainly should. But even this mechanism will be inadequate if it is not clarified, by the court or the legislature, so that when an individual makes his or her desires known in such a document, and designates a proxy through a durable power of attorney to carry them out, no further involvement by courts, consultants, or Ombudsmen is *required* to obligate the attending physician to honor the patient's wishes. These procedures should be *optional*, and used only by those who believe, usually because they are unsure of the patient's wishes or condition, that they must have legal immunity before withdrawing or withholding medical treatment.

Second, the arbitrary distinction between treatment decisions made in nursing homes and those made in hospitals cannot stand. If mandated intervention by the Ombudsman is appropriate, it must be appropriate in both settings. It makes more policy sense for the Ombudsman law to remain as it was before this opinion: the office *available* to be called on by anyone suspecting elder abuse. This is the model we employ in the child abuse setting, and it seems the appropriate one here as well. The *Conroy* scheme will require a lot of Ombudsmen who will waste a lot of time on cases that require no investigation or governmental review. It may also apply to nursing home residents who are temporarily in hospitals, and at least creates confusion in this area.

Finally, because the procedure is so universally required, it is likely to be ignored. If physicians are as worried about legal liability as this court portrayed them in *Quinlan*, they will continue to treat their incompetent patients even though their patients may not have wanted such treatment, or it is not in their best interests. This overtreatment is likely to be every bit as abusive as undertreatment.

First *Quinlan*, and now *Conroy*, demonstrate that the "New Jersey approach" of declaring broad-ranging substantive rights, and then attempting procedurally to restrict their application to very narrow categories of patients, fails to promote the autonomy of once-competent patients. The court has splendidly succeeded in articulating the right of competent adults to refuse treatment, but has frustratingly failed to provide any meaningful way to require proxies to exercise this right on behalf of incompetent patients.*

---

*Procedural isues were further discussed in three cases decided by the court in mid-1987, *Jobes, Farrell*, and *Peter* (529 A.2d 404). *See* Annas, G. J., "In Thunder, Lightning or in Rain: What Three Doctors Can Do," *Hastings Center Report*, **17**: 28–30, Oct./Nov. 1987.

Prisoner in the ICU

The Tragedy of William Bartling

In June 1984, a California court sentenced William Bartling to spend the rest of his life in an intensive care unit. His crime seems to have been that he suffered from fatal diseases instead of "terminal" ones, and that his desire to live seemed at odds with his desire to have the mechanical ventilator that sustained his life removed. His case is important to all who are concerned with respecting the autonomy of patients in hospitals.

The case illustrates how fear of liability can cause a hospital to alter its traditional role of offering services to willing patients, into one of forcing treatment on unwilling patients. It also illustrates how physicians, hospital administrators, and even judges can see themselves as responsible for the actions of a competent patient, and how their ambivalence about the patient's decision can cause them to compromise or abdicate their social roles to the patient's profound detriment.

Bartling's Condition

At the time of his current admission to the hospital, April 8, 1984, William Bartling was seventy years old, had been in failing health for eight years, and had been admitted to the hospital six times in the previous year. He suffers from at least five potentially fatal diseases and disorders: chronic obstructive pulmonary emphysema, which severely restricts his ability to breathe; diffuse atherosclerotic cardiovascular disease (hardening of the arteries); obstructive arteriosclerosis of the coronary arteries; an abdominal aneurysm; and inoperable adenocarcinoma of the left lung.

His lung cancer was diagnosed by biopsy. During this unnecessary procedure Bartling's lung collapsed, necessitating his prompt transfer to the Medical Intensive Care Unit, the insertion of a chest tube, and the use of a mechanical ventilator to sustain his breathing. The ventilator caused Bartling significant discomfort, distress, and pain; he repeatedly asked to have it removed, and on many occasions removed it himself. After a time, his hands were tied down to the sides of his bed by cloth cuffs to prevent further attempts to remove the ventilator.

Bartling's attorney, Richard Scott, made arrangements with the hospital and doctors to have Bartling's wishes honored. His physician, according to Attorney Scott, had no objections to this course of action, and agreed to remove the ventilator if the hospital administrator concurred. The hospital administrator reportedly agreed, provided legal counsel for the hospital approved. Legal counsel did not.

Instead of doing what almost all hospitals do in such a situation—go to court to get a declaration of rights and immunity—the hospital took the position that it could not countenance the removal of Bartling's ventilator because he was neither "terminally ill" nor in a persistent vegetative state nor "brain dead." At this point the hospital took an adversarial position in regard to its patient; instead of seeking legal means to follow the patient's wishes, it sought to use the law to oppose them. Scott sought judicial relief.

Bartling's Wishes

The evening prior to the judicial hearing, Bartling's deposition was taken and videotaped in his intensive care room. In addition to the lawyers for each side, present in an adjoining room were Mike Wallace and his *60 Minutes* crew. The story they were filming was aired on September 23, 1984. Viewers of that telecast saw the entire deposition. Bartling's attorney asked him three questions (the ventilator was attached by tracheotomy so Bartling could only mouth words and shake his head to respond):

"Do you want to live?" [Indicates yes]

"Do you want to continue to live on the ventilator?" [Indicates no]

"Do you understand that if the ventilator is discontinued or taken away, you might die?" [Indicates yes]

Brief cross-examination dwelt primarily on whether Bartling was satisfied with the nursing care. He was.

The Hearing

The hearing was held the next morning, June 22, before Los Angeles Superior Court Judge Lawrence Waddington.[24] The judge opened the hearing by expressing his "tentative inclination." He characterized Bartling's prognosis for recovery as "guarded and cautious, but optimistic"; assumed "he is competent in the legal sense"; and viewed Bartling's request as one for a "mandatory injunction," that is, requiring the affirmative act of ending treatment. He noted that such injunctions were disfavored under California law, and said he could find no case from any jurisdiction where a person applying for injunctive relief was not terminally ill or "in a comatose, vegetative or brain-dead state." On this basis he concluded that relief should not be granted.

Scott responded by noting several non-California cases in which courts permitted the withdrawal or withholding of treatment for patients who were not terminal, comatose, or brain-dead. He argued that the key to the case was the patient's competence. If Bartling was legally competent (and there was no dispute on this point), his wishes had to be respected; the patient in charge of his own body is the status quo, and the court had an obligation to restore that: "It is the patient's decision. The doctors can recommend, but the patient must decide."

The judge then asked a number of questions, including one on competence, which led to this response: "He understands the nature and consequences of the act he proposes.... He absolutely understands that if this ventilator is disconnected, there is a high chance that he would die. I asked him that. He nodded affirmatively...." Finally, when it was clear that the judge would not order any physician to withdraw the ventilator, Scott asked him to order that Bartling's hands be untied.

Arguing for the hospital, William Ginsburg attempted to distinguish the previous cases by arguing that unlike them: "Bartling eats ice cream. He watches LA Express games and Angels games. He communicates with nurses. And as Mr. Scott quite correctly states, he is very much with us..."

But his most important comments were on the issue of competence. Although Ginsburg did not dispute Bartling's competence, he characterized his actions as indicating "ambivalence." Specifically, he contrasted "I don't want to die" with "I don't want to live on the respirator," arguing that these positions are inconsistent, and evidence of ambivalence. In his words, "I don't think he really fully spiritually or emotionally understands what it is he is talking about ... There is a strong possibility that this man can be restored to useful life." Scott responded that there was no ambivalence or vacillation:

> He would prefer to live, but he does not prefer to live with that illness and with the necessity of his every breath being sustained by a ventilator, by his trachea being suctioned every two hours around the clock, confined to an ICU bed for the rest of his life, watching television and eating ice cream.

As for depression, Scott argued it was real and situational: his serious illnesses, his restraints in the ICU, the constant surveillance, and connection to a respirator all "give him very little cause to be cheerful." The judge was not persuaded, and so did not change his tentative decision, nor would he order that Bartling's hands be untied.

A Case That Should Not Have Gone to Court

This is a good example of a case that should never have gone to court, and may even be an example of a case in which a functioning ethics committee could have proven helpful and decisive. For some reason the physician felt he could not act without the approval of the hospital administrator, and the hospital administrator felt, in turn, he could not act without the approval of his attorney. And the attorney, apparently, did not understand either the question put to him by the hospital or the law of California. While we may never know the whole story, since the physicians involved are not talking, it appears that fear of criminal liability for either homicide or assisting suicide led all of the actors involved to defer to others rather than defer to the patient.

According to briefs they filed with the trial court, for example, the hospital attorneys believed treatment refusal by a competent nonterminal individual was suicide if death would result, and potential murder for the physicians involved. These beliefs apparently were fostered by an inability to distinguish *Bouvia** or to understand *Barber*.[25],** At the hearing Ginsburg insisted: "The issue of suicide

*Discussed in two essays at pp. 290–301.

**_Barber_ involved a homicide prosecution brought against two physicians, Robert Nejdl and Neil Barber, for terminating intravenous feeding on Clarence Herbert in August, 1981. After a number of hearings, a Superior Court Judge dismissed the complaint, and an appeals court affirmed the dismissal, stating that the family members should be able to decide to withdraw artificial feeding if continued treatment is "futile" or has "proved to be ineffective."

has been raised in the *Bouvia* case....And if that wasn't enough, the issue of homicide, murder, was raised in the *Barber* case...." The analytical issue in the *Bartling* case is the inability to equate the morality and legality of stopping an ongoing activity with that of not starting the same activity.

Counsel for the hospital also had the non-California cases backwards and was able to at least confuse the judge about their relevance. Far from restricting the right to refuse treatment to comatose, totally incompetent, and brain dead individuals, cases like *Quinlan* do the opposite: they hold that *even* comatose and incompetent patients must be provided with a mechanism to refuse treatment *because* if they were competent, they would have that right, and to deny it to them simply because they could no longer personally exercise it would devalue their lives. Likewise, although the California Natural Death Act limits its application to the terminally ill, this has never been a common law or constitutional limitation on the right to refuse treatment.[26] This argument, along with the "eating ice cream" argument, uses the fact that Bartling knows about his situation, and consciously suffers pain and despair, against him. We can only do this when we are more concerned with our own suffering than that of the patient.

Consistent with this view, the trial judge redefined the case as one in which *he* was being asked to order a physician to terminate Bartling's wanted life, not his unwanted treatment. Instead of seeing his role as protector of the liberty of an individual who was being held and treated against his will, the judge seemed to see himself as making a medical decision. Medically, the judge determined that Bartling's prognosis was "optimistic" and that he was evidencing "ambivalence." The judge seems also to have redefined the legal issue of competence in the case into a medical one to avoid taking any personal responsibility for a decision that might result in Bartling's death.

This case is a personal tragedy for William and Ruth Bartling; a dismal failure for physicians trying to administer humane care; and a disgrace for the judiciary. For the lawyers it is a throwback to the post-*Saikewicz* days in Massachusetts when lawyers instructed physicians that they could not write DNR orders or stop treating any patient unless court approval was sought and obtained. Such advice was legally inaccurate and resulted in immeasurable human suffering.[27,*]

A similar situation exists in California today in a post-*Barber* era. Some physicians are convinced, although *Barber* holds exactly the

---

* See discussion in "No Fault Death: After *Saikewicz*", at p. 267.

opposite, that termination of treatment on a competent adult with that person's informed consent could be aiding suicide or murder. This position is indefensible as a matter of law, and attorneys who give their health care provider clients such advice should be held accountable for its foreseeable consequences on the lives of the patients of those clients. Unless lawyers are at risk for negligent, uninformed advice to doctors and hospitals, their incentive will often be to delay and take cases to court. Their helath-provider clients will pay a high price in terms of transforming their social roles from serving patients to treating them against their wills; and patients will pay the ultimate price: they will be forced by ambivalent judges "to bear the unbearable and tolerate the intolerable."[28],*

---

*This case was reversed on appeal in late December, 1984, two months after Bartling died, still attached to the ventilator in the ICU. *Bartling v. Superior Court*, 163 Cal. App. 3d. 186 (1984). In mid-1986, dismissal of Ruth Bartling's civil suit for damages sustained by her as a result of the treatment of her husband was affirmed. In late 1987, attorney fees were awarded to Bartling's lawyers, and paid by the hospital and doctors in the amount of $155,000.

Government Regulation

Introduction

The essays in this section are devoted to areas in which the legislative or executive branch of the government has affirmatively and directly intervened to affect policy in the life sciences area. The first two deal with the two major federal commissions on bioethics of the past decade: the National Commission, and the President's Commission. Both were efforts by Congress to heighten the consciousness of medical researchers and others, and to develop useful and meaningful regulation of human experimentation and other areas of medical research and treatment.

Three unrelated issues involving life forms have reached the courts, and all are covered in the next two essays. The first involves an issue that recently reached the US Supreme Court: the teaching of "creationism" in the schools. The second, protecting endangered species, and the third, patenting novel life forms, are dealt with together. Both of these involve the interpretation of federal legislation by the US Supreme Court.

The section concludes by examining two large institutional systems that are run primarily based on state statutes and regulations: nursing homes and prisons. The nursing home essay involves the rights of nursing home residents to a hearing before they can be transferred out of a facility because the state has decided it should be closed. The second involves the limitations a state can reasonably place on prison hunger strikes.

The first of these essays, "Good as Gold," originally appeared in *Medicolegal News* (now *Law, Medicine and Health Care*), in December, 1980, and inspired more direct response and controversy than any other essay in this collection. All of the others are from the *Hastings Center Report*. "The President's Bioethicists" was published in February, 1979; "Creationism" in April, 1982; "Life Forms" in October, 1978; "Transfer Trauma" in December, 1980; and "Prison Hunger Strikes" in December, 1982.

Good as Gold

Report on the National Commission

In Joseph Heller's satirical account of federal commissions, Bruce Gold, the hero of *Good as Gold*, is appointed to one as an introduction to the Washington scene.[1] The first meeting of the Commission consists exclusively of adjournment. At the opening of the second meeting the chairman begins: "Let's quit. We've already spent more time on these problems than I think they deserve." After Gold protests that the Commission has not done anything, a more seasoned member responds: "And in record time, too. I was once on a Presidential Commission that took almost three years to do nothing, and here we've accomplished the same thing in only two meetings."

It would be both unfair and inaccurate to so characterize the work and the record of the National Commission for the Protection of Human Subjects of Biomedical and Behavioral Research, which produced a flood of documents, reports, and recommendations from 1974 to 1978. Indeed, almost every commentator on the Commission has treated it as if it were sacred, and criticism seems to amount to blasphemy.[2] This may be explained by the fact that, compared to most previous work on the subject by governmental entities, its work was exceptional; that the Commission provided a source of income to many experts in the field; or that most commentators actually believe that the "hands off" policy toward research regulation advocated by the Commission is a valid one.

There is, however, a minority opinion about federal commissions in general, and this Commission in particular, that deserves at least a brief hearing. This essay outlines the major reasons why the Commission called for continued reliance on a researcher-dominated local institutional review board (IRB) system to provide protection for research subjects, and indirectly comments on why such protection is inadequate. My hope is that if any of the points made are persuasive, they will be taken into account by the National Commission's successor, the President's Commission for the Study of Ethical Problems in Medicine and Biomedical and Behavioral Research,[3] and by others who want to protect research subjects.

The Commission's primary problem was structural, and this led it to three questionable premises on which almost all of its recommendations were based. The structural problem has to do with its composition; the three premises concern the value of research to progress, the risks to human subjects, and the appropriateness of researcher control of subject protection.

Commission Structure

The enabling legislation called for a Commission of eleven members, five of whom "shall be individuals who are or who have been engaged in biomedical or behavioral research involving human subjects."[4] Most commentators have been pleased that the majority of the Commission was "nonresearcher." But domination by the five researchers was almost complete since they effectively represented their industry, whereas the other six members, from a variety of diverse backgrounds, were relegated to minority status. This became evident almost immediately when the Commission elected as its chairman one of the five researchers. In my view, this was inevitable, as was the likelihood that a commission so composed would endorse only proposals looked upon with favor by the biomedical research community.

It should be noted that the argument is not a Marxist one, i.e., that elite groups define and organize practices in such a way as to benefit themselves (researchers) to the detriment of the powerless (subjects), but a much less doctrinaire one: individuals are likely to act in ways they think are in their own best interests, and members of groups are likely to act in ways that they believe will benefit their group.

The President's Commission departs from this model in two respects: only three of its eleven members must be researchers, and the chairman must be appointed directly by the President.[5] The fact that President Jimmy Carter chose a nonresearcher as chairman is, I believe, significant and makes the prospect of changing the current system much more likely, though by no means guaranteed.*

---

*See the following essay, "All the President's Bioethicists," for a discussion of the legislation that established the President's Commission.

One can reasonably question why any researchers at all should be on a commission concerned with the protection of human subjects. We do not expect members of the nuclear power industry to be on a commission to decide the risks of nuclear power; or employees of McDonnell-Douglas to be members of a commission to determine the safety of the DC-10; or drug company officials to be on a commission to determine the safety and efficacy of drugs. Expertise is not the answer; expertise can be supplied by consultants. The only satisfactory answer seems to be that we as a society tend to trust physicians and biomedical and behavioral researchers. Although we know they get paid for what they do and are rewarded based on what they "discover," society does not seem to place them in the same category as businessmen or lawyers.

Premise I: Research is Good

This is a premise almost no one will argue against. On the other hand, it has recently been accompanied by a significant number of caveats. We have seen the dangers of air, water, and land pollution from desirable activities; serious injury to workers and others from hazardous chemicals and pesticides; and risks of cancer from the almost universal indulgence, smoking. There are limits to new activities that we do not always recognize. There are also values we do understand. For example, near the beginning of the race to the moon, NASA could have quite quickly put a man on the moon, but we would have been unable to bring him back. Putting a man on the moon before Russia would have been (and eventually was) a tremendous scientific achievement. Nonetheless, the cost of the life of the first man on the moon was almost universally thought to be far too high a price to pay for this "scientific achievement." There are limits that society would place even on the acquisition of new knowledge, and most of the scientific community conceded this during the recent debate over research on recombinant DNA.[6]

Nevertheless, we all have a tendency to think that we (and others like us) can be trusted. We all want less government regulation and control of ourselves and our own activity. This is natural. On the other hand, beginning with the premise that research is good, and putting the burden of proof on those who argue that particular research is bad, is a way to help promote research, not a way to promote the rights of research subjects. It is a premise that leads one to examine carefully not the subjects of research, but the kinds of research done on subjects. It is, in short, the type of bias that is perfectly appropriate for a group promoting research on human subjects, but not at all appropriate for a group charged with "protecting human subjects." For that purpose, the group should be composed of individuals who have shown some concern for the protection of individual citizens against the arbitrary actions of private and public individuals and agencies.

One example may clarify this point: the Commission's recommendations concerning psychosurgery. Although the Commission admitted it could find "no evidence" that psychosurgery was ever beneficial to children, their recommendations still permitted this experimental procedure on children because the Commission did not want to deprive them of the benefits of this "new" modality.[7] Again using the psychosurgery example, the Commission purposely and strongly stepped outside of its Congressionally mandated role and affirmatively recommended that the Secretary of Health, Education and Welfare (HEW), now Health and Human Services (HHS), "...conduct and support studies to evaluate the safety of specific psychosurgical procedures and the efficacy of such procedures in relieving specific psychiatric symptoms and disorders." The Commission made this striking recommendation even though nothing in its report supports the concept that psychosurgery should be on HEW's priority list, and none of their studies of the multiple types of procedures being used by the more than 140 surgeons engaged in this type of experimentation indicated that psychosurgery would be beneficial to subjects.[8] Both of these recommendations are incomprehensible from a commission protecting human subjects, but perfectly logical from one that is actually promoting medical investigations. HEW firmly rejected both of these recommendations.[9]

The argument should not be misunderstood. It is not that research is bad. It is that the fundamental premise of the Commission should have been: people must decide for themselves whether or not to participate in ethically acceptable human experimentation.

Premise II: Experimentation Is Almost Never Harmful to Subjects

This is a recurrent theme of Commission reports and the commentary on them. It is of the genre of the nuclear power industry's claim (before the Three Mile Island accident); the tobacco industry's claims about smoking; and artificial sweetener manufacturers' claims about saccharin. Even Robert Veatch seems convinced. He writes, "It is getting harder to find examples of serious harm done to research subjects or serious violations of their rights."[10] The questions are, "More harm than what?" and "Serious to whom?" We have not seen anything like Germany during World War II, although Henry Beecher did publish a series of clearly unacceptable research protocols done in the United States long after the Nuremberg Code had been promulgated.[11] It is also true that the thalidomide research in the United States (involving literally thousands of physicians), the Jewish Chronic Disease Hospital case, the Tuskegee syphilis study case, and the Willowbrook experiments, to name some of the ones that come most quickly to mind, are all pre-Commission cases.

Nonetheless, the Commission's conclusion on this score seems again a product of their pro-researcher bias; a more subject-concerned group could have come up with much more persuasive and definitive data. It could even have been data that supports the little harm premise, but I remain unconvinced. The Commission bases its conclusion that research is a relatively "safe" activity primarily on studies in which investigators themselves were asked how many of their subjects had experienced harmful effects from what the investigators had done to them.[12] This is a highly unusual way of obtaining accurate data. We do not ask physicians how many times they have been negligent in their care of patients to obtain an incidence rate for medical malpractice; and we would not expect to study automobile accidents in the United States by asking drivers to identify and describe all the accidents they have caused. Reporting, even in the absence of the so-called "malpractice crisis" that existed during the data collection phase, is likely to exclude injuries with very serious consequences, and medical investigators are unlikely to be critical of the reports of other medical investigators.

This can again be illustrated fruitfully by the Commission's own report on psychosurgery. Its sample of cases was taken "exclusively" from patients and patient records "volunteered" by four surgeons. On the basis of these 61 patients, the Commission concluded that there was "...at least tentative evidence that some forms of psychosurgery can be of *significant* therapeutic value..." (emphasis added) and the Commission found no serious side effects from these procedures.[13] The evidence supports neither conclusion. The sample is inherently biased in favor of surgery, since its sole source is the surgeons themselves. The claim that there were no serious or harmful side effects can only be made by discounting such matters as seizure disorders, multiple brain operations, and the psychological aftermath of surgery experienced by some of these patients. The psychosurgery report also fails to mention any of the civil suits brought by former psychosurgery patients who feel strongly enough that they have been seriously injured to actually sue.

Other examples can be mentioned, although they were ignored by the Commission. The first is the entire area of drug research in the United States. Although the Food and Drug Administration (FDA) has primary authority over this area, it is human experimentation and necessitates IRB approval. Nevertheless, a 1976 report to Congress by the Comptroller General of the United States indicated that the FDA "...is not adequately regulating new drug testing to insure that human test subjects are protected and test data is accurate and reliable." This report noted that as of June 1974, there were 4600 active investigational new drugs involving more than a quarter of a million test subjects, 1200 sponsors, and 5000 clinical investigators.[14] In September 1979, officials of the FDA testified before Senator Edward Kennedy's Health Care Subcommittee that they were actively investigating researchers at a number of institutions who were involved in the fraudulent reporting of research data.[15] In cancer studies, for example, this could mean that consent to take part in experiments was obtained under false pretenses, since if data is falsified, the results of the experiments will be meaningless to others. This is even worse than the Jewish Chronic Disease Hospital case, and is something that the FDA believes is continuing in the United States on a significant scale. It is fatuous to argue that these research subjects are not being harmed; and equally absurd to think that such investigators would respond honestly to any questionnaire sent to them by the National Commission about "harm" to their subjects.

Two other examples merit attention. The first is a 1979 study of articles published in 1973–1975 by the obstetrics and gynecology faculty members of four New York City medical schools which, the author contends, shows that "one out of eight published papers revealed serious ethical problems involving the exposure of human subjects to unnecessary harm."[16]

Secondly, the great swine flu experiment deserves mention, since, as will be discussed in the next section, the National Commission itself was involved in writing the consent form for it. Although we still do not know how many "research subjects" were harmed, there are currently more than 3000 pending claims against the US Government asking for a total of more than three billion dollars in compensation.[17]

In conclusion, the reason no harm to subjects was found is that none was looked for. The Commission trusted people like themselves to be honest, but it is not the honest researcher we have to worry about. Moreover, the real magnitude of the problem of harm to subjects became apparent only in early 1979 after HEW issued a regulation requiring institutions to state in their consent forms whether or not subjects would be compensated if they were harmed. The cry of foul was heard across the land and almost no institution in the country guaranteed it would compensate victims of human experimentation for the harms that might befall them, harms the institutions and the Commission had been arguing up to this time were almost nonexistent and extremely unlikely.

Premise III: Faith in IRBs

The Commission found it easy to have faith in IRBs, since IRBs are by definition dominated by investigators themselves. Both existing and proposed regulations require such bodies to have five members, four of whom may be investigators from the very institution in which the proposed research will be performed. It is traditional peer review; the kind that makes researchers and physicians very comfortable, but does little to satisfy the average patient or potential research subject.

Of the three premises, this is the most destructive, since most of the Commission's recommendations depend almost exclusively on the local IRBs for their implementation. Thus, if the IRB does not function well to protect human subjects, the Commission's elaborate scheme to protect subjects in general, and children, mental patients, and prisoners in particular, collapses and we are left in a position substantially identical to that which we were in before the Commission began its work in 1974.

Unfortunately, the evidence is all but conclusive that IRBs as currently constituted do not protect research subjects, but rather protect the institution and the institution's investigators. A study of 293 research institutions published in 1973 by Bernard Barber and his colleagues indicated very little impact of the review process on actual practice: only 30 percent of those responding indicated they had ever turned down more than one proposal and only 16 percent required revisions in more than 10 percent of the projects submitted.[18] The National Commission did a more in-depth study of 61 research institutions, and found only 18 percent of the IRB-approved consent forms in use contained the elements actually required by federal law; fewer than 20 percent described available alternatives; statements of subjects' rights were often missing; and fewer than 7 percent were in language as simple as that found in *Time* magazine[19] Similarly, in an interview study by Bradford Gray of individuals actually participating in approved research, 40 percent did not even realize they were in a study, and 41 percent believed falsely that there were no risks involved.[20]

In addition, the almost all-researcher IRB is an anomalous body to determine societal values or even to determine the types of risks reasonable non-researchers might be willing to take, or how these risks can be made understandable to them. As Robert Veatch has expressed it:

> The problem is not one of lacking confidence or trust in the goodwill of the scientists. Trusting scientists to decide about the value of knowledge is a bit like trusting a rabbi friend to pick a good Easter ham. He is to be treated as a person of goodwill, but he is simply not the appropriate person to ask.[21]

Yet even in view of these condemning data, the National Commission continued to base its entire program on substantially the same IRB structure that was in place when it was sworn in. Any outside observer with an elementary knowledge of sociology or human nature would guess that an organization composed of investigators from one particular institution would tend to support research done by other investigators at the same institution. The guess would be correct. One would also guess that in cases where deadlines are pressing, or prominent members of the institution are involved, or large sums of money are in question, these factors will be terribly influential. This is harder to prove, but happily the Commission itself provides a public example.

During the congressional debate over the swine flu program, Congress decided that the consent form should be reviewed by the National Commission, much the way the National Commission believes consent forms should be reviewed by IRBs. The plan was to expose 150 million citizens to a new vaccine (40 million eventually got the drug). Liability was a prime concern, and a good consent form was needed. The legislation required that the Centers for Disease Control (CDC) ask the National Commission for a review and consultation on the consent forms.The head of the CDC, David Sencer, is quoted as having said, "I'll consult if they tell me I have to and then I'll do just what I want."[22] This is precisely what Sencer did. Sixty million consent forms were already printed, and the CDC was anxious to avoid further costs or delays. The National Commission was unable to improve the consent form significantly; indeed it may have made it worse by adding another form to the existing one and having the two stapled together. The result has been charitably described as "a messy product, hard to follow."[23]

It is unfair to suggest that the National Commission was sitting as a "blue ribbon IRB" and had the opportunity to practice what it preached concerning the efficacy of such bodies. But it is not unfair to say that in the one instance the National Commission had to act like the IRBs it puts so much trust in, it failed miserably. John Robertson has made ten modest proposals to improve IRBs; this is the least that should be done.[24]

Conclusion

We are not precisely where we were in 1974, but there is little doubt that if we are to take the interests of subjects more seriously than the interests of researchers, a new approach is needed—one that critically examines the three premises the National Commission took on faith. In this regard the President's new Commission may merit support: there is still much to do to protect human subjects.

Given its composition, we can be reasonably assured that it will do better than Gold's ill-fated Presidential Commission. When it concluded its affairs at its second meeting, Gold was instructed to write its final report:

> Make it short ... and make it long. Make it clear and make it fuzzy. Make it short by coming right to each point. Then make it long by qualifying those points so that nobody can tell the qualifications from the points or ever figure out what we're talking about.[25]

All the President's Bioethicists

Presidential commissions perform diverse functions. President Truman, for example, appointed the Hoover Commission to suggest ways the structure of the federal government could be improved. President Johnson appointed the Warren Commission to investigate the assassination of President Kennedy. Before February 1979, fulfilling a Congressional mandate, Jimmy Carter will appoint a Presidential Commission to ponder the perplexing issues of biomedical ethics. Has medical ethics come of age in Washington?

The Commission's Predecessors

Noting that law and social policy often lagged behind biomedical and technological advances, and were thus generally reactive rather than constructive or channeling forces, Senator Walter Mondale first proposed such a commission in 1968. He introduced legislation to create a National Advisory Commission on Health, Science, and Society "to consider and study the ethical, social, and legal implications of advances in biomedical science and technology." Hearings were held, but the proposal never reached the Senate floor. Senator Mondale continued to introduce his proposal, and in 1974 a modified version was passed as part of the National Research Act of 1974, sponsored by Senator Edward Kennedy.

Among other things, that act established the National Commission for the Protection of Human Subjects of Biomedical and Behavioral Research and gave it a mandate to identify the basic ethical principles that should underlie research on human subjects, develop guidelines for such research, and make recommendations to the Secretary of Health, Education and Welfare (now Health and Human Services) for new regulations that might be required to protect human subjects in general and prisoners, mental patients, and children in particular. The National Commission was also charged with doing a "special study" on the "ethical, social and legal implications of advances in biomedical and behavioral research and technology," a task directly taken from the Mondale proposal, and with making recommendations on fetal research and psychosurgery. The National Commission has completed its work and its statutory existence ended in October 1978.*

Although there have been disagreements with some of the National Commission's proposals, its general success and popularity prompted Senator Kennedy to introduce legislation to prolong its life (by transforming it into a Presidential Commission), and to broaden its mandate to include other research areas. In introducing his bill,[26] he called it "one of the most significant pieces of health legislation" ever debated in the Senate. He cited the recognized abuses of research subjects in the past that led to the National Commission's creation, and stressed that the National Commission had had a profound effect on public policy without having independent regulatory authority.

Kennedy noted that the National Commission's influence stemmed from the high intellectual quality of its work, which provided a vehicle and a forum for the discussion of the issues. Arguing that the need for such a commission was even greater in 1978 than 1974 because of the great expansion in research activities, he sounded a note of urgency and pointed out that the National Commission's work would soon stop and that most of its staff had already left or were leaving.

With his encouragement, the bill easily passed the Senate on June 26, 1978. This version, however, was poorly drafted and its intention inartfully articulated. It expanded the jurisdiction of the National Commission from HEW to encompass virtually all federal agencies and departments, including the Armed Forces and the CIA. It expanded the commission's mandate to include resource allocation in health care delivery, a uniform definition of death, and ethical issues in genetic counseling. Present members of the National Commission were to continue to serve on the President's commission until their successors were chosen. But this was a new mission, and arguably required an entirely new type of commission. Familiar with the ways of a Congress desperately wanting to adjourn for pre-election campaigning, many observers predicted privately that the Kennedy bill would die in

---

*The work of the National Commission is discussed in the preceding essay, "Good as Gold."

the House. On the last day of the session, October 14, 1978, this fate seemed likely. However, on that date the House passed a revised version of the bill. Later, after the House had actually voted to adjourn, the Senate adopted the House's amended version. On November 9, 1978, Jimmy Carter, the self-proclaimed anti-commission President, signed the bill into law.[27]

The New Commission's Mandate

Last-minute compromises and changes often produce flawed legislation. However, most of the House changes strengthened the bill and clarified the commission's role. First, the name was changed to indicate it was an "ethics" commission rather than solely a human research commission. Its formal name is The President's Commission for the Study of Ethical Problems in Medicine and Biomedical and Behavioral Research. Second, its duties were clearly outlined to include studies of the ethical and legal implications of:

(A)   The requirements for *informed consent* to participation in research projects and to otherwise undergo medical procedures
(B)   The matter of *defining death*, including the advisability of developing a uniform definition of death*
(C)   Voluntary testing, counseling, and information and education programs with respect to *genetic diseases and conditions*, taking into account the essential equality of all human beings, born and unborn
(D)   The differences in the *availability of health services* as determined by the income of residence of the persons receiving the services
(E)   Current procedures and mechanisms designed (i) to safeguard the *privacy of human subjects* of biomedical and behavioral research, (ii) to ensure the *confidentiality* of individually identifiable *patient records*, and (iii) to ensure *appropriate access* of patients to information contained in such records
(F)   Such other matters relating to medicine or biomedical or behavioral research as the President may designate for study by the Commission. (emphasis added)

---

*The Commission's 1981 Report on *Defining Death* was to become its most universally well-received product. *See* "Defining Death: There Ought to be a Law" at p. 365.

The priorities and order of these studies are to be determined by the President's Commission. In addition, the commission may undertake to investigate or study any other appropriate matter which relates to medicine or biomedical research "consistent with the legislation."* The jurisdiction of the Commission encompasses all federal agencies (including the Defense Department and the CIA), but its authority to "review in detail any program" of any agency has been deleted and replaced with more general language that would make such reviews the exception rather than the rule.

The mandate of the President's Commission is extremely broad, and could encompass most if not all of the subject matter of "medical ethics." Its powers are persuasive only, although recommendations for specific agency action must be published in the *Federal Register* and commented on by the affected agency within 180 days. The commission is given four years to do its work, and will terminate on September 30, 1982. It can hire a staff, and has an authorized budget of $5 million annually.

HEW's Ethics Advisory Board, which was established by Secretary Joseph Califano in the spring of 1978, and which began meeting in September to consider whether the moratorium on federal funding for in vitro fertilization would be lifted, will continue as a permanent body.** However, legislation authorizing a permanent National Advisory Council for the Protection of Human Subjects of Biomedical and Behavioral Research was explicitly repealed.

The President's Commission will prepare a biennial report on the protection of human subjects in research conducted or supported by federal agencies and make recommendations for further protection when appropriate. It will also report each year to the President, Congress, and appropriate federal agencies on its activities and prepare a list of all recommendations made to various agencies and the agencies' responses to them.

The Composition of the Commission

Those who like "blue ribbon" committees will like this one. Those who are generally cynical of such entities are likely to be skeptical of

---

*It was under this authority that the Commission prepared its most far-reaching report, *Decisions to Forego Life-Sustaining Treatment*, in March, 1983.

**The Ethics Advisory Board issued its own Report on IVF in 1979, but has been dormant since the election of President Ronald Reagan.

this one as well. This skepticism is warranted, since while the name change and enlargement of its mandate were both appropriate, the composition of the President's Commission (although altered slightly from the original Kennedy bill) remains its primary weakness. One might expect that a commission whose primary responsibility is to study "law and ethics" would be composed primarily of legal scholars, philosophers, and ethicists. But this is not the case. The majority of the Commission is made up of representatives from medicine and research, the fields likely to be most affected by its recommendations. Just as the National Commission, although almost evenly balanced between researchers and nonresearchers, essentially endorsed the existing researcher-dominated institutional review board (IRB) system as the subjects' primary protection mechanism, so this commission is likely to endorse only those policies supported by organized medicine and research.

Though unlikely, under the language of the law all eleven Commissioners could be physicians. Three must be distinguished in biomedical or behavioral research; three distinguished in the practice of medicine or provision of health care; and five distinguished "in one or more of the fields of ethics, theology, law, the natural sciences (other than biomedical or behavioral science), the social sciences, the humanities, health administration, government, and public affairs." The chairman and members of the commission are to be named by the President; the chairman with the advice and consent of the Senate.

Health care providers and researchers have an assured majority on the President's Commission. If the remaining five are all distinguished in "ethics, theology, and law" the Commission may provide a forum for important discussions. If, as appears likely, the other five members are spread out among the nine general areas listed in the qualifying language, the impact of ethics or law on this "ethics" commission with an "ethical and legal" mandate is likely to be negligible. It also seems highly desirable, but unlikely, that the chairman be a respected social or legal philosopher.*

It is premature to condemn the legislation as holding out the promise of intensive ethical studies on the one hand, and ensuring that the implications of such studies will not be vigorously pursued on the

---

*In fact the President did name a distinguished attorney, Morris Abram, to this post; and a bright and energetic legal scholar, Alexander Capron, was hired as Staff Director. This combination enabled the Commission to do wide-ranging and in-depth work on a variety of bioethics topics.

other. Nevertheless, both the President and the Commissioners he appoints should realize that the Commission has by its very composition a serious credibility problem, which, if it is to serve its purpose, it must strive to overcome.*

*The President's Commission did not propose any major changes in IRBs, and none have been made. IRBs are probably adequate at reviewing routine, low-risk research; but have demonstrated an almost total inability to deal with novel and high-risk experiments, such as embryo research, genetic engineering, brain implants, the artificial heart, and xenografts. *See* the essays on Barney Clark (p. 391), and Baby Fae (p. 384) for a further discussion of the role of IRBs in research review.

Creationism

Monkey Laws in the Courts

When the "moral majority" attained political prominence, it was reasonable to expect attempts to legislate a new (or old) morality. One tactic has been to resurrect early twentieth century fears that the teaching of evolution in schools would lead to atheism and a breakdown in traditional moral values. Statutes requiring a "balanced" treatment of evolution and "creation science" (evidence for the biblical creation story) were recently passed in Louisiana and Arkansas, and are pending in various forms before 20 state legislatures. The Arkansas statute has been struck down by a lower federal court on the grounds that creation science is not science at all, but religion.[28] But this judicial opinion may not close the books on the issue. Those religions that view evolution as a dangerous heresy will continue to oppose its teaching in any way they feel reasonable; and even if their statutes are consistently declared unconstitutional, some may still introduce them, and legislatures may still pass them, simply to keep the issue in the public eye. The resurgence of antievolution legislation provides us with a useful opportunity to review the history of this issue, and to put the current dispute in its First Amendment context.

The first clause of the First Amendment to the United States Constitution is that "Congress shall make no law respecting the establishment of religion or prohibiting the free exercise thereof ..." The theory of the amendment is that the state and organized religion will survive only if they keep to their own realms. As the US Supreme Court has expressed it, "A union of government and religion tends to destroy government and degrade religion";[29] and, "The law knows no heresy, and is committed to the support of no dogma, the establishment of no sect."[30]

Over the years, three basic antievolution legislative strategies have been attempted in the Southern states. The first was simply to outlaw teaching anything inconsistent with the biblical creation story. The second was to forbid only the teaching that man descended from lower animals. And the most recent is to require a "balanced" treatment in the classroom between evolution science and "creation science."

In March 1925, the Tennessee legislature passed a statute that made it a crime for anyone in the public schools to "teach any theory that denies the story of the divine creation of man as taught in the Bible and to teach instead that man has descended from a lower order of animals." That summer, Joseph Thomas Scopes was tried and convicted of violating the law. On the second day of the trial, his defense counsel, Clarence Darrow, summarized the First Amendment issue as he saw it:

> If today you can take a thing like evolution and make it a crime to teach it in the public schools, tomorrow you can make it a crime to teach it in the private schools ... At the next session you may ban books and newspapers. Soon you may set Catholic against Protestant, and Protestant against Protestant, and try to foist your own religion upon the minds of men.

H. L. Mencken described the trial as an attack on an alleged "conspiracy of scientists...to break down religion, propagate immortality, and reduce mankind to the level of the brute." In this view, Scopes was an agent of Beelzebub, "once removed." As to the jury, Mencken said, "It would certainly be spitting into the eye of reason to call it impartial." The guilty verdict was expected.

On appeal, the Tennessee Supreme Court refused to rule the statute unconstitutional, finding "no unanimity among the members of any religious establishment as to this subject."[31] Instead, the court reversed the conviction on a technicality: the judge had fined Scopes $100, and under Tennessee law, any fine in excess of $50 had to be assessed by the jury. The court did, however, encourage the Attorney General not to try the case again, saying, "We see nothing to be gained by prolonging the life of this bizarre case."

Citizens of Arkansas read the Tennessee opinion carefully, and in 1928 passed a statute they believed would be even stronger against a First Amendment attack. It made it a crime "to teach the theory or doctrine that mankind ascended or descended from a lower order of animals..." That "monkey law" rested undisturbed (and apparently undisturbing) until the mid-1960s, when its constitutionality was challenged by a young high school biology teacher. She sought an injunction barring the application of the statute, and a lower court granted it. On appeal, the Arkansas Supreme Court reversed in a two-sentence opinion, which found the act a "valid exercise of the state's power...." This set the stage for the first and only time an antievolution statute has been reviewed by the US Supreme Court. Justice Fortas wrote the Court's opinion, which struck down the statute as contrary to the First Amendment on the basis that it furthered no secular purpose, only a religious one:

> The overriding fact is that Arkansas law selects from the body of knowledge a particular segment when it proscribes for the *sole reason* that it is deemed to conflict with a particular religious doctrine: that is, with a particular interpretation of the Book of Genesis by a particular religious group... *Government* in our democracy, state and national, *must be neutral* in matters of religious theory, doctrine, and practice.[32] (emphasis added)

This is the only major case in which the Supreme Court relied exclusively on the secular purpose test to find a statute unconstitutional on First Amendment grounds. To refute an attack on a statute based on the establishment clause, three related tests must usually be met: the statute must have a secular purpose, a primary secular effect, and an absence of excessive governmental entanglement in religion.[33]

All three tests were used in early 1982 when the remodeled Arkansas monkey law came under the scrutiny of US District Court Judge William Overton. Seeking to avoid the constitutional problems of requiring or outlawing Bible stories or evolution, the new Arkansas statute required that "public schools...give balanced treatment to creation-science and to evolution-science." Unlike 1925, when a young teacher stood alone and accused of a crime by the state, in 1981 a distinguished group of citizens, including Catholic, Episcopal, and Methodist bishops, Southern Baptist, Presbyterian, and Methodist ministers, with the support of the American Civil Liberties Union, challenged the constitutionality of the statute under the First Amendment.

Judge Overton began his analysis by recounting the belief of fundamentalists in the literal meaning of the Bible. To combat growing secularism, some fundamentalists developed the notion of "scientific creationism" to bolster Genesis. A major group attempting to validate Genesis on the basis of science is the Creation Research Society. All its members must believe that the Bible is "historically and scientifically true," that all basic types of living things, including man, were made by direct creative acts of God during creation week as described in Genesis"; that "the Great Flood...was an historical event..."; and that "Jesus Christ is our Lord and Savior."

The drafter of the Arkansas act is of the opinion that evolution "is forerunner to many social ills, including Naziism, racism, and abortion." He sent the bill to Reverend W. A. Blount, a biblical literalist from Little Rock. At Blount's request, the Evangelical Fellowship unanimously adopted a resolution that it be introduced into the Arkansas legislature. Senator James L. Holstead, a self-described "born again" Christian fundamentalist, introduced the bill into the legislature. No hearings were held in the Senate, and only a 15 minute one was held in the House. The bill passed and was signed by the Governor, who indicated that he did not read it. The court found no evidence that the principals involved were motivated "by anything other than their religious convictions when proposing its adoption or during their lobbying efforts in its behalf." On this basis the judge concluded that the act "was simply and purely an effort to introduce the Biblical version of creation into the public school curricula."

To show that the act has a primary religious, not secular, effect, one must show that "creation science" is religion, not science. This the plaintiffs did. The act defines creation science as:

> ...the scientific evidences and related inferences that indicate: (1) sudden creation of the universe, energy, and life from nothing; (2) the insufficiency of mutation and natural selection in bringing about development of all living kinds from a single organism; (3) changes only within fixed limits of originally created kinds of plants and animals; (4) separate ancestry for man and apes; (5) explanation of the earth's geology by catastrophism, including the occurrence of a world wide flood; and (6) a relatively recent inception of the earth and living kinds.

The judge had no trouble finding that the source of this definition was Genesis. In his words, the ideas of the definition "are not similar to the literal interpretation of Genesis; they are identical and parallel to no other story of creation."

The plaintiffs also produced an impressive array of scientific experts to refute the major tenets of creationism. They argued first that creation science could not be considered science at all, since it lacked all of science's essential characteristics: its conclusions are not tentative, testable, or falsifiable. Its assertions need to be taken on faith and are absolute. For example, creationists put the age of the earth at 6000 to 20,000 years, based on the genealogy of the Old Testament. They refuse to acknowledge paleontological evidence that dates back to 4.5 billion years, or to acknowledge the possibility that the Bible may be in error on this point. When pressed, the best the defenders of creationism could do was to counter with the assertion that, though their method was not scientific, neither was evolution a scientific doctrine. And this may expose the real strategy of many of the creationists: since it is impossible to give a "balanced" presentation of evolution and creationism in the schools, they assume that schools would simply stop teaching evolution, their goal for more than half a century.

The only scientist the creationists called to testify who had not signed the Creation Research Society pledge (acknowledging the belief in the literal interpretation of the Bible) was Welsh astronomer Chandra Wickramasinghe. He testified that life may well not have originated on earth at all, but came from outer space. His theory is based on what he believes is the impossibility of the spontaneous origin of life from nonlife, and the existence of microorganisms and genes in outer space. A Creator, according to him, dispersed microorganisms throughout interstellar space, and they were able to seed planets like earth when they arrived in the tails of comets. New genes rain down periodically and create evolutionary jumps. Regardless of one's view of Wickramasinghe's hypothesis, he helped destroy the creationists' credibility when he answered the following question in the negative: "Could any rational scientist believe that the earth is less than one million years old?"

The judge dealt only briefly with perhaps the central issue in First Amendment religious cases: government entanglement in religious affairs. The law would necessarily involve government deeply in religion through the textbook selection committee, school boards, and administrators who must constantly monitor materials and teaching. This type of government activity is prohibited by the amendment.

Fundamentalists who are unalterably committed to treating the Bible as a repository of not only spiritual but also scientific truth will not be successful in legally mandating that their creed be made part of public school curricula. Judge Overton's opinion is sound, and his conclusions will be followed in other cases where similar statutes are challenged.

The law cannot, of course, inform us on the matter of how life began on earth or how humans evolved, but the First Amendment guarantees us all that religious convictions on these matters cannot be made part of the public school curriculum by statute. In our nontheocratic democracy, the law can play no part in measuring religious truth. Thank God!*

*In the summer of 1987 the US Supreme Court, in the case of *Edwards v. Aguillard*, finally had occasion to examine the constitutionality of the Louisiana creation science statute. It overwhelmingly found it unconstitutional on substantially the same grounds used by Judge Overton in the Arkansas case. *See* Annas, G. J., "Of Monkeys, Man, and Oysters," *Hastings Center Report*, **17**: 20–22 (Aug. 1987).

Life Forms
The Law and the Profits

Although the purposeful creation of new life forms captures the public's imagination, the negligent destruction of existing life forms continues steadily. A recent study of the Worldwatch Institute predicts that by the year 2000, thousands of unique life forms may be destroyed by pollution, destruction of natural habitats, excessive hunting, and population growth.

The law's role in this process has usually been indirect and not carefully reasoned. Two judicial decisions, however, give the US Congress an opportunity to consider the issue anew. Both cases involve the interpretation of statutes passed by Congress, and neither rests on constitutional issues. Thus, if Congress is not satisfied with the way the courts have decided these cases, it can amend the laws in question to nullify or alter the decisions. One case was recently decided by the US Supreme Court, the other is pending before it.

The Tellico Dam

In the case involving the Tellico Dam, the Court found that the language of the Endangered Species Act of 1973 required that a virtually completed dam be prohibited from opening on the basis of a finding by the Secretary of the Interior that its operation would eradicate the snail darter, an endangered species.[34] About $110 million has been spent on the Tellico Dam, a multipurpose project of the Tennessee Valley Authority. In 1973, during a court battle to halt construction, a University of Tennessee ichthyologist, David A. Etnier, found a previously unknown species of perch, the snail darter, *Percian imostoma tanasi*. One of approximately 130 known species of snail darter, it is three inches in length and tannish colored. The total 1973 population of about 15,000 snail darters lived in a habitat that would be destroyed by the operation of the Tellico Dam.

346

Four months after this discovery, but unrelated to it, Congress passed the Endangered Species Act, which provided that no federal action be taken that would "jeopardize the continued existence of such endangered species [as found by the Secretary of Interior] and threatened species or result in the destruction or modification of habitat of such species which is determined by the Secretary to be...critical." In November 1975, the Secretary of the Interior formally listed the snail darter as an endangered species, and found further that the operation of the Tellico Dam "would result in total destruction of the snail darter's habitat," and was thus a "critical habitat."

The Secretary's determinations were not challenged, but the Supreme Court was asked to find that the Endangered Species Act of 1973 was not meant to apply to a project that was virtually completed when the snail darter was discovered. In refusing to do so, Chief Justice Warren Burger, writing for the majority, held that the language of the statute "admits of no exception," and reviewed the legislative history in detail to support this conclusion. Included is this language from a House Committee Report:

> As we homogenize the habitats in which these plants and animals evolved, and as we increase the pressure for products that they are in a position to supply (usually unwilling) we threaten their—and our own—genetic heritage. The value of this genetic heritage is, quite literally, incalculable... From the most narrow possible point of view, it is in the best interests of mankind to minimize the losses of genetic variations. (emphasis added)

From this review, Justice Burger concluded that "Congress was concerned about the 'unknown uses' that endangered species might have and about the 'unforeseeable' place such creature may have in the chain of life on this planet." Finding that Congress viewed the value of an endangered species as "incalculable," he rejected the invitation to weigh the value of the snail darter against the more than $100 million spent on the project, noting that he had no authority "to make such fine utilitarian calculations." His opinion concludes by noting that the Court is not determining that Congress did the "right" thing in passing a statute with this result, but only that it was the judiciary's duty to interpret statutes, not to judge them on their merits. Justice Powell wrote the major dissent (two other Justices also dissented) arguing, among other things, that the result was nonsensical since it would permit the destruction "of even the most important federal project in our country" by a finding that "the continuation of the project would threaten the survival or critical habitat of a newly discovered species of water spider or amoeba."

Patenting Life

In the second case, the Court of Customs and Patent Appeals decided that an application to patent a biologically pure culture of a new microorganism should not have been rejected by the patent office, which had done so on the grounds that it was "a living organism."[35] Under consideration was a patent application by The Upjohn Company for a "Process for Preparing Lincomycin," which used a new microorganism, *Streptomyces vellosus*, in the production.

The question presented was whether a living organism could be patented under the provisions of the US Patent Act:[36]

> Whoever invents or discovers any new and useful process, machine, manufacture, or composition of matter, or any new and useful improvement thereof, may obtain a patent therefor, subject to the conditions and requirements of this title.

The patent office determined that the organism was not patentable under the statute. The Appeals Board affirmed: "If we were to adopt a liberal interpretation of 35 USC 101 new types of insects, such as honeybees, or new varieties of animals produced by selective breeding and crossbreeding would be patentable."

The Court of Customs and Patent Appeals reversed, making a number of points about the new microorganisms:

- It does not exist in, is not found in, and is not a product of nature
- It is man-made and can be produced only under carefully controlled laboratory conditions
- It is alive, and useful because of this
- It is an industrial product used in an industrial process
- Biologically pure cultures like this one "are much more akin to inanimate chemical compositions such as reactants, reagents, and catalysts than they are to horses and honeybees or raspberries and roses."

The fact that the organism was alive, unlike chemical compounds, was, the court concluded, a "distinction without legal significance," and therefore an insufficient reason to reject the patent application. The court further characterized as "far-fetched" the fear that their ruling would "make all new, useful, and unobvious species of plants, animals, and insects created by man patentable."

The decision was three to two. The dissenting judges argued that there is a "fundamental" difference between living things and inanimate chemical compositions. They noted that the microorganism in question, being alive, able to reproduce, and able to act on its surroundings, was more like a honeybee than a chemical. Thus, at least until the US Supreme Court acts on this case,* or Congress changes the law in relation to either case, the law is that no federal activity may destroy a designated endangered species or its habitat, and that new, man-made living organisms are patentable.

Destroying Species

The destruction of an entire species of life is an incredibly arrogant and usually equally ignorant act. There are currently about 1.4 million species of animals and 600,000 species of plants in the world. As many as 10 percent, or 200,000, are either endangered or threatened. It is ironic to note that before the 1970s most of the conservation of animals and critical habitats was funded by hunting and fishing license fees. "Wildlife management" is an old phrase among hunters and fishermen; and that group, which in the nineteenth century was responsible for the indiscriminate slaughter of many species of game, is now often found in the forefront of protection. Natural habitats are now being destroyed by environmental pollution and indiscriminate expansion. No one denies that we should do what we can to ensure the survival of as many diverse species as possible on this planet. This policy can be pursued for either purely selfish reasons (the species may turn out to be valuable to man in some currently unknown way) or out of a conviction that all species have a "right to survive."

Should the prohibition against the destruction of species be relative or absolute? Justice Burger refused to deal in "utilitarian" terms be-

---

*The US Supreme Court eventually affirmed, holding that certain life forms are patentable. *Diamond v. Chakrabarty*, 447 US 303 (1980). Congress has taken no action to amend the Patent Act.

cause he thought it both unauthorized and impossible. But legislation sponsored by Senator John C. Culver was recently passed in the Senate calling for the establishment of a "Review Board" that could consider such situations as the Tellico Dam project and grant exemptions under the Act if the benefits of the project "clearly outweigh" the value of the species. In short, the review board would be empowered to authorize the destruction of an entire species of living creatures for short-term technological advance.

One inherent problem that such a review board will face will be the varying appeal different species have for humans. Deer, eagles, and whooping cranes, for example, arouse keen concern; no federal project is likely to be permitted that would endanger them or a similarly "attractive" species. Cockroaches, spiders, bats, and rats, on the other hand, are not likely to have spirited defenders. Decisions are likely to be made not on any rational basis, but on the public's general emotional reaction to the species in question.

Since a specific project will almost never destroy the *only* place on the planet (or in our country) that a species could live, a rule that no federal project could destroy a critical habitat until one at least as useful to the species was developed in a federally protected area could serve as a workable compromise. The Tellico Dam is the only federal project to date where such a compromise could not be reached, and new habitats for the snail darter apparently can be developed. Such a policy is preferable to a review board with life-and-death authority over other species.

The Law and the Profits

The patentability of new forms of life is arrogance of a different nature. It raises the vexing vista not of destroying life for technological advance or profit, but of controlling life for personal or corporate gain. One assumes that we would not permit the patenting of a human clone, even if this meant patenting at the one-cell level, because of the potential for development into a human being. It may only be our provincial view of time that compels us to protect the "freedom" from commercial exploitation of one type of cell that can produce a human being in nine months, rather than another that might evolve into a thinking life-form in two billion years or less. In both cases, unfortunately, short-term profits rather than well-thought-out philosophical principles are likely to dictate federal policy.

As a society, we are still blessed with options. We can accelerate the modification of the earth to meet human needs, using our technology to clean up the environment, control population growth, and develop new food and energy sources. We can also accept an environment that will become increasingly polluted and difficult to live in, and embark on a genetic engineering task aimed at adapting humankind to this changing environment by developing pollution-breathing, cancer-resistant, grass-eating human beings. Currently we seem to be hedging our bets and traveling along an intermediate path. This path can, however, only be traveled for a few more decades before the second option becomes the only viable one for the human species.*

*Congress did not act to exclude living organisms from the Patent Act. In 1987, both the Board of Patent Appeals and the Commissioner of Patents and Trademarks ruled that the US Supreme Court's language in *Chakrabarty* (interpreting the Patent Laws to include "anything under the sun that is made by man") applied to all living nonhuman entities. The Patent Office says it currently has about fifteen patents pending involving animals. *See* Annas, G. J., "Of Monkeys, Man and Oysters," *Hastings Center Report*, **17**: 20–22 (Aug. 1987).

Transfer Trauma
and the Courts

The role of medical literature in court decisions is little known. In many cases courts cite medical literature to demonstrate that their position is consistent with current medical practice or ethics. But a more important use occurs when judges actually take the findings of medical and sociological studies as the foundation of their opinions. The 1973 abortion decision, *Roe v Wade*,* is such a case. There the majority based its finding about the safety of first trimester abortions on medical literature demonstrating that the risks to a woman were higher if she carried the pregnancy to term than if she chose abortion.

Two recent cases on the subject of involuntary transfer of patients from intermediate care facilities illustrate further how judges can use medical literature to bolster, and even determine, their decision. Both cases involve "transfer trauma," the physical and psychological injuries suffered by a patient involuntarily transferred from a residential care setting.

The Hale Mohalu Lepers

The first case involves a Hawaiian leprosarium. Since the late 1940s Hawaii had supplemented its large leprosarium on the island of Molokai with a smaller residential facility at Hale Mohalu, near Pearl City on the island of Oahu. This facility was established so that those leprosy patients in need of sophisticated care could live near the Honolulu hospitals. Although the facility was originally established by the federal government, in 1977 title to the eleven acres on which it is located reverted to the state of Hawaii, which shortly thereafter began proceedings to close it.

*Discussed at pp. 144–146.

Because of advances in the treatment of leprosy, only the worst cases still need residential care. Therefore most of the residents at Hale Mohalu were among the most elderly, afflicted, and handicapped. In January 1978, the facility was officially closed, but a number of the residents chose to remain. The state continued to provide them with water, electricity, telephone, food, and medicine until September 1978, when these services were terminated. Four days later the residents filed suit to compel the state to resume its services. The lower court refused to issue a preliminary injunction, but the Court of Appeals for the Ninth Circuit thought more information was needed before a decision could be reached.

The court found two possible reasons why the residents might be entitled to a hearing prior to forced transfer. The first was the possibility that the facility was a Medicaid "intermediate care facility," in which case the regulations required the residents be given a fact-finding hearing as to the cause of their transfer, since they were entitled not to be transferred except "for medical reasons or for their own welfare or that of other patients, or for nonpayment...." The second was the patients' interest in life and liberty. In the court's view, Hawaii owed a duty of care to these patients and it could not "act so as to reduce these services to the point of imperiling life or imposing other severe hardships without affording the recipients a predetermination hearing." Without specifically citing medical or sociological studies, the court noted: "There has been considerable judicial and scientific recognition of the phenomenon known as 'transfer trauma ...' characterized by physical and emotional deterioration as well as by increased rates of mortality." To determine whether transfer trauma might occur at Hale Mohalu, the court sent the case back for further proceedings.[37]

The Pennsylvania Nursing Home Patients

The second case is a US Supreme Court opinion dealing with 180 elderly residents of a nursing home in Philadelphia. The Department of Health, Education and Welfare (now Health and Human Services) found in 1977 that the nursing home failed to meet federal standards to qualify as a skilled nursing facility, and accordingly did not renew its Medicare provider status. Three days later, the state of Pennsylvania terminated its Medicaid provider status. The question presented was whether the "residents" of the nursing home have a constitutional right to a hearing before a state or federal agency may revoke its Medicare or Medicaid provider status. Justice John Paul Stevens, writing for the majority of a seven-to-one Court, found that they did not.[38]

This decision reversed a lower court, which had concluded that a hearing was required because the decertification of a home inevitably forced the transfer of all residents receiving Medicare and Medicaid benefits—triggering the residents' right to a hearing on the issue of whether there was an adequate basis for the transfer. Justice Stevens disagreed. He noted first that while Medicare and Medicaid recipients were entitled to receive services, they were not entitled to receive them from an unqualified provider. And even though the impact of a decertification decision is transfer of patients, that impact "is an *indirect or incidental result* of the Government's enforcement action, and does not amount to a deprivation of any interest in life, liberty, or property." (emphasis added)

Their position under these circumstances would be comparable to that of members of a family who have been dependent on an errant father, they may suffer serious trauma if he is deprived of his liberty or property as a consequence or criminal proceedings, but surely they have no constitutional right to participate in his trial or sentencing procedures.

Justice Stevens was willing to concede the existence of transfer trauma, and assume that some patients might suffer "both emotional and physical harm" or "severe hardship" from the transfer. But he concluded that since such results were only indirectly caused by the government, the patients had no right to a hearing prior to their transfer.

Blackmun on Transfer Trauma

As he did in his majority opinion in *Roe v Wade*, Justice Harry Blackmun went to the medical literature to bolster his position. In his concurring opinion, he terms the majority opinion "simplistic and unsatisfactory" and undertakes a lengthly and cogent analysis of principles of administrative law applicable to the case.

He labels the adjective "indirect" a platitude that submerges the analytical complexities of this case, noting that transfer is not only the inevitable consequence of decertification, it is also a "basic purpose of decertification." The government often believes patients will be better off in a facility that meets its standards. But this argues for a hearing, since "a basic purpose of affording a hearing in such cases is to test the Government's judgment that its action will in fact prove to be beneficial." However, since the nursing home had the opportunity to make the very arguments the patients were likely to make, Blackmun does not believe due process requires their participation in the hearing.

Blackmun's next argument on transfer trauma relies almost exclusively on the medical literature. The residents had argued that since they could be deprived of "life and liberty" by the transfer, a hearing was required. The majority, as noted, found it unnecessary to examine this contention because it found such effects "indirect." Blackmun disagrees. Chiding the Court for implying that "regardless of the demonstrated risk that widespread illness or even death" may result from transfer, no hearing is required. He terms this holding "heartless" and concludes that "a governmental decision that imposes a high risk of death or serious illness on identifiable patients must be deemed to have an impact on their liberty." But, based on the medical literature, Blackmun concludes that any danger of transfer trauma is speculative. In his words, "Substantial evidence suggests that 'transfer trauma' does not exist, and many informed researchers have concluded at least that this danger is unproved. Recognition of a constitutional right plainly cannot rest on such an inconclusive body of research and opinion."

He supports this conclusion by citing one article published in *The Gerontologist* that found that though six studies had concluded that transfer resulted in increased mortality rates, twelve studies had found no significant effect on mortality rates and one had actually found a significantly lower mortality rate.[39] The study does support Blackmun's conclusion (the authors argue that "Since 85.7% of all studies utilizing the experimental control design and the majority [75%] of all studies found similar mortality rates for both relocated and nonrelocated groups, the data overwhelmingly support the premise that relocation does not influence mortality"). However, it seems somewhat extreme to base a judicial decision primarily on one article that is not even cited or discussed by the other justices. The article may well be authoritative, but its mere appearance in *The Gerontologist* does not make it so. Indeed, Blackmun also cites two earlier articles from this journal that came to a different conclusion.

Conclusion

The law of the Pennsylvania nursing home case is that patients do not have a constitutional right to a hearing before the government terminates the Medicare or Medicaid provider status of a nursing home, or, arguably, therefore, before it closes one, like Hale Mohalu, that it operates itself. This rule may be too harsh, and at least for cases in which the patients allege that the termination of their services will cause them severe injury, and possibly death, a hearing should be required. Although there is no right to live in an unqualified home at government expense, patients should retain the right to demonstrate that no safe and reasonable alternatives exist. And if patients, like those at Hale Mohalu, can demonstrate the existence of "transfer trauma" in their own cases, the government should not be permitted to close the facility unless it can demonstrate a compelling state interest for doing so.

Proof of the likelihood of such harm, however, will apparently require better documentation of transfer trauma than currently exists in the medical and sociological literature. But since all but one member of the US Supreme Court is willing to concede that transfers can be traumatic, it need not require the impossible. Indeed, the Second Circuit in a post-*O'Bannon* case has already ruled that in an instance of *direct* transfer, nursing home residents should be provided the opportunity to demonstrate the likelihood of such harm. They must rely on their own situation, not on general conclusions from the medical literature. In the words of that court, "The record in this case contains ample evidence that transfer of elderly patients, even when it does not pose increased risks of morbidity, carries the undeniable possibility of emotional and psychological harm—at least in the case of many individuals."[40]

This is the proper approach. Instead of routinely denying hearings on the basis of indirect effect, or relying on studies showing that the phenomenon of transfer trauma has not been conclusively proven, courts should focus attention on specific individuals who may suffer specific harm. Though there are some practical reasons for denying patients a constitutional right to a hearing where the government terminates a program or a provider, the patients affected should at least have access to the courts in cases where they allege a serious threat to "life and liberty."

Prison Hunger Strikes
Why Motive Matters

When the Irish hunger strikers in Maze prison focused world attention on the situation in Northern Ireland, the British Government decided not to force feed them. In a statement to the House of Commons, Humphrey Atkins, the Secretary for Northern Ireland, said:

> We do not want any prisoners to die; but if they persist in their hunger strike they will not be forcibly fed. If they die, it will be from their own choice...we shall not let the way we run the prisons be determined by hunger strikes or any other threat.

The statement sums up many of the most important issues at stake when a prisoner decides to go on a hunger strike. What are the prisoner's motivations? Do they matter? Is the prisoner attempting to commit suicide? Does the state have an obligation to intervene to prevent death? What difference does the prison setting make? Hunger strikes differ from other forms of "suicidal" behavior because the death is a very slow one. This route is usually chosen because its very slowness gives others a chance to meet the political or personal demands of the hunger striker. Two other responses are, of course, possible. The striking prisoner could be force fed, or permitted to die.

The length of hunger strikes has ranged from a few days to almost two months. Death from dehydration will occur in only a few days if both food and water are refused; but a fast can last much longer if water or other liquids are consumed. After depleting the body's fat, muscle protein is used to produce glucose. Strikers can suffer blindness and brain damage before they ultimately starve to death.

In the United States, hunger-striking prisoners have been routinely force fed, and until recently no prison official or prisoner found it necessary or appropriate to go to court to obtain either an injunction against force feeding or permission to stop taking nourishment. Not long ago (1982), however, three appellate decisions were handed down. In each case the hunger-striking prisoner relied on court decisions involving the right to refuse medical treatment,* to argue that he had a constitutional right to refuse force feeding.

Three Hunger Strikes

Ted Anthony Prevatte was a "sane and rational" prisoner who began his hunger strike to obtain transfer out of the Georgia prison system to North Carolina. He felt he would be killed by other prisoners if he remained in any Georgia prison. Prevatte refused even medical examinations to determine his condition, and the prison officials sought an order authorizing them to examine him and, if necessary, force feed him to prevent his death. The state said that it had a duty to keep prisoners safe from harm and to render medical assistance when necessary. Prevatte countered that he had a constitutional right to control his own body and that the state could demonstrate no compelling interest in interfering with his exercise of this right.

A lower court denied the state's petition, concluding that the state may incarcerate and even execute a prisoner, but "has no right to destroy a person's will by frustrating his attempt to die if necessary to make a point." In a one-paragraph discussion, the Supreme Court of Georgia agreed. It based its conclusion on Prevatte's mental competence and his right of privacy, and cited three cases involving refusal of treatment by competent patients as authority.[41]

John Lennon's killer, Mark David Chapman, was transferred to a psychiatric unit after he began a hunger strike and expressed his intention to take his life by starvation. The prison system sought a court order authorizing force feeding of Chapman either intravenously or by a nasogastric tube. Testimony indicated that Chapman was competent and desired to die.

Although recognizing Chapman's right to privacy, the court found a compelling state interest in preventing suicide. It distinguished the cases involving the "right to refuse radical surgery or similar medical treatment" although it did not discuss its reasoning in any more than a conclusory fashion: "Even a superficial comparison...illustrates their essential dissimilarity."[42]

*See p. 259 et seq.

Jesse White is a murderer serving a life sentence without mercy in West Virginia. He began a hunger strike to protest conditions in the Moundsville State Penitentiary. After prison officials announced their intention to force-feed him, he sought an injunction forbidding this action. The court rejected his request. It disagreed with the Georgia court, saying it "failed to consider compelling reasons for preserving life, not the least being civility." Though recognizing that "competent, rational patients" have a right to refuse treatment, the court concluded that "West Virginia's interest in preserving life is superior to White's personal privacy...."[43]

The Right to Privacy

All three courts cite a series of right-to-refuse-treatment cases and make their decision based on either uncritically accepting them as determinative (Georgia) or uncritically rejecting them (New York and West Virginia). It thus seems reasonable to ask whether refusal of food by a prisoner should be considered the same as refusal of medical treatment by a patient.

In each instance the individual's primary argument against medical intervention is the right to privacy in the sense of autonomy. It is not a question of whether the patient or prisoner has such a right, but whether the state can enunciate a "compelling interest" that justifies interfering with its exercise.

Courts have traditionally listed four such interests: (1) preservation of life; (2) protection of innocent third parties; (3) prevention of suicide; and (4) protection of the ethical integrity of the medical profession. Do these operate differently in the context of a hunger strike than in a refusal-of-medical-treatment case?

In both contexts the preservation-of-life argument seems almost identical. As expressed by the Massachusetts Supreme Judicial Court:

> The constitutional right to privacy...is an expression of the sanctity of individual choice and self-determination as fundamental constituents of life. The value of life as so perceived is lessened not by a decision to refuse treatment, but by the failure to allow a competent human being the right of choice.

In examining this issue, courts have also considered the prognosis and the degree of invasion involved in treatment. The degree of invasion necessitated by force feeding may seem minimal compared with the prognosis; but as Lawrence Altman has pointed out, force feeding is likely to involve "dragging" the prisoner to the site of the feeding, sedating him, and using physical restraints to prevent the removal of the feeding tube. Placing the tube may also be "tricky" and could result in fatal complications should it enter the trachea. The procedure may have to be repeated daily.[44]

Protecting innocent third parties, like minor children, should also be treated the same with regard to both hunger strikes and refusal of treatment, since in both cases the children's interest in the survival of their relative would be similar.

Prevention of suicide is more problematic. Courts have concluded that refusal of treatment in the face of an inevitable death is not suicide for one of a variety of reasons: (1) the person may not have the specific intent to die; (2) even if he or she does, death occurs by natural causes and not because the person put the death-producing agent in motion; and (3) suicide involves irrational self-destruction, and not a "competent, rational decision to refuse treatment." The major difference here, of course, is that the prisoner is neither sick nor terminally ill. On the other hand, unless he expressly intends to kill himself by starvation (like Chapman), one can argue that he does not desire death, but, like the Irish hunger strikers, has made a rational decision to attempt to obtain specific objectives knowing that the method could lead to death.

A prisoner under a death sentence would be in a more analogous position to a terminally ill patient. It may seem anomalous to refuse to permit such an individual to commit suicide or starve himself to death (as, for example, Gary Gilmore attempted). The primary justification is the state's right to punish the offender and wreak vengeance for society—an issue not at stake outside the criminal setting. Thus the state may indeed have more of an interest in preventing the suicide of a prisoner than of a patient.

The ethical integrity of the medical profession has never proven decisive in a case involving refusal of treatment. Medical ethics generally does not conflict with judicial decisions in these areas; but if it did, the right of self-determination of the individual would probably be considered more important.

So, except in the case of a prisoner who is intent on self-destruction, the state interests in interfering with an individual's right to privacy are superficially similar, whether one is dealing with a refusal of medical treatment or a hunger striker. The essential difference has to do with the nature of imprisonment.

The Nature of Imprisonment

We circumscribe the rights of prisoners in many more ways than those of other citizens, and denying them the "right" to refuse force-feeding may be just one more way. Perhaps the most persuasive arguments have been advanced by the Massachusetts Supreme Judicial Court in another context: Kenneth Myers was a prisoner who, needing kidney dialysis to live, refused dialysis in an attempt to get a transfer to another correctional facility. Myers had consented to a kidney transplant before the case reached the state's highest court, but the court decided to rule on the case because he could change his mind.

The court found that because of the magnitude of the invasion involved in kidney dialysis, Myers would have a right to refuse dialysis if he had been a free man. But because he was a prisoner attempting to manipulate the system, he did not. This was because of a fifth state interest: upholding orderly prison administration. The Corrections Commissioner argued that Myers's action could prompt others to "attempt similar forms of coercion" and his death would generate a possibly "explosive" reaction. The court wrote:

> Correctional needs in a case such as this are urgent and ought to be given considerable weight, especially when the prisoner's refusal of life-giving treatment is predicated on an attempt to manipulate his placement within the prison system.

The court concluded that the state's interest in maintaining order in prisons could be compelling enough to coerce even invasive treatment if the prisoner's motives in refusing it were to manipulate the system to his own advantage.[45]

This result seems correct. In the two cases discussed above, both prisoners ate rather than be force fed. Jesse White voluntarily ended his fast and later went to work as the prison's chief cook. Within four months he regained 50 of the 100 pounds he had lost during his hunger strike. Chapman resumed taking liquid nourishment rather than be force fed via a nasogastric tube.

Motivation is the most crucial distinction between patients who refuse treatment and prisoners who refuse to eat. Because the latter generally seek either to manipulate the prison system for their own benefit or to commit suicide, prison officials seem justified in seeking court orders for force feeding to maintain order and security. On the other hand, when motivations are similar—as when the life of a prisoner has become intolerable to him or her because of the nature of an invasive medical treatment—the prisoner should have the same right to refuse treatment as a nonprisoner.

In analyzing right-to-refuse-treatment cases, the context of the refusal is extremely important. The social context will determine how we balance the rights of the individual versus the rights of the state. This is the issue the Georgia court completely overlooked. In failing to take context into account, it failed to arrive at a reasoned conclusion. We restrict the rights of prisoners in many ways. Force feeding them, rather than permitting them to starve themselves to death, is probably one of the most benign.

Transplants
and Implants

Introduction

It seems fitting to close this collection of essays with a series on transplants and implants because the legal and ethical issues debated today are almost the identical ones that were at the center of bioethics and legal debates twenty years ago. The players are new, but transplants and implants have been around a long time, and the legal issues they have spawned have not yet been adequately dealt with.

Perhaps this is because powerful medical technologies, like heart transplants and implants, do not just change what we as human beings can do to each other, they also change how we think about each other. One example is our concept of death. Heart transplants would involve a double homicide unless we had an alternative to declaring death on the basis of irreversible heart stoppage. "Brain death" has become this alternative, and by adopting this concept to permit organ transplantation, that technology has succeeded in actually changing the way we think about death itself.

This section provides a fitting conclusion for another reason: transplantation has again become a major media event, and discussion of it has been used to open the entire area of resource allocation in health care. It is an area that I had only been peripherally involved in until 1983, when I was named Chairman of the Massachusetts Task Force on Organ Transplantation, a multidisciplinary group that was charged with developing a policy for the introduction of heart and liver transplant services into the Commonwealth of Massachusetts. This experience gave me a somewhat broader perspective on the subject than I had previously had, and gave me the incentive to think about headline-grabbing experiments on artificial hearts and xenografts in a way not previously open to me.

My writing and speaking on the artificial heart, for example, led to an appearance before an FDA panel charged with reviewing the permanent artificial heart experience (December, 1985), and before a Congressional panel reviewing the "Status of the Artificial Heart Program" (February, 1986). On both occasions I urged a moratorium on further permanent and temporary implants. Prior to this time, in August, 1985, William DeVries was kind enough to invite me to spend three days with him in Louisville, reviewing his procedures, reading the protocol for permanent artificial hearts, and letting me talk with members of the IRB, as well as Juanita and Murray Hayden, and Margaret and William Schroeder. Although most people seem to agree that it is time to halt permanent implants and return to the lab for further work, temporary implants have become almost fashionable. The final two essays in this section are about this phenomenon, and concentrate on Jack Copeland's championing of temporary implants in the press.

But none of the essays in this collection struck a deeper and more responsive chord than the essay on the Baby Fae experiment. In retrospect it is much easier to see that this operation was primarily an exercise in "faith healing," with almost no scientific foundation or preparation. Indeed, a *Journal of the American Medical Association* editorial, commenting on the published case report, termed it "wishful thinking," while Jack Provencha, Loma Linda's bioethicist, termed it simply, "a leap of faith." These descriptions seem accurate, and this unscientific xenograft now seems little more than an atavistic sideshow.

All but the second essay in this section, "The Prostitute, the Playboy and the Poet" (which won the 1985 *Wall Street Journal* award for the "catchiest title by an academic" and was published in the *American Journal of Public Health* in February, 1985) are from the *Hastings Center Report*. "Defining Death" was published in February, 1983; "The Pursuit of Organ Sales" in February, 1984; "Baby Fae" in February, 1985; "Consent to the Artificial Heart" in April, 1983; "The Phoenix Heart" in June, 1985; and "No Cheers for Artificial Hearts" in October, 1985.

Defining Death
There Ought to Be a Law

Una Loy Clark asked her husband if, now that he had an artificial heart, "he might not love" his family any longer. The heart is our symbol of love and life; and its demise quite naturally seems the end of both. Nontheless, Barney Clark's was only the most spectacular in a series of technological advances demonstrating that life can continue even after one's own heart is removed. As a society we have come to accept this, and agree that death of the brain is a surer indication of biological finality than death of the heart.

Even though his heart was "dead" and his circulation was maintained artificially, Clark was obviously alive. Turning off the air compressor that powered his artificial heart without his prior approval would have been murder. And no statute was needed to declare him dead when his brain ceased to function (a good thing, since Utah had no brain death statute) even though his artificial heart could continue to circulate blood through his body for additional weeks or even years.

The Proposed Statute

"There ought to be a law" is a common phrase in third year law school seminars; it is also increasingly suggested as a way for physicians to deal with problems that make them uncomfortable. In the early 1960s, for example, physicians called for the passage of "Good Samaritan statutes" to protect them from a perceived danger of liability suits if they rendered emergency aid. More recently physicians have called for legislation to define their rights regarding termination of treatment and writing orders not to resuscitate. The American Medical Association (AMA) has even changed its policy on brain death statutes. It now supports uniform legislation to define death, presumably on the assumption that this will relieve physicians of legal liability for pronouncement of death based on brain death criteria.

365

Legislation has many advantages, not the least of which is clarification of the law on a particular subject. But it is a serious error to assume either that in the absence of legislation there is no law, or that legislation will solve the myriad of personal and emotional factors that control the way both physicians and nonphysicians deal with difficult issues. Perhaps nowhere is this as clear as in the debate on brain death legislation. Currently, the AMA has been joined by the American Bar Association, the National Conference of Commissioners on Uniform State Laws, and the President's Commission for the Study of Ethical Problems in Medicine and Biomedical and Behavioral Research in suggesting that all states adopt the following statute:

> An individual who has sustained either (1) irreversible cessation of circulatory and respiratory functions, or (2) irreversible cessation of all functions of the entire brain, including the brain stem, is dead. A determination of death must be made in accordance with accepted medical standards.

It is hard to oppose such a statute, and I do not. Nevertheless, it should be clearly recognized that this statute merely codifies the common law. For those who are confused by the current law, it brings clarification; for those who are not, it is unlikely to help in their day-to-day decisionmaking. To put it another way: the law has always been that an individual is dead when a doctor, using accepted medical criteria, declares the individual dead. This law has consistently been interpreted by courts to include the utilization of total brain death. Though society might want to take a stance that this is unjustified, no state legislature has so acted; thus physicians may utilize accepted brain death criteria in every state, whether or not it has adopted (or ever does adopt) legislation on this topic.

Brain Death Criteria

Uneasiness about brain death is probably a product of its genesis and the physical appearance of an individual on a ventilator who meets accepted brain death criteria. When Harvard's Ad Hoc Committee on the Definition of Death published its report in the *Journal of the American Medical Association* in 1968, it gave two reasons for proposing a new definition: (1) Improved methods of resuscitation and support of individuals whose brains were damaged irreversibly permitted their hearts to continue beating, at great emotional and financial cost; and (2) It was difficult to obtain organs for transplant using the traditional definition of death. These reasons remain the basis for using "brain death" criteria to determine death when technological means are used to prolong respiration and circulation in a body whose brain has ceased to function.

The committee further suggested that since the courts had always accepted the definition of death adopted by the medical profession as valid for all legal purposes, legislation to redefine death was unnecessary as long as the medical community agreed upon brain death criteria.

Fending Off Lawsuits

This conclusion was correct, and nothing has changed since 1968 to render it obsolete. Nonetheless, since that time there seems to have been a shift in physician attitudes toward the law, as seen not only in the medical malpractice insurance crisis of the mid-1970s, but also in the series of cases involving discontinuance of medical treatment.* Many physicians have apparently returned to the early 1960s mentality that a lawsuit is an omnipresent threat, and that physicians should not be expected to do what they believe is right and in the best interests of their patients, even with their patients' consent, if there is any possibility at all that they might be sued.

Although such an attitude is, perhaps, understandable, it is based primarily on ignorance. In the "Good Samaritan" example, mentioned previously, no physician has ever been successfully sued for stopping and rendering emergency assistance outside his usual working hours. Likewise, although there have been some legal challenges to the notion that doctors can declare a person dead using brain death criteria, most involved criminal defendants charged with homicide who alleged that it was the physician who discontinued the respirator, and not the defendant's act, that caused the death of the victim. In these, and in every single case in which physicians have applied brain death criteria, courts have ultimately accepted it as a legally satisfactory definition of death.

*These cases are extensively reviewed in "Death, Dying, and Refusing Medical Treatment," p. 259 *et seq.*

In the Absence of Statute

What if the quest for uniform legislation is not successful? Will this raise the prospect (one that worries many commentators) that a person may be legally dead in one state, but all of a sudden become legally "alive" when the ambulance crosses into another? Although a theoretical possibility, such a result is unlikely. All current statutes permit death to be pronounced when the patient's total brain function has ceased irreversibly. A "brain death" pronouncement also can be made in the absence of a statute, since this practice has been accepted by the medical community and the courts. There is also no meaningful distinction between "medical death" and "legal death."

In retrospect, it is easy to agree with the initial assessment of the Harvard committee that legislation was unnecessary. But we can't turn back the clock. We already have brain death statutes in a majority of the states and will likely get them in most others. No harm is likely to come from this exercise. Nevertheless, it is critical for the "deceased" and their families that physicians understand that they can, both now and in the future, use accepted medical criteria for determining death, and the failure of their state to adopt the uniform act will not affect this.

Contrary arguments are unpersuasive. Some see physicians who declare patients dead on the basis of accepted medical criteria in states that do not have a statute as playing "a game of Russian roulette with prosecutors or with disgruntled relatives of a person declared dead by brain-based criteria." They argue that assuring physicians that their best clinical judgment in declaring death will be legally accepted is "similar to the assurance one might derive from the fact that no American physician has ever been convicted for mercy killing."[1]

What can this mean? It cannot mean that the prosecutor, the family, or the judge or jury can declare someone who is dead alive again. Nor can it seriously be argued that affirmatively killing an individual is the same (or even analogous to) diagnosing and declaring brain death. A majority of legislatures and every court that has ever studied the matter has sanctioned the use of brain death criteria as standard medical practice. In contrast, no court or legislature has ever sanctioned mercy killing; its legal prohibition is clear, and it has never been accepted medical practice. It is simply a play to paranoia to suggest that any physician will ever be convicted of homicide for terminating treatment after pronouncement of brain death based on accepted medical standards.

Of course, physicians may not declare death on whatever criteria they choose, and society does retain the ultimate right to decide what criteria should be used in the event any criteria other than whole brain death criteria is used (e.g., neocortical death). However, when the criteria physicians actually do use comport with irreversible biological death (like the current brain death criteria) there is no need for additional legislation to protect physicians from criminal prosecution or civil lawsuits based on utilization of these criteria.

Nor is legislation likely to make those physicians who feel uncomfortable about pronouncing death on the basis of brain death any more comfortable; just as "Good Samaritan" legislation has had no measurable effect on the number of physicians who stop and render aid. Physicians refuse to stop for a variety of reasons (they don't want to get involved; they don't know emergency medicine) of which the law may be the least important. Likewise, many physicians will continue to feel uneasy about brain death regardless of the existence of a state statute.

Consider the case of Korean boxer Duk Koo Kim, who was beaten to death in a Las Vegas ring in November 1982. Nevada has a brain death statute, and this diagnosis was made shortly after the fight ended. Nevertheless, Duk Koo Kim was not immediately removed from the ventilator. Instead efforts were made to determine the identity of his wife, and to have him treated by a Korean physician, who said, "In the culture of the Korean people, Kim is still alive." His attending physician is quoted as having said he would make the decision to disconnect the respirator consistent with the "law, medical ethics, and the wishes of the family." Later he called in a judge, to "protect" himself, and it was only when the patient was actually declared dead with the concurrence of the judge and the patient's family that the ventilator was finally removed. Public attention obviously was a more important determiner of action than Nevada's brain death statute.

Difficult problems cannot be legislated away. Nor should problems which have already been dealt with be the subject of protracted legislative battles. Such battles tend to confuse more than educate. If the quest for a uniform statute is simply a quest for clarity, that goal can be obtained as easily by repealing all the existing brain death statutes and relying on the common law. The result would be the same as if all fifty states adopted the Uniform Determination of Death Act.

The Prostitute, the Playboy, and the Poet

Rationing Schemes
for Organ Transplantation*

In the public debate about the availability of heart and liver transplants, the issue of rationing on a massive scale has been credibly raised for the first time in US medical care. In an era of scarce resources, the eventual arrival of such a discussion was, of course, inevitable.[2] Unless we decide to ban heart and liver transplantation, or make them available to everyone, some rationing scheme must be used to choose among potential transplant candidates. The debate has existed throughout the history of medical ethics. Traditionally it has been stated as a choice between saving one of two patients, both of whom require the immediate assistance of the only available physician to survive.

More recently national attention was focused on decisions regarding the rationing of kidney dialysis machines when they were first used on a limited basis in the late 1960s. As one commentator described the debate within the medical profession:

> Shall machines or organs go to the sickest, or to the ones with most promise of recovery; on a first-come, first-served basis; to the most "valuable" patient (based on wealth, education, position, what?); to the one with the most dependents; to women and children first; to those who can pay; to whom? Or should lots be cast, impersonally and uncritically...[3]

*Originally prepared as a background paper for the *Report of the Massachusetts Task Force on Organ Transplantation*, Dept. of Public Health, Commonwealth of Massachusetts, Boston, MA, Oct., 1984 (Annas, G.J., Chairman)

In Seattle, Washington, where the dialysis shunt was developed, an anonymous screening committee was set up to pick who among competing candidates would receive life-saving kidney dialysis. One lay member of the screening committee is quoted as saying:

> The choices were hard...I remember voting against a young woman who was a known prostitute. I found I couldn't vote for her, rather than another candidate, a young wife and mother. I also voted against a young man who, until he learned he had renal failure, had been a ne'er do-well, a real playboy. He promised he would reform his character, go back to school, and so on, if only he were selected for treatment. But I felt I'd lived long enough to know that a person like that won't really do what he was promising at the time.[4]

When the biases and "selection criteria" of the committee were made public, there was a general negative reaction against this type of arbitrary device. Two experts reacted to the "numbing accounts of how close to the surface lie the prejudices and mindless cliches that pollute the committee's deliberations," by concluding that the committee was "measuring persons in accordance with its own middle-class values." The committee process, they noted, ruled out "creative nonconformists" and made the Pacific Northwest "no place for a Henry David Thoreau with bad kidneys."[5]

To avoid having to make such explicit, arbitrary, "social worth" determination, the Congress, in 1972, enacted legislation that provided federal funds for virtually all kidney dialysis and kidney transplant procedures in the US.[6] This decision, however, simply served to postpone the time when identical decisions will have to be made about candidates for heart and liver transplantation in a society that does not provide sufficient financial and medical resources to provide all "suitable" candidates with the operation.

There are four major approaches to rationing scarce medical resources: the market approach, the selection committee approach, the lottery approach, and the "customary" approach.

The Market Approach

The market approach would provide an organ to everyone who could pay for it with their own funds or private insurance. It puts a very high value on individual rights, and a very low value on equality and fairness. It has properly been criticized on a number of bases, including that the transplant technologies have been developed and are supported with public funds, that medical resources used for transplantation will not be available for higher priority care, and that financial success alone is an insufficient justification for demanding a medical procedure. Most telling is its complete lack of concern for fairness and equity.[7]

A "bake sale" or charity approach that requires the less financially fortunate to make public appeals for funding is demeaning to the individuals involved, and to society as a whole. Rationing by financial ability says we do not believe in equality, but believe that a price can and should be placed on human life and that it should be paid by the individual whose life is at stake. Neither belief is tolerable in a society in which income is inequitably distributed.

The Committee Selection Process

The Seattle Selection Committee is a model of the committee process. Ethics Committees set up in some hospitals to decide whether or not certain handicapped newborn infants should be given medical care may represent another.[8] These committees have developed because it was seen as unworkable or unwise to explicitly set forth the criteria on which selection decisions would be made. But only two results are possible, as Guido Calabresi has pointed out: either a pattern of decisionmaking will develop or it will not. If a pattern does develop (e.g., in Seattle, the imposition of middle-class values), then it can be articulated and those decision "rules" codified and used directly, without resort to the committee. If a pattern does not develop, the committee is vulnerable to the charge that it is acting arbitrarily, or dishonestly, and therefore cannot be permitted to continue to make such important decisions.

In the end, public designation of a committee to make selection decisions on vague criteria will fail because it too closely involves the state and all members of society in explicitly preferring specific individuals over others, and in devaluing the interests those others have in living. It thus directly undermines, as surely as the market system does, society's view of equality and the value of human life.

The Lottery Approach

The lottery approach is the ultimate equalizer which puts equality ahead of every other value. This makes it extremely attractive, since all comers have an equal chance at selection regardless of race, color, creed, or financial status. On the other hand, it offends our notions of efficiency and fairness since it makes no distinctions among such things as the strength of the desires of the candidates, their potential survival, and their quality of life. In this sense it is a mindless method of trying to solve society's dilemma which is caused by its unwillingness or inability to spend enough resources to make a lottery unnecessary. By making this macro spending decision evident to all, it also undermines society's view of the pricelessness of human life. A first-come, first-served system is a type of national lottery since referral to a transplant program is generally random in time. Nonetheless, higher income groups have quicker access to referral networks and thus have an inherent advantage over the poor in a strict first-come, first-served system.[9]

The Customary Approach

Society has traditionally attempted to avoid explicitly recognizing that we are making a choice not to save individual lives because it is too expensive to do so.[10] As long as such decisions are not explicitly acknowledged, they can be tolerated by society. For example, until recently there was said to be a widely held understanding among general practitioners in Britain that individuals over age 55 suffering from end-stage kidney disease not be referred for dialysis or transplant. In 1984, however, this unwritten practice became highly publicized, with figures that showed a rate of new cases of end-stage kidney disease treated in Britain at 40 per million (versus the US figure of 80 per million) resulting in 1500–3000 "unnecessary deaths" annually.[11] This has, predictably, led to movements to enlarge the National Health Service budget to expand dialysis services to meet this need, a more socially acceptable solution than permitting the now publicly recognized situation to continue.

In the US, the customary approach permits individual physicians to select their patients on the basis of medical criteria or clinical suitability. This, however, contains much hidden social worth criteria. For example, one criterion, common in the transplant literature, requires an individual to have sufficient family support for successful after-care. This discriminates against individuals without families and those who have become alienated from their families. The criterion may be relevant, but it is hardly medical.

Similar observations can be made about medical criteria that include IQ, mental illness, criminal records, employment, indigency, alcoholism, drug addiction, or geographical location. Age is perhaps more difficult, since it may be impressionistically related to outcome. But it is not medically logical to assume that an individual who is 49 years old is necessarily a better candidate for a transplant than one who is 50 years old. Unless specific examination of the characteristics of older persons make them less desirable candidates is undertaken, such a cut off is arbitrary, and thus devalues the lives of older citizens. The same can be said of blanket exclusions of alcoholics and drug addicts.

In short, the customary approach has one great advantage for society and one great disadvantage: it gives us the illusion that we do not have to make choices; but the cost is mass deception, and when this deception is uncovered, we must deal with it either by universal entitlement or by choosing another method of patient selection.

A Combination of Approaches

A socially acceptable approach must be fair, efficient, and reflective of important social values. The most important values at stake in organ transplantation are fairness itself, equity in the sense of equality, and the value of life. To promote efficiency, it is important that no one receive a transplant unless the person wants one and is likely to obtain significant benefit from it in the sense of years of life at a reasonable level of functioning.

Accordingly, it is appropriate for there to be an initial screening process that is based *exclusively* on medical criteria designed to measure the probability of a successful transplant, i.e., one in which the patient survives for at least a number of years and is rehabilitated. There is room in medical criteria for social worth judgments, but there is probably no way to avoid this completely. For example, it has been noted that "in many respects social and medical criteria are inextricably intertwined" and that therefore medical criteria might "exclude the poor and disadvantaged because health and socioeconomic status are highly interdependent." Roger Evans gives an example. In the End Stage Renal Disease Program, "those of lower socioeconomic status are likely to have multiple comorbid health conditions such as diabetes, hepatitis, and hypertension making them both less desirable candidates and more expensive to treat."[12]

To prevent the gulf between the haves and have nots from widening, we must make every reasonable attempt to develop medical criteria that are objective and independent of social worth categories. One minimal way to approach this is to require that medical screening be reviewed and approved by an ethics committee with significant public representation, filed with a public agency, and made readily available to the public for comment. In the event that more than one hospital in a state or region is offering a particular transplant service, it would be most fair and efficient for the individual hospitals to perform the initial medical screening themselves (based on the uniform, objective criteria), but to have all subsequent nonmedical selection done by a method approved by a single selection committee composed of representatives of all hospitals engaged in the particular transplant procedure, as well as significant representation of the public at large.

As this implies, after the medical screening is performed, there may be more acceptable candidates in the "pool" than there are organs or surgical teams to go around. Selection among waiting candidates will then be necessary. This situation occurs now in kidney transplantation, but since the organ matching is much more sophisticated than in hearts and livers (permitting much more precise matching of organ and recipient), and since dialysis permits individuals to wait almost indefinitely for an organ without risking death, the situations are not close enough to permit use of the same matching criteria. On the other hand, to the extent that organs are specifically tissue and size-matched and fairly distributed to the best matched candidate, the organ distribution system itself will resemble a natural lottery.

When a pool of acceptable candidates is developed, a decision about who gets the next available, suitable organ must be made. We must choose between using conscious, value-laden, social worth selection criteria (including a committee to make the actual choice), or some type of random device. In view of the unacceptability and arbitrariness of social worth criteria being applied, implicitly or explicitly, by committee, this method is neither viable nor proper. On the other hand, strict adherence to a lottery might create a situation where an individual who has only a one-in-four chance of living five years with a transplant (but who could survive another six months without one) would get an organ before an individual who could survive as long or longer, but who will die within days or hours if he or she is not immediately transplanted. Accordingly, the most reasonable approach seems to be to allocate organs on a first-come, first-served basis to members of the pool, but permit individuals to "jump" the queue if the second-level selection committee believes they are in immediate danger of death (but still have a reasonable prospect for long-term survival with a transplant) and the person who would otherwise get the organ can survive long enough to be reasonably assured that he or she will be able to get another organ.

The first-come, first-served method of basic selection (after a medical screen) seems the preferred method because it most closely approximates the randomness of a straight lottery without the obviousness of making equity the only promoted value. Some unfairness is introduced by the fact that the more wealthy and medically astute will likely get into the pool first, and thus be ahead in line, but this advantage should decrease sharply as public awareness of the system grows. The possibility of unfairness is also inherent in permitting individuals to jump the queue, but some flexibility needs to be retained in the system to permit it to respond to reasonable contingencies.

We will have to face the fact that should the resources devoted to organ transplants be limited (as they are now and are likely to be in the future), at some point it is likely that significant numbers of individuals will die in the pool waiting for a transplant. Three things can be done to avoid this: (1) medical criteria can be made stricter, perhaps by adding a more rigorous notion of "quality" of life to longevity and prospects for rehabilitation; (2) resources devoted to transplantation and organ procurement can be increased; or (3) individuals can be persuaded not to attempt to join the pool.

Of these three options, only the third has the promise of both conserving resources and promoting autonomy. Although most persons medically eligible for a transplant would probably want one, some would not—at least if they understood all that was involved, including the need for a lifetime commitment to daily immunosuppression medications, and periodic medical monitoring for rejection symptoms. Accordingly, it makes public policy sense to publicize the risks and side effects of transplantation, and to require careful explanations of the procedure be given to prospective patients *before* they undergo medical screening. It is likely that by the time patients come to the transplant center they have made up their minds and would do almost anything to get the transplant. Nonetheless, if there are patients who, when confronted with all the facts, would voluntarily elect not to proceed, we enhance both their own freedom and the efficiency and cost-effectiveness of the transplantation system by screening them out as early as possible.

Conclusion

Choices among patients that seem to condemn some to death and give others an opportunity to survive will always be tragic. Society has developed a number of mechanisms to make such decisions more acceptable by camouflaging them. In an era of scarce resources and conscious cost containment, such mechanisms will become public, and they will be usable only if they are fair and efficient. If they are not so perceived, we will shift from one mechanism to another in an effort to continue the illusion that tragic choices really don't have to be made, and that we can simultaneously move toward equity of access, quality of services, and cost containment without any challenges to our values. Along with the prostitute, the playboy, and the poet, we all need to be involved in the development of an access model to extreme and expensive medical technologies we can live with.

Life, Liberty, and the Pursuit of Organ Sales*

The media attention focused on a Virginia physician's macabre proposal to set up an "International Kidney Exchange" to broker the sale of kidneys between "consenting adults" is testimony to the plan's journalistic value rather than its intrinsic merits. There has been almost universal revulsion at the notion of a market in organs that would inevitably lead to the poor selling their body parts to the rich, and federal legislation has been proposed to outlaw such sales. Inspired by images of ghouls, vampires, and cannibals, most of the discussion has taken place on an emotional plane, with scant attention to the social or legal context.

The State of the Law

Under present law, transfers of vital organs, such as the heart and liver, would constitute suicide on the part of the vendor and homicide on the part of the harvesting surgeon. Accordingly, organ sales by living vendors (sometimes referred to as "donors" because of the source of the organ rather than the motivation for transfer) must be limited to nonvital organs, such as one kidney in an individual who has two healthy kidneys. Society currently tolerates the sale of blood and sperm, but unlike kidneys, these body products are renewable and can be transferred with almost no medical risk to the seller.

There have been no reported cases of kidney sales in this country, although no state or federal law explicitly prohibits such a transaction.

*Adapted from testimony presented before the US House Subcommittee on Investigations and Oversight of the Committee on Science and Technology, Nov. 9, 1983. Federal legislation outlawing the sale of organs was enacted in 1984.

The question of whether such a sale and transfer is legal or not rests upon the interpretation of the state's criminal law related to mayhem, and the criminal and civil law of assault and battery.

Historically, mayhem derives from the feudal system; it involved rendering an individual less able "in fighting, either to defend himself, or to annoy his adversary." Mayhem was criminal because the king had the right to the fighting services of all his subjects. Consent was no defense. Cutting off an ear or nose, or simply disfiguring an individual without weakening him, did not constitute mayhem. Most commentators do not believe courts should find mayhem involved in an operation in which the donor receives a significant psychological benefit (by donating an organ to a close family member or friend), although this proposition is far from certain.

When the only benefit is financial, the question is more difficult and may involve a precise calculation of the harm done to the vendor versus the benefit received and the social value of the activity. The sale of an eye could be considered a classic "maim" and subject the physician who removes it from a living seller to criminal prosecution. Further, if the consent of the seller is not valid (because it is not voluntary or informed) the physician removing a kidney could be sued both civilly and criminally for assault and battery, and civilly for negligence. The possibilities of these suits, added to the general reluctance of transplant physicians to use nonrelated donors, make it unlikely that many physicians will be involved in transplantation using vendor kidneys in the absence of a statute permitting such transactions.

There does not seem to be any constitutional right implicit in the "right to privacy" to sell one's nonvital organs. It does seem likely, however, that a court would find a constitutional right for a parent (or perhaps any close relative) to donate a nonvital organ in an effort to save the life of that person's child. The current state of the law is uncertain, and state legislation either to permit (with or without regulation) or outlaw the sale of nonvital organs would be constitutional. Which action is most desirable?

Arguments in Favor of Sale

There are two major policy arguments in favor of permitting living individuals to sell their nonvital organs: (1) it could increase the supply; and (2) individuals should have the liberty to use and dispose of their bodies as they see fit.

An increase in supply (even if obtained), however, is not in itself sufficient justification, since supply could also be increased by drafting individuals at random to serve as organ donors, or by using the residents of our mental institutions and homes for the retarded as donors—alternatives that most people would find more repellent than organ sales. Indeed, if the primary issue is supply, increased efforts to make the present procurement mechanisms more efficient should be exhausted before we embark on a radically new and controversial course.

The second argument is more compelling, but we limit individuals' liberty to do what they want with their bodies in many different ways. Mayhem is only one. Others include laws prohibiting prostitution, limits on third trimester abortions, and occupational health and safety laws. Even where we permit individuals to risk their lives and health, we often limit the degree of danger. For example, boxers cannot fight unless they meet the standards of a state boxing commission; football players cannot play without helmets; and eventually none of us will be able to drive in cars not equipped with passive restraints. All these actions can be viewed as governmental paternalism. The issue is not whether the state can act paternalistically (it obviously can), but whether it can justify its action to the satisfaction of its citizens (or, if a fundamental constitutional right is at issue, whether it can demonstrate a "compelling state interest" in interfering with its exercise).

In upholding a statute under which a boxing commission prohibited a 51-year-old prize fighter who had fought more than 300 bouts from risking injury in another fight, the court found that the risk was not "purely personal," but also of interest to the state: (1) to prevent certain "brutal and degrading" features of boxing; and (2) to promote contests conducted within "the legitimate limits of the sport."[13] A statute outlawing sales could likewise prevent the potential "degrading" features of organ sales and protect legitimate organ donations.

We might outlaw sales and yet permit donations for at least two reasons: (1) we believe donors obtain a priceless psychological reward by acting as "heroes"; and (2) we generally limit such donations to close family members (and the law traditionally respects "private" family decisions, at least where the effort is to save life). On the other hand, if we permit a father to donate his kidney to his wife or child, it is difficult to logically justify not permitting him to trade his kidney for another (directly or through a brokerage firm) in the event that his tissue is not a close enough match for the transplant, since the quality of the consent, the risks undertaken, and the motivation will all be identical.

Arguments Against Sale

There are many arguments against a market in organs, including a dilution of altruism, a less-dependable product, the pricing of a priceless gift, the tendency to view human body parts as commodities that may lead us to view human beings as things, and the general unsavoriness of a market in body parts. All have some merit. From a legal perspective, however, there is really only one major argument against permitting a competent adult to sell his or her nonvital organs; sale is an act of such desperation that *voluntary* consent is impossible. In the words of Bernard Dickens:

> The policy behind legislation prohibiting body-material sales [which exists in a number of Canadian Provinces] is ethical and humane; scientific research and medical education should no longer be pursued at the physical cost of the economically and socially deprived, nor should it exploit the needs of the vulnerable...Equally, the poor should not be induced by money to offer the material resources of their bodies, and to convert their own health into a salable commodity in the market place of human replacement parts.[14]

Dickens concedes that we do permit the poor to enroll in disproportionate numbers in the armed forces, and to take on essential but uncomfortable or dangerous occupations for money. But in these cases, benefits accrue to society as a whole, and although the employment choices may be limited, choices are not overtly coerced.

An analogous approach is to ask what risks we would permit an individual to consent to in human experimentation. We require voluntary consent to human experimentation, and generally prohibit monetary inducement that is so high as to be potentially coercive. We know that the amount that may induce one person to participate may not move another, depending upon relative wealth and present need. There are surely some experiments that we would forbid outright, such as permitting a person to volunteer to have his liver removed so surgeons could experiment with a new method of reimplanting it.

But are there valid experiments that we would approve if no compensation were given to volunteers, but disapprove if a large payment was offered? Imagine, for example, a study designed to test a new method of preserving organs from living donors that required 100 kidneys to be biopsied from healthy donors. Would we approve the study if the donors weren't compensated, but disapprove the same study if each donor–subject was to be paid $10,000 for undergoing the potentially risky biopsy? If we would, it must be because we believe the $10,000 payment would be a coercive inducement that vitiates the validity of the consent. In analogous situations we no longer permit prisoners to "volunteer" for risky medical experiments because we do not believe that they can give voluntary consent.

Of course, coerciveness may be equal or greater in the case of a family-member donor. But society does not create or endorse the coerciveness of that setting, and denying a relative the right to rescue only adds an additional burden to the family facing an already tragic situation.

The argument that sale will only sanction the transfer of organs from the poor to the rich can be at least partly answered by having a governmental agency be the sole purchaser and distributor of organs, with distribution based on a criterion other than wealth or social worth. The poor, however, will still likely serve disproportionately as sources of kidneys, a situation that would change only with a more equitable income distribution.

State or Federal Law?

The law governing medical treatment, consent, definition of death, donation of organs, autopsy, burial, and the disposition of dead bodies is exclusively state law. Unless there is a compelling justification, it seems extreme for the federal government to legislate in this area. However, since essentially all kidney procurement in this country is directly financed by the federal government, it would seem appropriate for the Congress to forbid any reimbursement for procurement or transplantation of an organ obtained by paying the vendor ("donor").

The strongest argument against a market in nonvital organs is society's instinctive reaction against permitting individuals to directly sacrifice themselves and their health for monetary rewards. Perhaps this revulsion originates in a primitive perception of organ transfers as ghoulish and somewhat cannibalistic; it is animated by a more sophisticated refusal to accept a symbolic and highly visible form of exploitation of the poor.

If this is so, the organ sale debate could provide an opportunity to educate the public about organ donation and procurement, and open educate the public about organ donation and procurement, and open the broader discussion of social justice and equity in income distribution and access to health care. Outlawing such sales by statute seems excessive before there is any reasonable likelihood of organ selling becoming a reality, and trivial when compared to the broader social justice issues.*

---

*In 1988, discussion of organ sales is confined to a fringe element that believes that everything has (and should have) a market price. The field has been taken over by the "required request" strategy, in which hospitals are required (by state statute, Medicare regulation, and/or JCAH accreditation standards) to inform the relatives of all potential organ "donors" about the option of organ donation.

Baby Fae

The "Anything Goes" School of Human Experimentation

Was Baby Fae a brave medical pioneer whose parents chose the only possible way to save her life, or was she a pathetic sacrificial victim whose dying was exploited and prolonged on the altar of scientific progress? To answer this question we need to examine the historical context of this experiment, together with the actions and expressed motives of the parents and physicians.

In an exclusive interview in *American Medical News* ten days after he had transplanted the heart of a baboon into Baby Fae, Leonard Bailey described James D. Hardy as "my silent champion." Speaking of Hardy's transplant of a chimpanzee heart into a human being in 1964, he said, "He's an idol of mine because he followed through and did what he should have done...he took a gamble to try to save a human life."[15]

James Hardy, of the University of Mississippi, did the world's first lung transplant on a poor, uneducated, dying patient who was serving a life sentence for murder. John Richard Russell survived the transplant for seventeen days, and died as a result of kidney problems that were expected to kill him in any event. Less than seven months later, in January 1964, Hardy performed the world's first heart transplant on a human being, using the heart of a chimpanzee. The recipient of the chimpanzee heart, Boyd Rush, did not consent to the procedure. Like Russell, he was dying and poor. Although not a prisoner, he was particularly vulnerable because he was a deaf-mute. He was brought to the hospital unconscious and never regained consciousness. A search for relatives turned up only a stepsister, who was persuaded to sign a consent form authorizing "the insertion of a suitable heart transplant" if this should prove necessary. The form made no mention of a primate heart. In later written reports Hardy contended that he had discussed the procedure in detail with *relatives,* although there was only one. Rush survived two hours with the chimpanzee heart.

Hardy's justifications for using the chimpanzee heart were the difficulty of obtaining a human heart and the apparent success of Keith Reemtsma in transplanting chimpanzee kidneys into Jefferson Davis at New Orleans Charity Hospital. Davis was a 53-year-old poor black man who was dying of glomerulonephritis. Davis described his consent in this transcript of a conversation with his doctors after the operation:

> You told me that's one chance out of a thousand. I said I didn't have no choice...You told me it gonna be animal kidneys. Well, I ain't had no choice.[16]

The operation took place on November 5, 1963; the patient was doing well on November 18 when he was visited by Hardy. On December 18 he was released to spend Christmas at home. Two days later he was back in the hospital, and on January 6, 1964, he died.

Whatever else one wants to say about these transplants, it is doubtful that anyone would seriously attempt to justify either the consent procedures or the patient selection procedures. Both experimenters took advantage of poor, illiterate, and dying patients for their own research ends. Both seem to have violated the major precepts of the Nuremberg Code regarding voluntary, competent, informed, and understanding consent; sufficient prior animal experimentation; and no *a priori* reason to expect death as a result of the experiment.

The parallels are striking. Like Russell, Rush, and Davis, Baby Fae was terminally ill; her dying status was used against her as the primary justification for the experiment. We recognize that children, prisoners, and mental patients are at special risk for exploitation, but the terminally ill are even more so, with their dying status itself used as an excuse to justify otherwise unjustifiable research. Like these previous subjects, Baby Fae was also impoverished; subjects in xenograft experiments have traditionally been drawn from this population. Finally, as a newborn, she was even more vulnerable to exploitation. Three issues merit specific discussion: (1) the reasonableness of this experiment on children; (2) the adequacy of IRB review; and (3) the quality of the consent.

The Reasonableness of the Experiment

Though different accounts have been given, it seems fair to accept the formulation by immunologist Sandra Nehlsen-Cannarella: "Our hypothesis is that a newborn can, with a combination of its underdeveloped immune system and the aid of the anti-suppressive drug, cyclosporine, accept the heart of a baboon if we can find one with tissue of high enough comparability." Questions that need answers are: Is there sufficient animal evidence to support this "underdeveloped immune system" hypothesis as reasonable in the human? Does the evidence provided give any reason to anticipate benefit to the infant? And is there any justification for experimenting on infants before we experiment on adults who can consent for themselves? The answer to all three questions seems to be no.

Only two new relevant scientific developments have occurred since the 1963-64 experiments of Reemtsma and Hardy: better tissue-matching procedures* and cyclosporine. Both of these, however, are equally applicable to adults. Only the "underdeveloped immune system" theory, which posits that transplants are more likely to succeed if done in infants with underdeveloped immune systems, is applicable to newborns, and this could be tested equally well with a human heart. Without this type of prior work, we are engaged, as one of my physician colleagues put it, in "dog lab" experiments, using children as means to test a hypothesis rather than as ends in themselves. Without adult testing, there could be no reasonable anticipation of benefit for this child; the best that could be hoped for is that the parents would bury a very young child instead of an infant. There should be no more xenografts on children until they have proven successful on adults.

The Adequacy of IRB Review

Since the Loma Linda IRB seems to have dealt with these concerns inadequately, we must question whether the IRB mechanism is able

---

*At the time of the Baby Fae xenograft Bailey said, "We can tissue-type our baboons for comparativity to a potential human recipient more thoroughly than we can tissue-type most human donors." Approximately one year later it was disclosed that Bailey had failed even to match the blood type of Baby Fae (type O) with that of the donor baboon (type AB). Whether this was simple ignorance or deceit on the part of Bailey remains uncertain.

to protect human subjects involved in first-of-their-kind organ transplants. The record is not very good. The Utah IRB failed to protect Barney Clark from being used as a means to promote the artificial heart.* Likewise, the Humana Heart Institute IRB seems to have been more interested in promoting its own institutional concerns through advertising than in protecting William Schroeder. For example, its consent form requires the subject to sign over all rights he or his heirs or other parties might have in "photographs, slides, films, video tapes, recordings or other materials that maybe used in newspaper, magazine articles, television, radio broadcasts, movies or any other media or means of dissemination."** Very little is known about the Loma Linda IRB and its process. According to its chairman, Richard Sheldon, the 23 member IRB first received the protocol in August 1983 and approved it later that year. Bailey was told to present any changes in it to the IRB when a suitable candidate was available. These were presented and approved by a nine to seven vote, two days before Baby Fae's transplant.

Some general observations about IRBs may explain their failure in these cases. First, IRBs are composed primarily (sometimes almost exclusively) of employees and staff of the research institute itself. When that institute, in addition to its basic research mission, has another common set of beliefs, based on a shared religion like Mormonism or Seventh Day Adventism, or a secular belief in the profit motive, there is a disturbing homogeneity to the IRB. This is likely to lead to approval of a project by a researcher who shares the same belief system.

Second, IRBs are way over their heads in this type of surgical innovation. There is no history of successful IRB review of first-of-their-kind kidney, liver, or heart transplants. Ross Woolley has described the Utah IRB that approved the Barney Clark experiment as a "bunch of folks who get together and stumble around and do our thing." More courteously, Albert Jonsen described the plight of the same IRB as akin to being "asked to build a Boeing 747 with Wright Brothers parts." Homogeneous IRBs without experience in transplant innovation are no match for surgical "pioneers."

**See: The Lion and the Crocodiles* at p. 391.

**Indeed, when Schroeder died, Humana sent its Public Relations Director, not a physician, to appear on ABC's *Nightline* with me and Denton Cooley to discuss the future of artificial hearts (August 6, 1986).

The Consent Process

On day ten after Baby Fae's transplant Bailey said:

> In the best scenario, Baby Fae will celebrate her 21st birthday
> without the need for further surgery. That possibility exists.

This was, in fact, never a realistic or reasonable expectation, and raises serious questions both about Bailey's ability to separate science from emotion, and what exactly he led the parents of Baby Fae to expect. He seemed more honest when he described the experiment as a "tremendous victory" after Baby Fae's death. But this could only mean that the experiment itself was the primary end, and that therapy was never a realistic goal.

As of this writing, the Baby Fae consent form remains a Loma Linda Top Secret Document.* But the process is much more important than the form, and it has been described by the principals. Minimally, there should have been an independent patient selection committee to screen candidates to ensure that the parents could not easily be taken advantage of, could supply the child with sufficient stable support to make long-term survival possible, were aware of all reasonable alternatives in a timely manner, and were not financially constrained in their decisionmaking.

Baby Fae's parents had a 2-1/2 year old son, had been living together for about four years, had never married, and had been separated for the few months prior to Baby Fae's birth. Her mother is a high school dropout who was forced to depend on Aid to Families with Dependent Children at the time of the birth of Baby Fae. Baby Fae's father had three children by a previous marriage and describes himself as a middle-aged adolescent. He was not present at the birth of Baby Fae and did not learn about it until three days later. Both felt guilty about Baby Fae's condition, and wanted to do "anything" that might "save her life."

Bailey describes the crux of the consent process as a conversation with the parents from about midnight until 7 AM on October 20. In Bailey's words:

---

*The consent form was ultimately released by HHS in March 1985. The form can be charitably described as incomplete and misleading.

> Apparently, the parents had spent three or four hours in debate at home [before admitting the baby] and now, from midnight until well into the next morning, I spent hours talking to them very candidly and very frankly. While Baby Fae was resting in bed, I showed them a film and I gave them a slide show, explaining our research and our belief why a baboon heart might work.

This account, given slightly more than two weeks after the transplant, is in error. Apparently Bailey is following Hardy's precedent of exaggerating the number of "relatives" involved in the consent process. What really happened is recounted by the couple in their exclusive interview in *People* magazine. Present at the midnight explanation were not "the parents," but the mother, the grandmother, and a male friend of the mother who was staying at her home at the time of Baby Fae's birth. Baby Fae's father was *not* in attendance, although he says, "I would have been there at the meeting with Dr. Bailey if I'd known it was going to turn into a seven hour discussion." Nonetheless, even though he missed the explanations about what was going to happen to his daughter, "when it came time to sign the agreements, I was up there."[17]

It is unclear that either of the parents ever read or understood the consent form, but it is evident that the father was not involved in any meaningful way in the consent process.

Lessons of the Case

This inadequately reviewed, inappropriately consented to, premature experiment on an impoverished, terminally ill newborn was unjustified. It differs from the xenograft experiments of the early 1960s only in the fact that there was prior review of the proposal by an IRB. But this distinction did not make a difference for Baby Fae. She remained unprotected from ruthless experimentation in which her only role was that of a victim.

David B. Hinshaw, the Loma Linda spokesman, understood part of the problem. In responding to news reports that the hospital might have taken advantage of a couple in "difficult circumstances to wrest things from them in terms of experimental procedures," he said that if this was true, "The whole basis of medicine in Western civilization is challenged and attacked at its very roots."[18] This is an overstatement. Culpability lies at Loma Linda.

Some will find this indictment too harsh. It may be (although none of us can yet know) that the IRB followed the NIH rules on research involving children to the letter, and that the experiment *could* be fit into the federal regulations by claiming that Baby Fae's terminally ill status was sufficient justification for an attempt to save her life. But if the federal regulations cannot prevent this type of gross exploitation of the terminally ill, they must be revised. We may need a "national review board" to deal with such complex matters as artificial hearts, xenographs, genetic engineering, and new reproductive technologies. That Loma Linda might be able to legally "get away with" what they have done demonstrates the need for reform and reassertion of the principles of the Nuremberg Code.

As philosopher Alasdair MacIntyre told a recent graduating class of Boston University School of Medicine, there are two ways to be a bad doctor. One is to break the rules; the other is to follow all the rules to the letter and to assume that by so doing you are being "good." The same can be said of IRBs. We owe experimental subjects more than the cold "letter of the law."

The Loma Linda University Observer, the campus newspaper, ran two headline stories on November 13, 1984, two days before Baby Fae's death. The first headline read "...And the beat goes on for Baby Fae," the second, which covered an unconnected social event, could have more aptly captioned the Baby Fae story: "'Almost Anything Goes' comes to Loma Linda."

Consent to the
Artificial Heart
The Lion and the Crocodiles

When Christiaan Barnard performed the first human heart transplant in 1967, he did not obtain informed consent. He purposely led Louis Washkansky to believe that the procedure had an 80 percent chance of success, instead of telling him that the probability applied only to his surviving the operation. Barnard's justification: "He had not asked for odds or any details...he was at the end of the line....For a dying man, it is not a difficult decision....If a lion chases you to the bank of a river filled with crocodiles, you will leap into the water convinced you have a chance to swim to the other side. But you would never accept such odds if there were no lion."[19]

Denton Cooley expressed similar thoughts about the first recipient of a temporary artificial heart. He said of Haskell Karp, "He was a drowning man. A drowning man can't be too particular about what he's going to use as a life preserver. It was a desperate thing and he knew it."

Similarly, after implanting the first permanent artificial heart in Barney Clark, William DeVries argued that Clark unquestionably made the right choice. "He was too old for a transplant, there were no drugs that would help; the only thing that he could look forward to was dying."[20]

Barnard made no pretext of obtaining informed consent. Cooley did have Haskell Karp sign a consent form regarding the possible temporary use of a "mechanical cardiac substitute" in the event the operation to reconstruct his left ventricle was unsuccessful. But it was only the incredible conclusion of two federal courts (applying Texas law) that this first-of-its-kind procedure was primarily therapeutic,

391

not experimental, that helped him prevail in a lawsuit brought by Karp's widow for failure to obtain informed consent.

Clark's operation occurred more than 13 years after Karp's. During this period there have been highly touted advances in institutional review boards (IRBs) and the required documentation of informed consent for experimental procedures. One would expect the informed consent procedures used in the Clark case to be of a very high order. And so they have seemed in the press. Indeed, Morris Abram, chairman of the President's Commission for the Study of Ethical Problems in Medicine and Biomedical and Behavioral Research, has been quoted as saying that the informed consent procedures followed in the Clark case are "a perfect example of what we recommended. The patient was fully able to comprehend and give autonomous consent. And there seemed to be total revelation by the medical team of what it planned to do and the possible risks and effects."[21]

Although the consent procedure used in the Barney Clark case was an improvement over the earlier ones, it can hardly be characterized as "perfect."

The Consent Form

The FDA has put a temporary hold on further implants because the early fracture of one of the valves was unexpected. However, the agency has not questioned the consent process itself and sees consent as an IRB responsibility.

Almost everyone knows that Barney Clark signed an "11-page" consent form. Unfortunately the form is more notable for its length than its content. It is incomplete, internally inconsistent, and confusing. Consent is a process and not a form, but the form should reflect the process and advance the knowledge and autonomy of the subject. The form's most crucial shortcoming is that it assumes Clark will either continue to be competent and able to consent to further treatment, or that he will die. It takes no account of a "halfway success": survival coupled with severe confusion, mental incompetence, or coma. Two clauses deal with the predictable and likely probability that additional surgical procedures will be done following the initial implant. Obviously, if the IRB believes consent means anything in this context, these additional procedures would require it as well. Accordingly, one clause states that repair or replacement of the device would not be done unless "a new consent form [is] signed by me." The next clause contemplates testing the functioning of the heart by certain procedures, and guarantees that "each of these new procedures will have a consent form which must be signed before they are performed."

A number of surgical procedures have actually been required since the implantation, both to repair the artificial heart and to treat related complications (such as nosebleed). The consent form Barney Clark signed clearly contemplated that only he would be allowed to decide whether such additional procedures would be done, and that he would agree by signing additional consent forms. In fact, Barney Clark did not sign any new consent forms for any of these surgical procedures. Hospital officials considered him mentally incompetent; they sought his wife's signature on all subsequent consent forms, although he did verbally assent to the procedures. Nor are the forms his wife signed the specific forms mentioned in the original consent document, but rather the hospital's routine consent form. Neither Barnard nor Cooley faced this halfway success problem, but it should have been faced in Utah. Who would Clark have wanted to make decisions for him? On what basis should such decisions be made? And when should the experiment be ended from Clark's perspective?

This lack is especially ironic in view of the form's withdrawal clause: "I understand I am free at any time to withdraw my consent to participate in this experimental project, recognizing that the exercise of such an option after the artificial heart is in place may result in my death." Surely Clark must be afforded this option. But just as surely someone else whom he designated should have been empowered to exercise it for him, applying the criteria he would have wanted to apply.

There are at least two approaches to this quandary: the "living will" and the durable power of attorney. In a living will, Clark could have specified the conditions under which, should he become incompetent, he would want the experiment terminated. For example, how long would he want to continue if he remained incompetent and hospitalized? What if he became comatose? What if he required kidney dialysis to survive? These are predictable questions only Clark should answer.

The durable power of attorney also would be of use, but arguably only if the same type of instructions accompanied it. This instrument permits the designation of a proxy to make decisions for the individual that is valid even if the individual becomes incompetent. Though no court has yet ruled that such authority gives the proxy the right to make medical decisions, it should be recognized for this purpose provided such decisions are consistent with the prior expressed wishes of the person. Standing alone, however, with no indication of the person's actual desires concerning treatment, the durable power of attorney would not add much to our knowledge of Clark's own wishes.

It might be argued that no one knows better what Clark would want done than his wife (and this may be true), but unless Clark has made his wishes known to her and designated her as the person to carry them out, we can only speculate on this issue. Just as we would not have allowed Mrs. Clark to consent to the heart implant on behalf of her husband, we should not permit her to consent to its termination or to additional procedures related to it on his behalf.

The problem is one of ascertaining Clark's own wishes. His son, Stephen Clark, a surgeon, said on the day of the operation: "Even with his illness, he still maintained a great interest in life. If, however, he finds out that while he is alive he still has a lot of pain and a lot of nausea, he will be very disappointed. I don't think he would want that....As long as he is comfortable, he has enough things in life to enjoy."[22]

Other Problems

Other shortcomings in the form deserve at least passing mention, because they indicate a lack of attention to detail in the entire process. Though the form is entitled a "Special Consent for Artificial Heart Device Implantation and Related Procedures," the "artificial heart device" itself is variously designated "a mechanical heart," a "mechanical artificial heart," an "artificial heart," a "mechanical heart device," an "experimental total artificial heart device," and "the device." The clauses on alternatives are also ambiguous at best. One assures that "replacement of my natural heart is the only treatment"; another that "I have discussed alternative treatments with my physician [including drug treatment and surgical repair of my heart]."

Other clauses simply appear to have been cut and pasted from standard hospital consent forms:

> I consent to the administration of anesthesia as arranged by my physician with such assistants as may be designated, and to the use of such anesthetics as they may in their judgment deem advisable.

I hereby authorize the University of Utah Hospital, through its authorized agents, to dispose of or use for any purpose, any tissue or body parts which may be removed during the surgical procedures.

The first adds nothing to the consent (unless we assume that Clark thought the procedure could be performed without anesthesia), and the second is so overbroad in this context as to be macabre. The only "tissue or body part" being removed are Clark's left and right ventricles, and requiring Clark to agree that they could be used "for any purpose" (display in an art museum or in the Smithsonian? sold to the *National Enquirer*?) is bizarre. This clause is also in sharp distinction to one following it, which carefully limits the use of videotapes, photographs, and drawings of the operation.

Why the Consent Form Wasn't Better

Writing consent forms by committee is a difficult business. Nevertheless, there are no excuses for the quality of Clark's consent form, only explanations.

The first explanation is that the form's writers opted for a general form that would cover all candidates—even though it was clear from the outset (the FDA had approved seven implants) that unless the first implant was successful, there would be no more for some time. The only blank lines on the form are for the patient's diagnosis, and the members of the surgeon's evaluation committee. The rest is boiler plate.

The second explanation is that the protocol, and thus the consent form, were changed in May 1982, after initial IRB approval in February 1981. The change was needed to expand the pool of potential artificial heart recipients to include those patients, like Barney Clark, suffering from "chronic nonoperable, end-stage progressive congestive heart failure," because no patients qualified for the implant under the previous protocol.

Finally, the IRB itself seems to have focused on two specific consent issues, almost to the exclusion of others: life style and finances.[23] Both are dealt with adequately in the form, although the sections on finances (which require Clark to assume all financial responsibility) are probably the clearest in the entire document.

Lessons

From the press reports it seems that the consent process used with Barney Clark was superior to the form that only evidences it. It is also true that Clark received significantly more information about the permanent artificial heart than did the first recipients of a temporary artificial heart, an artificial heart valve, a primate heart, a human heart, a human kidney, and primate kidney.[24] Surgical procedures are seldom subjected to prior review by either IRBs or the FDA. Indeed, of this list of firsts, this is the first that involved such external review. But prior review of ethical and legal considerations took a back seat to the technical aspects of the implant, and the sloppy consent form is merely one piece of evidence supporting this. The debate over Clark's access to the "key" that he could use to turn off the air compressor is another. Both his ability to use the key should he remain competent, and the conditions under which others should use it if he did not, should have been dealt with *before* the implantation of the artificial heart.

The problem of readable, reliable, and useful consent forms is endemic to the human research enterprise. The fact that IRBs spend the vast majority of their time worrying and debating about the words used in consent forms should disturb those who look to the IRB to protect subjects. The IRB may take the words in consent forms very seriously, but as Charles Bosk has noted, "The quality of the consent obtained is not an issue that excites surgeons or affects their evaluation of each other."[25]

At least in cases of surgical intervention designed to prolong the life of a dying subject, the contents of the consent form seem to be treated as bureaucratic red tape. But forms can be more. Specifically tailored to Barney Clark, and setting forth actions to be taken under reasonably foreseeable developments, "forms" could have provided him a measure of self-determination and dignity now denied him. The opportunity has been lost for Clark, but not for those who will follow.*

---

*Unfortunately, neither the consent form, nor the consent process has improved. *See* Annas, G. J., "Death and the Magic Machine: Informed Consent to the Artificial Heart," **9** *Western New Engl Law Rev* 89 (1987). Clark lived for 112 days on the artificial heart.

The Phoenix Heart
What We Have to Lose

In the two years following the first human-to-human heart transplant by Christiaan Barnard, more than 150 heart transplants were performed around the world at 60 different centers. The results were almost uniformly disastrous for the patients. This "unseemly rush to climb on this glamorous bandwagon" club of heart transplanters is one of the darkest episodes in the annals of surgery.[26] Did we learn anything from it, or are heart surgeons around the world poised to repeat this exercise in premature human experimentation, this time using artificial hearts?

Repetition does not seem likely with permanent artificial hearts, since the results of experiments to date have been so devastating to the patients. Use of the artificial heart as a *temporary* measure, however, is gaining support; thoughtful planning will be needed to prevent another round of indiscriminate experimentation.

The Phoenix Heart Implant

On Tuesday morning, March 5, 1985, Jack Copeland, Chief of University Medical Center's Heart Transplant Team in Tucson, Arizona, performed a human heart transplant on Thomas Creighton, a 33-year-old divorced father of two. Copeland later explained that the donor, an accident victim who had been hospitalized for several days, "wasn't what we'd call an excellent donor candidate (but in view of the urgency) we elected to proceed with the transplant."[27] The procedure was not a success, because of rejection of the heart. At 3:00 AM, Wednesday a search for another human heart began; Creighton was placed on a heart-lung machine.

At 5:30 AM a call was placed to Cecil Vaughn of Phoenix, asking if he had an artificial heart ready for human use. Vaughn was scheduled to implant an experimental model developed by dentist Kevin Cheng into a calf late that day, but had never considered use of the device in a human. Nonetheless, he called Cheng, who told him, "It's designed for a calf and not ready for a human yet." Asked to think about it for ten minutes, Cheng recalls, "I knelt and prayed." When Vaughn called him back he said, "The pump is sterile, ready to go."[28]

The two flew by helicopter from the hospital to the airport, chartered a jet to Tucson, and then took another helicopter to the Tucson hospital. They arrived at 9:30 AM Wednesday. The implant procedure began at noon. Designed for a calf, the device was too large, and the chest could not be closed around it. The implant maintained circulation until 11:00 that night when, in preparation for a second heart transplant, it was turned off, and Creighton was put back on the heart-lung machine. By 3:00 AM Thursday, the second human heart transplant was completed. The next day Mr. Creighton died.

The press treated the story like a modern American melodrama. *USA Today* called the implantation of Cheng's device "the fulfillment of an American dream."[29] The *New York Times* editorialized that "the artificial heart has at last proved it has a useful role..."[30] *Newsweek* faulted the FDA, noting, "It's hardly fair to doctors, or their patients, to make them break the law to save a life."[31] The FDA initially termed the unauthorized experiment a violation of the law, but by week's end had done an about face and was flailing itself as "part of the problem."[32]

Melodrama calls on us to suspend our critical judgment, identify with the protagonists, and join emotionally in the drama. This may be an appropriate response to soap operas, but we need to take a rational view of the events in Arizona—not to judge the actors in that drama, but to decide what transplant policies we should now pursue.

Justifications for the Implant

The physicians and their supporters have given three basic justifications for the implant: (1) the "only other option was just to let him die" so "we had nothing to lose," (2) in an emergency, a physician can do anything to save the patient's life; and (3) FDA regulations do not apply to dying patients. None of these excuses can survive scrutiny.

A matter of life or death: Copeland has justified the incident primarily by saying, we had "nothing to lose" by trying the artificial heart.[33]   William DeVries also used this justification for Barney Clark's permanent implant.

But the choices were not just "live or die." The reality was closer to: "Accept the implant and you'll almost certainly die anyway; and if you do live, you could spend the rest of your life severely disabled, mentally and physically." When the possibility of a "halfway success," survival in a severely impaired state, is added to the equation, the patient has much to lose, including his self-determination and dignity as a human being.

The Arizona implant was not performed in isolation, but as a response to an unsuccessful human heart transplant. In planned procedures like transplants, only the patient should be permitted to decide whether to resort to extreme and experimental methods of maintaining life, like artificial hearts. That Thomas Creighton was not the person who made this decision raised the question of what he was led to expect from the heart transplant, what risks (including the risk of rejection) were explained to him, and what steps he had agreed to should any of the risks materialize. Proxy consent (reportedly obtained from the patient's mother and sister) is inadequate to justify extreme experimentation, unnecessary for emergency therapy, and irrelevant if quackery is involved.

Taken to its logical extreme, the "nothing to lose" excuse can justify *any* experiment on a dying patient "to save the patient's life." The "right to die with dignity" movement is only one reaction to this type of thinking. This excuse also embodies a "magical" myth: that the physician has the power to conquer death and that prolonging life is *always* a reasonable goal. As psychiatrist Jay Katz has noted in another context in his book *The Silent World of Doctor and Patient:*

> At such times, all kinds of senseless interventions are tried in
> an unconscious effort to cure the incurable magically through a
> "wonder drug," a novel surgical procedure, or a penetrating
> psychological interpretation...The doctors' heroic attempts to
> try anything...may turn out to be a projection of their own needs
> onto patient.[34]

The emergency justification: University officials justified the incident on the basis that physicians are privileged to do anything they believe is appropriate in a medical emergency. This is wide of the mark. The emergency rule is, "treat first, and ask *legal* questions later." Thus an emergency situation may justify a physician's decision not to review federal regulations prior to acting, but it can never justify a physician in not considering *medical* data before acting. The medical "reasonableness" of using the artificial heart in a true emergency is debatable. But was this a true, unanticipated, emergency?

Organ rejection is a known risk of all transplant procedures. Creighton was Copeland's third patient to suffer immediate heart rejection. He vowed after the second to do all he could to save the next such patient. Thus, not only is organ rejection a "reasonably foreseeable risk" of transplantation, Copeland knew of this risk firsthand and had ample opportunity to develop a plan to deal with the next case he encountered. Under these circumstances organ rejection is not an unanticipated "emergency" that justified rash, extreme interventions.

FDA approval: The press tended to use the Arizona incident as an illustration of how government regulation doesn't work; and, unfortunately, the FDA seems to agree, to the point of almost apologizing for its own existence. The life-or-death and emergency excuses would equally justify the use of unapproved drugs like laetrile on dying cancer patients by licensed or unlicensed practitioners, and could be used to justify an artificial heart implant by an unlicensed practitioner. The FDA cannot sanction these types of reckless behaviors.

As the US Supreme Court noted in upholding the FDA's authority over laetrile, "To accept the proposition that the safety and efficacy standards of the Food, Drug and Cosmetic Act have no relevance for terminal patients is to deny the Commissioner's authority over all drugs, however toxic or ineffectual, for such individuals...[the terminally ill deserve protection] from the vast range of self-styled panaceas that inventive minds can devise."[35]

Even if *all* physicians opposed reasonable regulation of the safety and efficacy of drugs and medical devices, the FDA should not apologize for the important role Congress has assigned it to protect the public against unsafe, untested, and useless medical devices. Copeland's argument that not to use the device would have made "the government his executioner"[36] cannot be taken seriously. Mr. Creighton would have died not from an FDA rule, but from an unsuccessful heart transplant. The FDA properly forbade the "emergency" use of devices that have not been approved even for human experimentation, and should continue to do so.

In short, the unplanned use of unapproved temporary artificial hearts is not justified. But the planned use of approved temporary artificial hearts is problematic as well.

Reconsidering the Queue

Before we commit ourselves to more experimentation with the artificial heart as a temporary device, it seems reasonable to anticipate how the use of such devices should fit into the broader area of heart transplantation and organ rationing.[37] *

The use of "medical urgency" as a justification for "jumping the queue" is problematic. If temporary artificial hearts are used in individuals who are *least* likely to survive human heart transplants and be rehabilitated, thus taking away organs from those who are most likely to benefit, this new technology will not "save lives." Instead, its use will indirectly lead to the deaths of individuals on waiting lists. In short, we must recognize that while the shortage of human hearts for transplant exists, temporary artificial hearts will likely do more harm than good.

We should deal with these issues directly; we all have a lot to lose from an incoherent introduction of temporary artificial hearts.

See "The Prostitute, The Playboy, and the Poet" at p. 370.

No Cheers for
Temporary Artificial Hearts

It is no accident that the press and hospitals alike have adopted the sports metaphor for artificial heart implants. Americans love the spectacle of sports, and the combination of life-or-death outcomes and desperate searches for organs from accident victims is almost irresistible. Both hospitals and newspaper writers gain a large and sympathetic audience: large because drama overwhelms reality; sympathetic because we all cheer the team that challenges death. The FDA has not effectively regulated the use of temporary artificial hearts, and we are on the verge of an unparalleled free-for-all in human experimentation.

The Sporting Life

Michael Drummond, the first recipient of an FDA-approved artificial heart as a bridge to transplant (the fourth "bridge" recipient in history), was referred to the University of Arizona Hospital for evaluation for possible heart transplant on Monday, August 26, 1985. He was listed as a candidate for a human heart transplant on Tuesday, when the possibility of the artificial heart was mentioned for the first time. Early Thursday morning, because Jack Copeland and others decided Drummond had only 48 hours to live, he and his family were reportedly given an option: wait for a human heart transplant and die if one did not become available, or go ahead with an artificial heart implant. They opted for the latter because, as his mother put it, "We had no choice."

The patient himself had no clear idea what he was consenting to. When he awoke from the procedure, his first words were, "What's this clunking" [in my chest]? The operator of the drive system for his artificial heart, who was present, said he doubted that Drummond "really understands totally the artificial heart and what it is" [referring to the tubes, drive lines, console and, he could have added, the generators].[38] In addition, after his human heart transplant, Drummond was enthusiastic about the artificial heart, but reiterated that there was "no alternative" and said he thought he "only had four days to live."[39]

All these statements are troublesome, but not surprising. The consent process was not well thought out at Arizona. The FDA-"approved" consent form, for example, is the most misleading, rudimentary, and confusing one yet used for a total artificial heart implant. It describes the five previous permanent implants as having been done "successfully" with no further explanation of what is meant by this term. It incorrectly states that the Symbion [Jarvik-7] heart is attached "to a portable external drive console which powers the heart with compressed air" when, in fact, the portable drive system was not available at Arizona. It incorrectly states that "it is the *only alternative* which is available to maintain life until a suitable donor heart is found." (emphasis added) And it makes *no* provision for how the patient will die if a human heart transplant is not feasible, or who will make decisions for the patient, and on what basis, if the patient is unable to make them.

Indeed, although the form acknowledges that complications may preclude a human heart transplant, and assures the patient that "you will be supported by the artificial heart as long as possible," it does not explain how this will be done, and Arizona has made no provisions for so supporting a patient (such as the apartment and use of a portable drive available to William Schroeder at Humana Heart Institute).

No wonder Drummond wanted to know where all the noise was coming from. We may never know if he had four days to live (as he apparently believed) or two days (as his physician apparently believed), but we do know that he did have an alternative (even if he and his family did not think they had one). That alternative was to have his urgency status on the human heart transplant registry upped to "category 9," a category that usually includes only those patients expected to live forty-eight hours or less. This strategy would have provided him with a reasonable chance (though no guarantee) to obtain a human heart initially, without having to undergo the artificial heart implant. There was a significant failure of communication if he and his family did not understand this option.

The Transplant

On September 8, two "sporting events" were covered on the front page of the Sunday *New York Times*. Hana Mandlikova defeated Martina Navratilova in the finals of the US Open, and Michael Drummond received a human heart. Reporter Lawrence K. Altman, who has been the chronicler of artificial heart implants, took full advantage of the sports metaphor. Both his opening sentence and the headline for the story inaccurately described a "new human heart" (it was, of course, a "used" one) brought to Arizona after doctors "raced for seven hours to and from Tyler, Texas."

The sole photograph accompanying the article shows a technician running with the heart in a cooler. Altman describes the scene at the finish line: "A team member carrying the cooler sprinted across the parking lot to the hospital and was greeted by cheers from surgeons and nurses in the operating room." After the procedure, there was also the thrill of victory, Robert Vaughn describing the experience as "the thrill of a professional lifetime." Altman reported hospital spokesperson Nina Trasoff as saying, "[Drummond] watched some of the tennis matches yesterday and some football, and he's anxious for the tennis and football to get underway this morning so he can watch some more."

Experimentation is being turned into a sporting event and a spectator sport. As a sporting event it has no rules (since the FDA and NIH have abdicated any role as referee). As a spectator sport it is the most gruesome and morbid type of entertainment.

What If It Works?

As long as there is a shortage of transplantable human hearts, temporary artificial hearts that predictably "work" can serve *no useful purpose,* and are potentially destructive of important human values. Most heart transplant surgeons realize this, and that is why they are not rushing to adopt this "new technology." But the pressure to adopt it from an uninformed public, easily impressed by flash and drama, may make the technology difficult to resist. Accordingly, it is important to understand the argument.

First, as long as there is a shortage of transplantable human hearts, *temporary* artificial hearts cannot increase the total number of human heart transplants performed; they can only change the identity of the individuals who obtain them. For example, assume hypothetically that we have 1000 individuals annually qualified for human heart transplants in the US and that 600 human hearts are available. The remaining 400 will not receive a heart and will therefore die. If before their deaths these 400 are put on artificial hearts, the "waiting pool" will increase to 1400 next year, although only 600 hearts will be available. If priority is given to those on artificial hearts, these 400 will get hearts, but only 200 of the 1000 remaining on the list will get them. If the remaining 800 are not to die, they will get temporary artificial hearts. But this will already outstrip the next year's supply of human hearts. Six hundred will get them, 200 will wait another year, and the new group of 1000 potential recipients will either all die, or will have to get "temporary" hearts that will actually be "permanent" for most of them.

If we could dramatically increase the total number of available human hearts, or decrease the number of candidates, this scenario could change, but there is *no* reasonable expectation of this happening. Until there is, this technology merely increases the total cost of doing heart transplants on the same number of patients. It also leaves us with a significant number of patients on artificial hearts that were believed to be temporary, but have become permanent. Since we have *no* policy or plan concerning what to do with them, or how to make decisions about complications or turning off their artificial hearts, it is unethical even to begin this process.

In addition to being useless to society as a whole, temporary artificial hearts are unfair to those other patients on the waiting list for a human heart, at least if temporary artificial heart recipients are given priority over all others. After the artificial heart implant, Drummond was placed in category 9, the most urgent, although his hospital admitted he did not meet the criteria for that category since he was not expected to live 48 hours or less without the human heart. According to a hospital spokesperson, the *only* reason he was thought to qualify for that category was "because of the history of the Jarvik [7], it is felt that he is in fact in imminent danger of death because of that potential for stroke."[40] It is possible that because Drummond obtained the human heart on the basis of his "category 9" status and the intense national publicity surrounding his case, someone else in category 9— someone who really was dying imminently—did die. Whether or not this actually happened in the Drummond case, there is no doubt that if more temporary hearts are used, and if the hospitals that use them manipulate the procurement system to have recipients placed at the top of the priority list, others will die.

We need a moratorium on permanent artificial hearts because of the devastating effects they have had on their recipients. We need a moratorium on temporary artificial hearts because there is no guarantee they will not be permanent, we have yet to develop an ethically acceptable method of allocating human hearts to those with artificial hearts, and there is no room in our health care system for an extreme and expensive medical technology that is useless because it does not increase the total number of human lives saved by heart transplants. If the referee won't call time out, the players themselves should.*

*There has, in fact, been a moratorium on permanent artificial heart implants since the fourth US implant was performed on Jack Burcham on April 14, 1985. The publication of an entire issue of *JAMA* devoted to the artificial heart in February, 1988 memorializes this moratorium, but uncoordinated use of temporary artificial hearts continues. *JAMA*, Feb. 12, 1988 (Artificial Heart Issue).

References

Patients Rights

The Hospital: A Human Rights Wasteland

1. Analysis and case law supporting each of these rights appears in Annas, G.J., *The Rights of Hospital Patients*, Avon, NY, 1975.
2. For a more detailed discussion of the Patients Rights Advocate proposal, *see* Annas, G.J. and Healey, J.M., "The Patients Rights Advocate: Redefining the Doctor–Patient Relationship in the Hospital Context," **27** *Vanderbilt Law Rev* 243 (1974).
3. For a more detailed discussion of medical malpractice, *see* Annas, G.J., Katz, B.F. and Trakimas, R.G., "Medical Malpractice Litigation Under National Health Insurance: Essential or Expendable?" **1975** *Duke Law J* 1335.
4. *See* Annas, G.J., "Medical Privacy and Confidentiality," (In) Day, S.B. & Brandejas, J.F. (eds.), *Computers for Medical and Patient Management*, Van Nostrand Reinhold, NY, 1982, pp. 102–124.
5. Almost 40 states now have living will statutes, but there are problems. *See* Legal Advisers, Concern for Dying, "The Right to Refuse Treatment: A Model Bill," **73** *Amer J Public Health* 918 (1983) and Annas, G.J. and Densberger, J.E., "Competence to Refuse Medical Treatment: Autonomy v. Paternalism," **15** *Toledo Law Rev* 561 (1984).

The Emerging Stowaway: Patients Rights in the 1980s

6. Rawls, J., *A Theory of Justice*, Harvard Univ. Press, Cambridge, MA, 1971.
7. Dworkin, R., *Taking Rights Seriously*, Harvard Univ. Press, Cambridge, MA, 1977.
8. Altman, J.H., et al., "Patients Who See Their Medical Record," *New Engl J Med* **302**: 169 (1980).
9. Lipsett, D., Editorial, *New Engl J Med* **302**: 169 (1980).
10. Burt, R., *Taking Care of Strangers: The Rule of Law in Doctor–Patient Relations*, Free Press, NY, 1979, at 43.
11. Stevens, D.P., Stagg, R., MacKay, I., "What Happens When Hospitalized Patients See Their Own Records," *Annals of Internal Med* **86**: 474,476 (1977).

For a more complete discussion of this issue, including additional studies of patients given routine access to their records, *see* Annas, G.J., Glantz, L.H., Katz, B.F., *The Rights of Doctors, Nurses and Allied Health Professionals*, Avon Books, NY, 1981 at 159–161.

12. Golodetz, A., Ruess, J., Milhous, R., "The Right to Know: Giving the Patient His Medical Record," *Arch Phys Med Rehabilitation* **57**: 78,80 (1976).

13. *See*, e.g., Laforet, E.G., "The Fiction of Informed Consent," *JAMA* **235**: 1579 (1976).

14. Cassileth, B.R., et al., "Informed Consent—Why Are Its Goals Imperfectly Realized?" *New Engl J Med* **302**: 896 (1980).

15. *See generally* chapter on "Informed Consent" in Annas, G.J., Glantz, L.H., Katz, B.F., *The Rights of Doctors, Nurses and Allied Health Professionals*, Avon Books, NY, 1981; and Annas, G.J., Glantz, L.H., Katz, B.F., *Informed Consent to Human Experimentation: The Subject's Dilemma*, Ballinger, Cambridge, MA, 1977.

16. Thomas, L., "On Magic in Medicine," *New Engl J Med* **299**: 461, 462 (1978).

17. Lear, M.L., *Heartsounds*, Simon & Schuster, New York, 1980 at 47.

18. Annas, G.J., "The Care of Private Patients in Teaching Hospitals: Legal Implications," *Bull NY Acad Med* **56**: 403–411 (1980).

Breast Cancer: The Treatment of Choice

19. Mass. G.L. c.111, sec.70E.

20. Katz, J., *Experimentation with Human Beings*, Russell Sage Foundation, NY, 1972, p. 775.

21. Henderson, I.C., and Canellos, G.P., "Cancer of the Breast" (part II), *New Engl J Med* **302**: 78,87 (1980).

22. Crile, G., *What Women Should Know About the Breast Cancer Controversy*, Macmillan, NY, 1973, pp. 92–98.

23. *Washington Post*, Jan. 2, 1980, p. A17.

24. Bosk, C.L., *Forgive and Remember*, University of Chicago Press, Chicago, 1979.

Beyond the Good Samaritan

25. *Wilmington General Hospital v. Manlove*, 54 Del. 15, 174 A.2d 135 (1961).

26. *Guerrero v. Copper Queen Hospital*, 112 Ariz. 104, 537 P.2d 1329 (1975).

27. *E. Kentucky Welfare Rights Org. v. Simon*, 506 F.2d 1278 (1974), *vacated on other grounds*, 426 US 26 (1976).

28. Annas, G. J., "Myths and Realities in the Medical School Classroom," 1 *Am J Law Med* 195,199 (1975).

29. *Lyons v. Grether*, 239 S.E. 2d 103, 105 (Va. 1977).

Adam Smith in the Emergency Room

30. Shem, S., *House of God*, Marek, NY, 1978.
31. *Mercy Medical Center v. Winnebago Co.*, 206 N.W.2d 198,201 (Wisc. 1973).
32. *See*, e.g., Himmelstein, D. U. et al., "Patient Transfers: Medical Practice as Social Triage," *Am J Public Health*, **74**: 494 (1984).
33. *Thompson v. Sun City Community Hospital*, 688 P.2d 605 (Ariz. 1984).
34. *St. Joseph's Hospital v. Maricopa Co.*, 688 P.2d 986,990 (Ariz. 1984).

Sex in the Delivery Room

35. *Backus v. Baptist Medical Center*, 510 F. Supp. 1191 (1981).

Conception

Redefining Parenthood

1. Dept. of Health and Social Security, *Report of the Committee of Inquiry into Human Fertilization and Embryology*, London, 1984.

Artificial Insemination

2. Curie-Cohen, H., Luttrell, L., and Shapiro, S., "Current Practice of Artificial Insemination by Donor in the United States," *New Engl J Med* **300**: 585 (1979).
3. Behrman, S.J., Editorial, *New Engl J Med* **300**: 592 (1979).

Contracts to Bear a Child

4. *Washington Post*, Feb. 11, 1980, p.1.
5. *Am Med News*, June 20, 1980, p. 13.
6. *People*, Dec. 8, 1980, p. 53.
7. *Doe v. Kelly*, 6 FLR 3011 (1980).
8. Op. Atty. Gen., 81-18.
9. Note, "Contracts to Bear a Child," **66** *Cal. Law Rev* 611 (1978).

The Baby Broker Boom

10. Kentucky Rev. Stat., 199.590 (1982).
11. *Surrogate Parenting Associates v. Kentucky*, 704 S.W.2d 209 (1986).
12. *Smith & Smith v. Jones & Jones* (85-532014 DZ, Detriot, MI, 3d Jud. Dist., March 14, 1986) (Battani, J.).
13. MCLA Sec. 722.711 *et seq.* (1968).

Surrogate Embryo Transfer

14. Jones, H.W., editorial, "Variations of a Theme," *JAMA* **250**: 2182 (1983).
15. Bustillo, M., Buster, J.E., Cohen, S.W., et al., "Nonsurgical Ovum Transfer as a Treatment in Infertile Women: Preliminary Experience," *JAMA* **251**: 1171 (1983).
16. Walters, L., editorial, "Ethical Aspects of Surrogate Embryo Transfer," *JAMA* **250**: 2183 (1983).
17. "Establish Clinics Nationwide and Commercialize Service?" *Am Med News,* Feb. 20, 1984, p. 5.
18. Brotman, H., "Human Embryo Transplants," *New York Times Magazine,* June 8, 1983, p. 47.
19. Cancila, C., "First Embryo Transfer Baby Born," *Am Med News,* Feb. 17, 1983, p. 38.
20. *Griswold v. Connecticut,* 381 US 479,485 (1965).

Pregnancy and Birth

Fetal Neglect

1. *Washington Post,* Oct. 2, 1986, p. A17.
2. *New York Times,* Oct. 9, 1986, p. A22.
3. *Cal.* Penal Code, Sec. 270 (West, 1986).
4. *People v. Yates,* 298 P. 961,962 (L.A. Supr. Ct. 1931).
5. Law, S., "Rethinking Sex and the Constitution," **132** *U Pa Law Rev* 955,1008 (1984).
6. Johnsen, D.E., "The Creation of Fetal Rights: Conflicts with Women's Constitutional Rights of Liberty, Privacy and Equal Protection," **95** *Yale Law J* 599, 618 (1986).

Medical Paternity and Wrongful Life

7. *Becker v. Schwartz,* and *Park v. Chessin,* 413 NYS 2d 895 (1978).
8. Annas, G.J., and Coyne, B., "'Fitness' for Birth and Reproduction: Legal Implications of Genetic Screening," **9** *Family Law Quarterly* 463,478 (1975).

Righting the Wrong of Wrongful Life

9. Capron, A.M., "The Wrong of 'Wrongful Life'" (In) Milunsky, A. & Annas, G.J. (eds.), *Genetics and the Law II,* Plenum, NY, 1980, p. 89.
10. *Curlender v. Bio-Science Laboratories,* 165 Cal. Rptr. 477 (Ct. App. 2d Dist, Div. 1, 1980).

11. *Berman v. Allan,* 80 NJ 421, 404 A.2d 8 (1979) (dissenting opinion).
12. *Gildiner v. Thomas Jefferson Univ. Hospital,* 451 F. Supp. 692 (E.D. Pa. 1978).
13. Shaw, M., "Preconception and Prenatal Torts," (In) Milunsky, A. & Annas, G.J. (eds.), *Genetics and the Law II,* Plenum, NY, 1980, pp. 225, 229.
14. Capron, *supra,* note 9.

Is a Genetic Screening Test Ready When the Lawyers Say It Is?

15. 519 P.2d 981 (Wash. 1974).
16. 60 F.2d 737, 740 (2d Cir. 1932).

Homebirth: Autonomy vs Safety

17. Gots, R.E., *The Truth About Medical Malpractice,* Stein and Day, NY, 1975, p. 48.
18. *Bowland v. Municipal Ct. for Santa Cruz City,* 134 Cal. Rptr 630,638,556 P.2d 1081 (1976).
19. *Kahn v. Suburban Comm. Hospital,* 340 N.E.2d 398 (Ohio 1976).

Forced Cesareans

20. Leiberman J.R. et al., "The Fetal Right to Live," *Obstetrics and Gynecology* **53**: 515, 1979.
21. *Jefferson v. Griffin Spalding Co. Hospital Authority,* 247 Ga. 86, 274 S.E.2d 457 (1981).
22. Bowes, W.A., and Selgestad, B., "Fetal v. Maternal Right: Medical and Legal Perspectives," *Am J Obstetrics & Gynecology* **58**: 209–214, 1981.
23. *Raleigh Fitkin-Paul Morgan Memorial Hospital v. Anderson,* 201 A.2d 537,538 (NJ 1964).
24. *Application of the President and Directors of Georgetown College,* 331 F.2d 1000 (1964).
25. *Application of the President and Directors of Georgetown College,* 331 F.2d 1010 (1964), *cert den.* 377 US 978 (1964).

Disconnecting the Baby Doe Hotline

26. *See* Fost, N., "Putting Hospitals on Notice," *Hastings Center Report,* Aug., 1982.
27. *Fed. Reg.,* March 7, 1983, pp. 9630–9632.
28. *Am Academy of Pediatrics v. Heckler,* 561 F. Supp. 395 (DC 1983)

The Baby Doe Redux: Doctors as Child Abusers

29. **48** *Fed. Reg.* 30846–30852.

Checkmating the Baby Doe Regulations

30. *Bowen v. American Hospital Association*, 476 US 610 (1986).
31. *United States v. University Hospital*, 729 F.2d 144 (1984).
32. For a discussion of infant care review committees, *see* Annas, G.J., "Ethics Committees in Neonatal Care: Substantive Protection or Procedural Diversion," *Am J Public Health*, **74**: 843 (1984).

Reproductive Liberty

Round II

1. *Roe v. Wade*, 410 US 113 (1973).
2. *Danforth v. Planned Parenthood of Missouri*, 428 US 52 (1976).
3. *New York Times*, June 22, 1976, p. 40.
4. See *Doe v. Rampton*, 366 F. Supp. 189 (Utah, 1973).
5. *Bellotti v. Baird*, 423 US 982 (1975).

Let Them Eat Cake

6. *Beal v. Doe*, 432 US 438 (1977).
7. *Maher v. Roe*, 432 US 464 (1977).
8. *Poelker v. Doe*, 432 US 519 (1977).
9. Paraphrased from *San Antonio School Dist. v. Rodriguez*, 411 US 1, 17 (1973).
10. *Roe v. Wade*, 410 US 113 (1973).
11. *Doe v. Bolton*, 410 US 179 (1973).

Parents, Children and the Supreme Court

12. *Planned Parenthood of Missouri v. Danforth*, 428 US 52 (1976).
13. *Bellotti v. Baird*, 443 US 622 (1979).
14. *Parham v. J.L. and J.R.*, 442 US 584 (1979).

The Irrelevance of Medical Judgment

15. Woodward, R., and Armstrong, S., *The Brethren*, Simon & Schuster, NY, 1980, p. 175.
16. *Harris v. McRae*, 448 US 297 (1980).
17. *Maher v. Roe*, 432 US 464 (1977).

Roe v. Wade Reaffirmed

18. *Roe v. Wade,* 410 US 113 (1973).
19. *Akron v. Akron Center for Reproductive Health,* 462 US 416 (1983).

Reaffirming Roe v. Wade

20. *Thornburgh v. ACOG,* 90 L. Ed. 2d 779, 476 US 747 (1986).
21. *Akron v. Akron Center for Reproductive Health,* 462 US 416 (1983).
22. *Danforth v. Planned Parenthood of Missouri,* 428 US 52 (1976).

Medical Practice

Medical Malpractice

1. The best book on the crisis of the mid-1970s is Law, S., and Polan, S., *Pain and Profit: The Politics of Malpractice,* Harper and Row, New York, 1978; *see also* Annas, G.J., Katz, B.F., and Trakimas, R.G., "Medical Malpractice Under National Health Insurance: Essential or Expendable?," **1975** *Duke Law J* 1335.
2. *Forbes,* March 10, 1986, p. 63.
3. *New York Times,* April 13, 1986, III, p. 3.
4. Danzon, P., *Medical Malpractice,* Harvard Univ. Press, Cambridge, MA, 1985.

Doctors Sue Lawyers

5. *Am Med News,* June 14, 1976, p. 1.
6. *Wolfe v. Arroyo,* 543 S.W.2d 11 (Tex. Ct. App. 1976).
7. *Drago v. Buonagurio,* 391 NYS.2d 61 (Sup. Ct. Schenectady Co., 1977).
8. *Spencer v. Burglass,* 377 So.2d 596 (La. Ct. App. 4th Cir. 1976).
9. *Am Med News,* August 1, 1977, p. 1.

Confidentiality and the Duty to Warn

10. *People v. Poddar,* 10 Cal. App. 3d 750, 518 P.2d 342 (1974).
11. *Tarasoff v. Regents of U. California,* 118 Cal. Rptr. 129, 529 P.2d 553 (1974).
12. *Tarasoff v. Regents of U. California,* 551 P.2d 334 (1976).
13. *Merchants National Bank v. US,* 272 F. Supp. 409 (D.N.D. 1967)
14. E.g., Flemming J. & Maximov B., "The Patient or his Victim: The Therapist's Dilemma," **62** *Cal Law Review* 1025 (1974).

15. Note, **37** *Pitt Law Review*, 155, 165 (1975).
16. *Supra*, note 14 at 1031.
17. Lambert T., *ATLA Newsletter*, Feb. 1975, p. 51.

Who to Call When the Doctor is Sick

18. Green, R.C., Carroll, G.J., and Buxton, W.D., *The Care and Management of the Sick and Incompetent Physician*, Charles C Thomas, Springfield, Ill., 1978.

CPR: The Beat Goes On

19. Standards and Guidelines for Cardiopulmonary Resuscitation (CPR) and Emergency Cardiac Care (ECC), *JAMA* **244**: 453; 1980.
20. *New York Times*, Jan. 27, 1981, p. A18.

CPR: When the Beat Should Stop

21. *JAMA*, **244**: 506, 1980.
22. "The Do-Not-Resuscitate Order in a Teaching Hospital," *Ann Intern Med* **96**: 660–664, 1982.
23. *Matter of Shirley Dinnerstein*, 380 N.E.2d. 134 (Mass. App. 1978).
24. *Jane Hoyt v. St. Mary's Rehabilitation Center*, No. 774555, 4th Judicial Dist., Hennepin Co., Mn., Jan. 2, 1981, Arthur, J.
25. *Custody of a Minor*, 385 Mass. 697, 710 (1982).

The Mentally Retarded and Mentally Ill Patient

Denying the Rights of the Retarded

1. *In re Phillip B.*, 156 Cal. Rptr. 48 (1st App. Dist., Division 4, 1979).
2. *See*, e.g., *In re Hofbauer*, 65 A.D. 2d 108, 411 NYS 2d 416 (App. Div. 1978), aff'd 419 NYS 2d 936 (1979).
3. *See* Annas, G.J., "Reconciling *Quinlan* and *Saikewicz*: Decision Making for the Terminally Ill Incompetent," 4 *Am J Law Med* 367 (1979).
4. Robertson, J., "Involuntary Euthanasia of Defective Newborns: A Legal Analysis," **27** *Stanford Law Rev* 213, n. 5 (1975).
5. *See* Dworkin, R., *Taking Rights Seriously*, Harvard Univ. Press, Cambridge, MA, 1978, pp. 204–205.

A Wonderful Case and an Irrational Tragedy

6. *Guardianship of Phillip Becker*, Superior Court of Cal., Santa Clara Co., No. 101981, Aug. 1981; 7 FLR 2648, *aff'd Heath v. Becker*, 8 FLR 2017 (1981).
7. Dickens, B., "The Modern Function and Limits of Parental Rights," **97** *Law Quarterly Review* 462, 465–466 (1981).
8. Goldstein, J., Freud, A., Solnit, A., *Before the Best Interests of the Child*, Free Press, New York, 1979.
9. *Id.*
10. Dickens, *supra* note 7.

Sterilization of the Mentally Retarded

11. *Buck v. Bell*, 274 US 200 (1927).
12. *Skinner v. Oklahoma*, 316 US 535 (1942).
13. *In the Matter of Lee Ann Grady*, 426 A. 2d 467 (NJ, 1981).
14. Baron, C., "Voluntary Sterilization of the Mentally Retarded," (In) A. Milunsky, G.J. Annas (eds.) *Genetics and the Law*, Plenum, NY, 1976, pp. 273–274.

Refusing Medication in Mental Hospitals

15. *Rennie v. Klein*, 462 F. Supp. 1131, 1145 (1978).
16. *Rogers v. Okin*, 478 F. Supp. 1342 (1979).
17. *Boston Globe*, November 21, 1979, p. 13.

The Incompetent's Right to Die: The Case of Joseph Saikewicz

18. Kindregan, C.P., "The Court as a Forum for Life and Death Decisions in Symposium—Mental Incompetents and the Right to Die," **9** *Suffolk Univ Law Rev* 919–973 (1977).
19. *Super. of Belchertown v. Saikewicz*, 370 N.E. 2d 417 (Mass. 1977).

The Case of Mary Hier

20. *In the Matter of Mary Hier*, 18 Mass. App. Ct. 200 (1984).
21. *Supra*, note 19, 370 N.E. 2d at 428.

Death, Dying, and Refusing Medical Treatment

The Case of Karen Ann Quinlan

1. "The Quinlan Decision: Five Commentaries," *Hastings Center Report*, Feb. 1976, pp. 8–19.

2. *In re Quinlan*, 70 N.J.10, 355 A.2d 647 (1976).
3. *John F. Kennedy Hospital v. Heston*, 58 NJ 576, 279 A.2d 670 (1971).
4. Teel, K., "The Physician's Dilemma: A Doctor's View," 27 *Baylor Law Rev* 6 (1975).
5. Milgram, S. *Obedience to Authority*, Yale Univ. Press, New Haven, CT, 1970.

No Fault Death: After Saikewicz

6. Rabkin, M.T., Gillerman, G. and Rice, N.R., "Orders Not to Resuscitate," *New Engl J Med* **295**: 364–369 (1976).
7. Knox, R., "Bay State Judges and the Right to Die," *Boston Globe*, Jan. 30, 1978, pp. 1–5.
8. Relman, A., "The *Saikewicz* Decision: Judges as Physicians," *New Engl J Med* **298**: 508–509 (1978).
9. Burt, R. "Authorizing Death for Anomalous Newborns," (In) Milunsky, A. and Annas, G.J. (eds.) *Genetics and the Law*, Plenum, NY, 1976, p. 445.

Brother Fox and the Dance of Death

10. *Eichner v. Dillon*, 73 AD 2d 431, 426 NYS 2d 517 (1980).

Help from the Dead: The Cases of Brother Fox and John Storar

11. *In the Matter of Philip Eichner, In the Matter of John Storar*, 52 NY 2d 363 (1981).

Quality of Life in the Courts: Earle Spring in Fantasyland

12. *In the Matter of Earle Spring*, 380 Mass. 629, 405 N.E. 2d 115 (Mass. 1980).
13. *In the Matter of Earle Spring*, 8 Mass. App. Ct. 831, 399 N.E. 2d 493 (1979).
14. *Minerva v. National Health Agency*, 40 US 2d 345 (2002) (Euterpe, J., dissenting) reprinted in **3** *Am J Law and Medicine* 59 (1977).

When Suicide Prevention Becomes Brutality

15. *Bouvia v. Co. of Riverside*, No. 159780, Sup. Ct., Riverside Co., Cal., Dec. 16, 1983, Tr. 1238–1250.
16. *Super. of Belchertown v. Saikewicz*, 370 N.E.2d 417, 426 (Mass. 1977).
17. *Id.* at 427.

Elizabeth Bouvia: Whose Space Is This Anyway?

18. *Bouvia v. Glenchur*, L. A. Superior Court, C 583828, Feb. 21, 1986.

19. *Cobbs v. Grant*, 502 P.2d 1 (Cal. 1972); *Barber v. Superior Court*, 195 Cal. Rptr. 484 (1983), *Bartling v. Superior Court*, 163 Cal. App. 3d 186 (1984).

Do Feeding Tubes Have More Rights than Patients?

20. *Patricia Brophy v. New England Sinai Hospital*, Norfolk Probate Ct., No. 85 E0009-G1, 1985.
21. *Super. of Belchertown v. Saikewicz*, 370 N.E.2d 417 (Mass. 1977).

The Case of Claire Conroy: When Procedures Limit Rights

22. *In the Matter of Claire Conroy*, 486 A.2d 1209 (NJ 1985).
23. *In re Conroy*, 188 NJ Super. 523 (Ch. Div. 1983) discussed in Annas, G.J., "Nonfeeding: Lawful Killing in California, Homicide in New Jersey," *Hastings Center Report*, Dec. 1983, pp. 19–20.

Prisoner in the ICU: The Tragedy of William Bartling

24. Quotations are from the Transcript, *Bartling v. Glendale Adventist Medical Center*, Case No. C 500 735, June 22, 1984, Superior Court, Los Angeles, CA (Dept. 86, Waddington J.).
25. *Barber and Nejdl v. Sup. Ct.*, 195 Cal. Rptr. 484 (Cal. App. 2 Dist 1983) discussed in Annas, G.J. "Nonfeeding: Lawful Killing in California, Homicide in New Jersey," *Hastings Center Report*, Dec. 1983, p. 19.
26. Annas, G.J. and Densberger, J.E., "Competence to Refuse Medical Treatment: Autonomy vs. Paternalism," **15** *Toledo Law Rev* 561 (1984).
27. Annas, G.J., "Reconciling *Quinlan* and *Saikewicz*: Decisionmaking for the Terminally Ill Incompetent," **4** *Am J Law & Med* 367 (1979).
28. *Foster v. Tourtellotte*, USDC No. CV 81-5046-RMT, Cent. Dist. CA (1981); atty. fees denied, 704 F.2d 1109 (1983).

Government Regulation

Good as Gold: Report on the National Commission

1. Heller, J., *Good As Gold*, Simon and Schuster, New York, 1979, p. 195.
2. *See*, for example, Levine, R. J., "Clarifying the Concepts of Research Ethics," *Hastings Center Report*, **9**(3): 21–26, June, 1979; and Neville, R., "On the National Commission: A Puritan Critique of Consensus Ethics," *Hastings Center Report* **9**(2): 22–27, April, 1979.
3. Established 42 USC 300v (subchapter xvi), Nov. 9, 1978.

4. National Research Service Award Act of 1974, PL 93-348, Title II, 201(b)(1), July 12, 1974.
5. 42 USC, (a)(1)(B),(C).
6. *See generally*, "Limits of Scientific Inquiry," *Daedalus*, Spring, 1978.
7. Annas, G.J., Glantz, L.H., Katz, B.F., *Informed Consent to Human Experimentation: The Subjects Dilemma*, Ballinger, Cambridge, Mass. 1977, 234–242.
8. *Id.*
9. *43 Fed. Reg.* 53242, Nov. 15, 1978.
10. Veatch, R. M., "The National Commission Recommendations on IRBs: An Evolutionary Approach," *Hastings Center Report* 9(1):22–28, Feb. 1979.
11. Beecher, H. K., "Ethics and Clinical Research," *New Engl J Med* 274: 1354–1360 (1966).
12. *National Commission for the Protection of Human Subjects of Biomedical and Behavioral Research, Report and Recommendations on Institutional Review Boards*, DHEW Pub. No. (OS) 78-008, Washington, DC (1978) pp. 61–62; and P. Cardon, et al., "Injuries to Research Subjects," *New Engl J Med* **295**: 650–654 (1976).
13. *National Commission for the Protection of Human Subjects of Biomedical and Behavioral Research, Report and Recommendations: Psychosurgery*. DHEW, Washington, DC (1977); and Annas, *Informed Consent, supra* note 7.
14. Controller General of the United States, *Federal Control of New Drug Testing is Not Adequately Protecting Human Test Subjects and the Public*, HR Doc. 76-96, July 15, 1976.
15. Holden, C., "FDA Tells Senators of Doctors Who Fake Data in Clinical Drug Trials," *Science* **206**: 432–433 (1979).
16. Rosenberg, K., "Human Experimentation: Adding Insult to Injury," *Health/PAC Bulletin* (1979), p. 2.
17. Letter Report of Comptroller General, HR Doc. 79-49 (Feb. 6, 1979). One could also discuss bone marrow transplant cases involving death from graft vs. host disease in unsuccessful cases. *See also*, J. M. McCartney, "Encephalitis and Ara-A: An Ethical Case Study," *Hastings Center Report* 8(6): 5–7, Dec. 1978.
18. Barber, B., Lally, J.J., Makarushka, J. L., Sullivan, D., *Research on Human Subjects: Problems of Social Control in Medical Experimentation*, Russell Sage Foundation, New York, 1973.
19. Appendix (bound separately) to *Report and Recommendations on Institutional Review Boards, supra* note 12.
20. Gray, B.H. *Human Subjects in Medical Experimentation: A Sociological Study of the Conduct and Regulation of Clinical Research*, Wiley, New York, 1975.
21. *Supra* note 10, p. 26.
22. Nuestadt, R.E., and Fineberg, H.V., *The Swine Flu Affair: Decision-making on a Slippery Disease*, U S Govt. Printing Office, Stock No. 017-000-00210-4, Washington, DC, 1978, p. 63.
23. *Id.*

24. Robertson, J., Ten Ways To Improve IRBs, *Hastings Center Report* 9(1):29–33, Feb. 1979.
25. *Supra* note 1, p. 198.

All the President's Bioethicists

26. S. 2579.
27. P.L. 95-622.

Creationism: Monkey Laws in the Courts

28. *McLean v. Arkansas Bd. of Education*, 529 F. Supp. 1255 (Ark. 1982) [opinion reprinted in *Science*, 215: 934–943, Feb. 19, 1982].
29. *Engel v. Vitale*, 370 US 421, 431 (1962); and "The law knows no heresy, and is committed to the support of no dogma, the establishment of no sect." *Watson v. Jones*, 80 US 679, 728 (1871).
30. *Id.*
31. *Scopes v. State*, 289 S.W. 363, 367 (Tenn. 1927).
32. *Epperson v. Arkansas*, 393 US 97, 103–104 (1968).
33. Tribe, L., *American Constitutional Law*, Foundation Press, NY, 1978, 14–80.

Life Forms: The Law and the Profits

34. *Tennessee Valley Authority v. Hill*, 437 US 153 (1978).
35. *Application of Bergy*, 563 F.2d 1031 (1977).
36. US Patent Act, 35 USC 101.

Transfer Trauma and the Courts

37. *Brede v. Department of Health*, 616 F.2d 407 (9th Cir. 1980).
38. *O'Bannon v. Town Court Nursing Home*, 447 US 773 (1980).
39. Borup et al., "Relocation and Its Effect on Mortality," *The Gerontologist* 19: 135, 1979.
40. *Yaretsky v. Blum*, 629 F.2d 817 (1980).

Prison Hunger Strikes: Why Motive Matters

41. *Zant v. Prevatte*, 248 Ga. 832, 286 S.E. 715 (1982).
42. *Von Holden v. Chapman*, 450 NYS 2d 623 (App. Div. 1982).
43. *White v. Natick*, 292 S.E. 2d 54 (W. Va. 1982).
44. *New York Times*, Jan. 20, 1981, p. C3.
45. *Commissioner v. Myers*, 399 N.E.2d 452 (Mass. 1979)

Transplants and Implants

Defining Death: There Ought to be a Law

1. *Am Med News*, August 27, 1982, p. 6.

The Prostitute, The Playboy, and the Poet

2. Calabresi, G, and Bobbitt, P., *Tragic Choices*, Norton, NY, 1978.
3. Fletcher, J., "Our Shameful Waste of Human Tissue," (In) *The Religious Situation*,, Cutler, D.R. (ed.), Beacon Press, Boston, 1969, pp. 223–252.
4. Quoted in Fox, R. and Swazey, J., *The Courage to Fail*, Univ. of Chicago Press, Chicago, 1974, p. 232.
5. Sanders & Dukeminier, "Medical Advance and Legal Lag: Hemodialysis and Kidney Transplantation," **15** *UCLA Law Rev* 357 (1968).
6. Rettig, R.A., "The Policy Debate on Patient Care Financing for Victims of End Stage Renal Disease," **40** *Law & Contemporary Problems* 196 (1976).
7. President's Commission for the Study of Ethical Problems in Medicine, *Securing Access to Health Care*, US Govt. Print Office, Washington, DC, 1983, p. 25.
8. Annas, G.J., "Ethics Committees and Neonatal Care: Substantive Protection or Procedural Diversion?," *Am J Public Health* **74**: 843 (1984).
9. Bayer, R., "Justice and Health Care in an Era of Cost Containment: Allocating Scarce Medical Resources," *Social Responsibility*, 9: 37–52, 1984.
10. Annas G.J., "Allocation of Artificial Hearts in the Year 2002: *Minerva v. National Health Agency,*" 3 *Am J Law Med* 59–76 (1977).
11. Commentary, "UK's Poor Record in Treatment of Renal Failure," *Lancet,* July 7, 1984, p. 53.
12. Evans, R., "Health Care Technology and the Inevitability of Resource Allocation and Rationing Decisions," Part II, *JAMA* **249**: 2008, 2217 (1983).

Life, Liberty and the Pursuit of Organ Sales

13. *Fitzsimmons v. NY*, 146 Supp. 117 (NY 1914).
14. Dickens, B., "Control of Living Body Materials," 27 *Toronto Law Rev* 142, 165 (1977).

Baby Fae

15. Breo, D.L., "Interview with Baby Fae's Surgeon: Therapeutic Intent was Topmost," *Am Med News*, Nov. 16, 1984, 1.
16. Thorwald, Jurgen , *The Patients*, Harcourt Brace Jovanovich, NY, 1972.

17. Hoover, E., "Baby Fae: A Child Loved and Lost," *People*, Dec. 3, 1984, 49–63.
18. *New York Times*, Nov. 15, 1984, A27.

Consent to the Artificial Heart: The Lion and the Crocodiles

19. Barnard, Christiaan, *One Life*, Macmillan, NY, 1969, 348.
20. *Newsweek*, Dec. 13, 1982, 35–36.
21. *Medical World News*, Jan. 10, 1983, 13–14.
22. *New York Times*, Dec. 4, 1982, 14.
23. Eichwald, E.J., "Insertion of the Total Artificial Heart," *IRB: A Review of Human Subjects Research*, August/September, 1981.
24. Annas, G.J., Glantz, L.H., and Katz, B.F., *Informed Consent to Human Experimentation: The Subject's Dilemma*, Ballinger, Cambridge, MA, 1977, 10–18.
25. C. Bosk, *Forgive and Remember*, Univ. of Chicago Press, Chicago, 1979, 218.

The Phoenix Heart: What We Have to Lose

26. Jennett, B., *High Technology Medicine*, London, 1984.
27. *Arizona Daily Star*, March 7, 1985, p. 2.
28. *New York Times*, March 19, 1985, pp. C1–C2.
29. *USA Today*, March 8, 1985, p. 1A.
30. *New York Times*, March 9, 1985, p. 22.
31. *Newsweek*, March 19, 1985, p. 86–88.
32. *New York Times*, March 17, 1985, p. 87.
33. *Arizona Daily Star*, March 8, 1985, pp. 1–2.
34. Katz, J., *The Silent World of Doctor and Patient*, Free Press, New York, 1984.
35. *US v. Rutherford*, 544 US 442, 1979.
36. *USA Today*, March 11, 1985, p. 10A
37. Annas, G.J., "Regulating the Introduction of Heart and Liver Transplantation," *Am. J. Public Health*, **15**: 93, 1985.

No Cheers for Artificial Hearts

38. *New York Times*, Sept. 2, 1985, p. 10.
39. *New York Times*, Sept. 15, 1985, p. 33.
40. *New York Times*, Sept. 7, 1985, p. 7.

Index

423